Optics Essentials

Optics Essentials

Edited by
Roderick Swayne

WILLFORD PRESS
www.willfordpress.com

Published by Willford Press,
118-35 Queens Blvd., Suite 400,
Forest Hills, NY 11375, USA

ISBN: 978-1-68285-579-9

Cataloging-in-Publication Data

Optics essentials / edited by Roderick Swayne.
    p. cm.
Includes bibliographical references and index.
ISBN 978-1-68285-579-9
1. Optics. 2. Light. 3. Physics. I. Swayne, Roderick.
QC355.3 .O68 2019
535--dc21

For information on all Willford Press publications
visit our website at www.willfordpress.com

WILLFORD PRESS

Contents

Preface

In my initial years as a student, I used to run to the library at every possible instance to grab a book and learn something new. Books were my primary source of knowledge and I would not have come such a long way without all that I learnt from them. Thus, when I was approached to edit this book; I became understandably nostalgic. It was an absolute honor to be considered worthy of guiding the current generation as well as those to come. I put all my knowledge and hard work into making this book most beneficial for its readers.

The field of optics is the study of light in its entire spectrum like high energy X-rays, ultraviolet, visible, infrared light, low energy radio waves, etc. and their individual interactions with matter. Optical phenomena are generally approached from the domains of physical and geometric optics. The classical electromagnetic description of light as well as the quantum mechanical approach to light together creates a comprehensive framework of optical dynamics. It has prominent applications in engineering, photography and medicine. The objective of this book is to give a general view of the different areas of study within the field of optics and their applications. It also presents researches and studies performed by experts across the globe. It is meant for students and researchers who are looking for an elaborate reference text on optics.

I wish to thank my publisher for supporting me at every step. I would also like to thank all the authors who have contributed their researches in this book. I hope this book will be a valuable contribution to the progress of the field.

Editor

Measurement to radius of Newton's ring fringes using polar coordinate transform

Ping An[1], Fu-zhong Bai[1]* , Zhen Liu[1]*, Xiao-juan Gao[1] and Xiao-qiang Wang[2]

Abstract

Background: Newton's ring method is often used to measure many physical parameters. And some measured physical quantity can be extracted by calculating the radius parameter of circular fringes from Newton's ring configuration.

Methods: The paper presents a new measuring method for radius of circular fringes, which includes three main steps, i.e., determination of center coordinates of circular fringes, polar coordinates transformation of circular fringes, and gray projection of the transformed result which along the horizontal direction. Then the radius values of each order ring are calculated.

Results: The simulated results indicate that the measuring accuracy of the radius under the effect of random noise can keep the degree of less than 0.5 pixels.

Conclusions: The proposed method can obtain the radius data of each order closed circular fringes. Also, it has several other advantages, including ability of good anti-noise, sub-pixel accuracy and high reliability, and easy to in-situ use.

Keywords: Newton's ring, Polar coordinates transform, Fringes pattern, Gray projection

Background

The parameter estimation of interference fringe patterns has been widely used in optical metrology, including holographic interferometry, electronic speckle pattern interferometry (ESPI) and fringe projection. Such optical techniques have been applied to measure physical parameters such as curvature radius, displacement, strain, surface profile and refractive index. The information regarding the measured physical quantity is stored in the radius parameter of the captured fringe pattern [1]. Some optical fringes, i.e., elementary fringes that have great importance in optical measurement (e.g., Newton's rings fringe patterns), have a quadratic (i.e., second-order polynomial) phase. Therefore, the fringes pattern is unequispaced fringe.

In general, Newton's rings method is used to measure physical parameters such as film thickness [2, 3], stain [4], and curvature radius [5] as well. In some application, phase demodulation needed to be done in Newton's ring

interference configuration. And the Fringe Center Method (FCM) [6, 7] or the Fourier transform [8, 9] are still an important inspection method to extract the character information of the fringes pattern. However, for example, the FCM Manual intervention is introduced to link the processes, such as the fringe patching and the assignment of the fringe orders.

In the measurement of curvature radius based of Newton' ring configuration, the radius of fringes is a key parameter and should be accurately obtained from fringes pattern. In the traditional method, the radius of the fringes is measured by observing the microscope and the scale with the eye. The disadvantage of the method is obvious, i.e., the visual field of microscope is small and hence make the fringe center difficult to observe. Additionally, scale is easy to misread due to the fatigue of human eye. Also, parameters of circular fringes can be retrieved with the Fourier transform via the estimation of the phase and its derivatives [9]. However, the required iterative procedure is a time-consuming approach. And it is error-prone because the procedure requires phase unwrapping and numerical differentiation

* Correspondence: fzbaiim@163.com; lz_water@sina.com
[1]College of Mechanical Engineering, Inner Mongolia University of Technology, Huhhot 010051, China
Full list of author information is available at the end of the article

operations [10]. The least squares method [11] is also developed to analyzed the circular fringes and estimate the parameter of optical fringes. However, it requires initial approximations for the fringe parameters to be determined.

With the development of digital image processing technology, it has been applied to the fields requiring non-contact, high speed, automatic processing and large dynamic range [12, 13]. It is especially suitable for the occasion that the traditional method is difficult to be applied. At present, the image processing technique used in analyzing the circular fringes includes several reprocessing steps, such as noise removal, fringe thinning, fringe patching, assignment of the fringe orders and so on [14–16]. For the Hough transform [17, 18] used to determine the parameters and the orders of circular fringes, the computational mount is heavy and the efficiency is low.

Especially aiming at the measurement of radius of plate-convex lens based on the Newton's ring configuration, the paper propose a new analyzing method of the ring fringes to improve automatic processing technique. Through transforming circular fringes to straight fringes with polar coordinates transform, the method carries out the measurement of radius of each order circular fringes. The principle of polar coordinates transform and the processing algorithm of Newton's ring interference pattern are introduced in the paper. Moreover, the accuracy of the method is analyzed and the experiment are done.

Methods
Principle of polar coordinate transform

The task of polar coordinate transform is that an image under the Cartesian coordinate (x - y) space is transformed to another image under polar coordinate (ϕ- r) space. The expression of polar coordinate transform is expressed as [19]

$$\begin{cases} r=\sqrt{x^2+y^2} \\ \phi= \arctan(y/x) \end{cases}. \tag{1}$$

The schematic diagram of polar coordinate transform is shown in Fig. 1.

In polar coordinate space, the meaning of r describes the distance of a point (x,y) to the origin position in Cartesian coordinate space, and ϕ discribes the angle of vector and its range is from 0 to 359°. Due to the origin symmetric of polar coordinate transform, the transform needs to be carried out in the range of 0° to 179°.

According to Eq. (1) and Fig. 1, one point under the Cartesian coordinate space corresponds uniquely to one point under the polar coordinate space. One circle in the Cartesian coordinate space whose center coincides with the origin, will corresponds to one line along ϕ-axis in the polar coordinate space, and the radius of the circle corresponds to the distance of this line to the origin in polar coordinate space.

Determination of center of circular fringes

Newton's ring interference fringes is composed of alternating light and dark stripes, and light and dark area are clear, as shown in Fig. 2(a). Through using the Otsu method [20] the fringes image is processed with threshold segmentation, and so a binary image B(x, y) can be obtained from the fringes image f(x, y) according to the following expression,

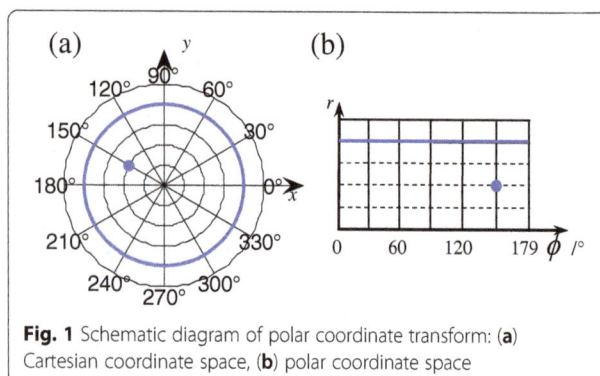

Fig. 2 Calculation of center ordinates of circular fringes: (a) Newton's ring fringes image, (b) binary image, (c) the first order ring, (d) circular region and center position

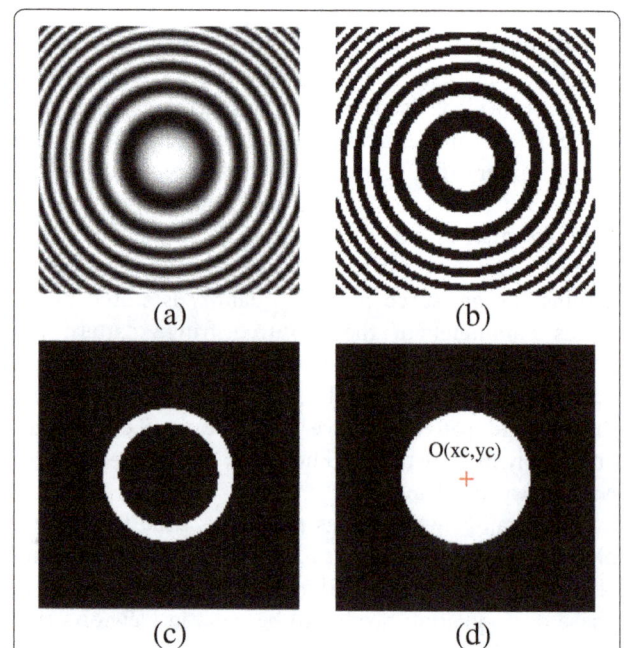

Fig. 1 Schematic diagram of polar coordinate transform: (a) Cartesian coordinate space, (b) polar coordinate space

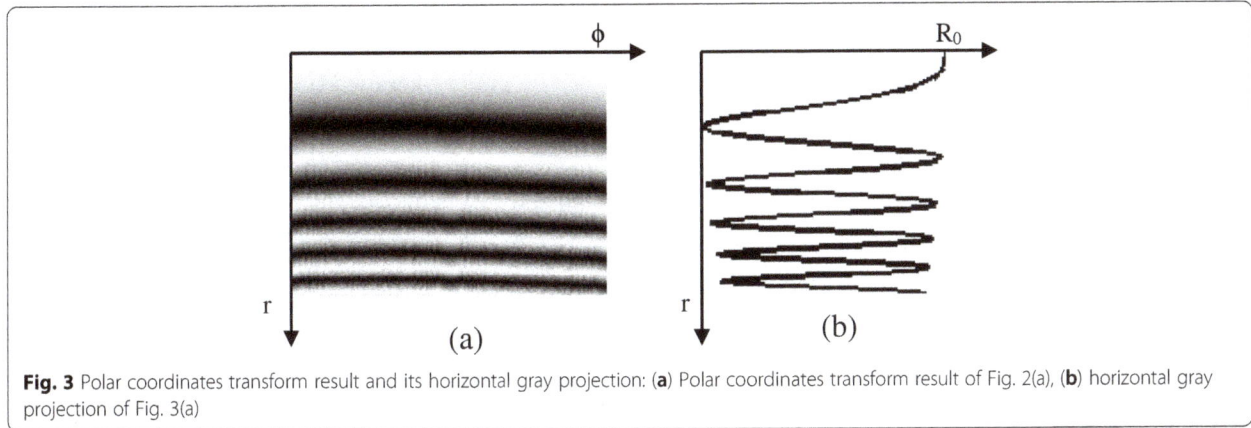

Fig. 3 Polar coordinates transform result and its horizontal gray projection: (**a**) Polar coordinates transform result of Fig. 2(a), (**b**) horizontal gray projection of Fig. 3(a)

$$B(x,y)=\begin{cases} 1, f(x,y)<T \\ 0, f(x,y)\geq T \end{cases}, \qquad (2)$$

$$\begin{cases} x_c=\frac{1}{n}\sum x_i \\ y_c=\frac{1}{n}\sum y_i \end{cases}, \qquad (3)$$

Here, assumed that light fringes are regarded as the target and the radius of light fringes will be calculated, and T is threshold value. The binary image of Fig. 2(a) is shown in Fig. 2(b).

Through using the connected component labeling algorithm [18] the first order ring from the binary image can be extracted, which is shown in Fig. 2(c). Then, the circular region is filled by the morphological operation, which is shown in Fig. 2(d). Furthermore, the edges of the target region is smoothed by using the opening operation, then the gravity ordinates (xc, yc) of circular region (i.e., the region with white gray-scale pixels) can be calculated according the following equation,

where, n is the number of the white pixels as shown in Fig. 2(d). Also, the center of circular fringes is marked in Fig. 2(d).

Calculation of circular fringes radius

Based on the calculated center the circular fringes are transformed to the polar coordinate space with polar coordinates transform method introduced in Section 2. Therefore, the circular fringes can be transformed to straight fringes. The transformed result of the original image as shown in Fig. 2(a) is shown in Fig. 3(a) and expressed as $p(r, \phi)$.

To calculate the radius of each order ring and eliminate immensely the effect of random noise, the straight fringes as shown in Fig. 3(a) is implemented the horizontal gray projection according to the following equation,

$$R_0(r) = \int p(r, \phi)d\phi. \qquad (4)$$

The projection curve of Fig. 3(a) is shown in Fig. 3(b). In the case, the r coordinate value corresponding to each peak position in the projection curve denotes the radius value of each order ring, and hence the method can

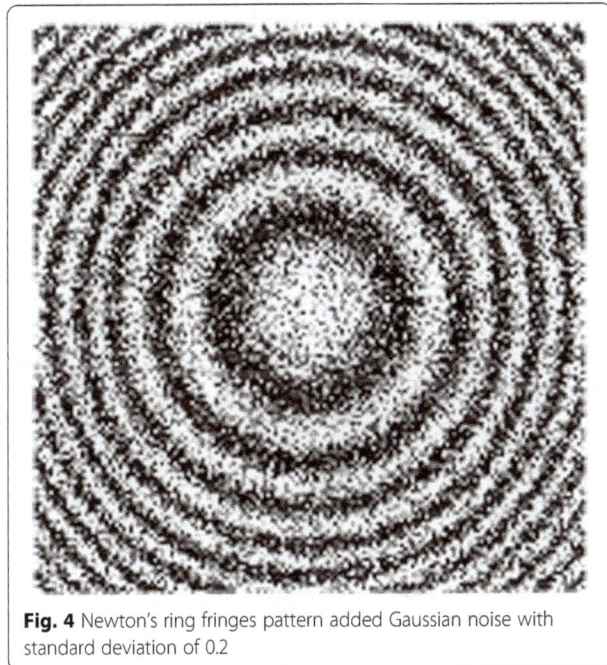

Fig. 4 Newton's ring fringes pattern added Gaussian noise with standard deviation of 0.2

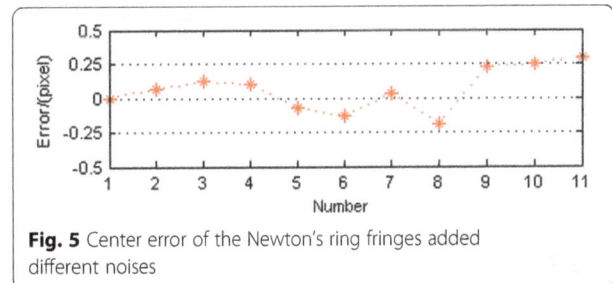

Fig. 5 Center error of the Newton's ring fringes added different noises

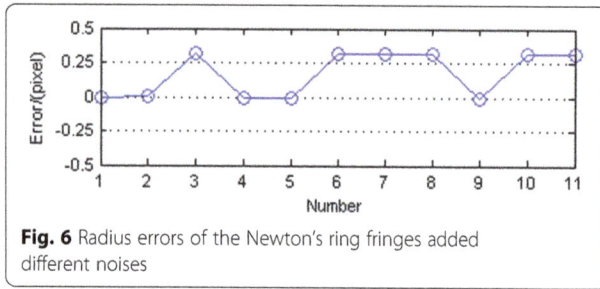

Fig. 6 Radius errors of the Newton's ring fringes added different noises

calculate respectively the radius parameter of each order ring from circular fringes.

Results
Center positioning accuracy from noise
The center of circular fringes is one of important parameters to circular fringes, and the center positioning accuracy is affected mainly by the noise. Therefore, it is necessary to analyze the effect of noise on the center positioning accuracy.

Simulated fringes pattern with noise is used to show the effectiveness of the proposed approach as shown in Fig. 4 for an image size of 255×255 pixels. The center of the simulated image is (128, 128). To investigate the effect of random noise on the center positioning, 11 frames of Newton's ring interference fringes containing four closed rings are generated by numerical method, which added to an independent Gaussian white noise with a mean value of zero and standard deviation varying from 0 to 0.2. The standard deviation of added noise in Fig. 4 is equal to 0.2.

The center coordinates of each frame of images are calculated by the proposed method introduced in Section 3.1, and the difference between the calculated center and the given center is obtained. The maximum value of the transverse and longitudinal coordinate error is seen as the error of center positioning for each image. The error curve of center positioning is shown in Fig. 5. It can be found from Fig. 5 that the positioning error

under the effect of random noise is not larger than 0.5 pixels with the proposed method.

Measuring accuracy of radius
Similarly, we generate 11 frames of Newton's ring interference fringe patterns, which contain four closed rings, and the size of simulated image is 255×255 pixels. Then different Gaussian noise with the standard deviation varying from 0 to 0.2 is respectively added to images. For each frame of fringes patterns, the radius values are calculated with the polar coordinates transform algorithm.

According to the horizontal projection curve, the radius of any order closed ring can be obtained. For simple analysis the radius error of the third order ring is seen as the radius measuring error. It can be obtained according to the difference between the measured value and the ideal value of radius. The radius errors of every frame of interference fringes with different random noises are shown in Fig. 6, and the measuring accuracy of the radius can also keep the degree of less than 0.5 pixels.

Experimental result
To illustrate the performance of the proposed method, this method is applied to an experimental fringes pattern. The Newton's ring experimental setup for recording circular interference fringes is shown in Fig. 7. The He-Ne laser with output wavelength of 632.8 nm is used as experimental light source, and lenses L1 and L2 and spatial filter are used as laser beam expander and collimation. The beam splitter is used to adjust the energy of the reference beam and the object beam. By moving a controlled shifting stage M driving by the computer, we can record different phase-shifting interferograms. The aperture A1 and A2 are used respectively to control the diameter of laser beam and to filter stray light. The 8-bit

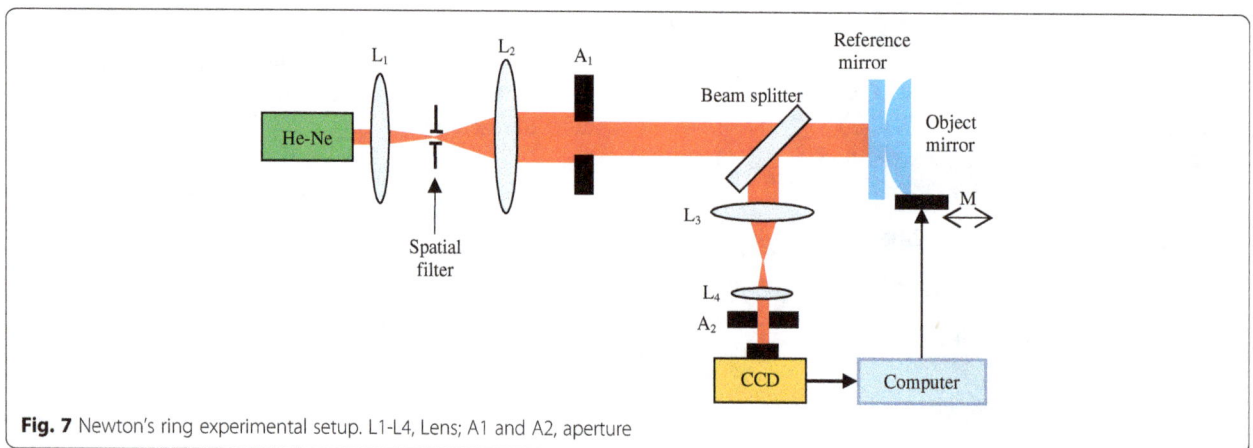

Fig. 7 Newton's ring experimental setup. L1-L4, Lens; A1 and A2, aperture

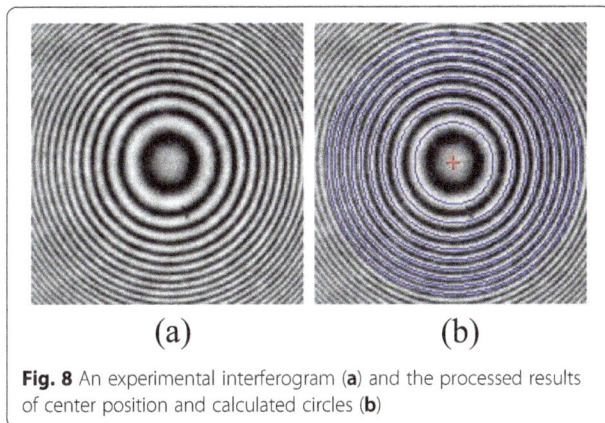

Fig. 8 An experimental interferogram (**a**) and the processed results of center position and calculated circles (**b**)

Basler CCD camera with pixel size of 4.4 μm × 4.4 μm is used to capture the interferograms.

The acquired experimental interferogram with the image size of 454 × 455 pixels is shown in Fig. 8(a). The center position and the circular fringes from 1th to 9th order as efficient closed-ring are calculated by the proposed method and are plotted in Fig. 8(b). The calculated radius values of each order fringes are shown in Table 1.

According to the measured results, the curvature radius of the lens may be calculated,

$$R = \frac{r_{m+j}^2 - r_{n+j}^2}{(m-n)\lambda}, \tag{5}$$

where, λ is the wavelength of the incident light, r_m and r_n are the radius of the mth and nth order bright fringes, respectively. If the curvature radius of the lens is known, the wavelength of the incident light can be calculated based on this method and optical setup, and the equation is expressed as

$$\lambda = \frac{r_{m+j}^2 - r_{n+j}^2}{(m-n)R}. \tag{6}$$

Conclusions

The paper proposes a method to analyze the Newton's ring interference fringes. With this method the radius of circular fringes can be determined, and the radius parameter of each order fringes can be obtained. Results of simulation and experiment show that this method hold performance of anti-noise, sub-pixel accuracy and high

reliability, and it is convenient to use in in-situ measurement of curvature radius of plate-convex lens. In the practical measurement, we generally use a monochromatic laser output as the incident light. As long as the two order fringes to be measured can be captured by the CCD pixels in the case of fulfilling the sampling theorem, the method is efficient and its measuring accuracy can be ensured. If the incoming light with certain spectral width incidents the Newton's ring configuration, the fringes pattern will show a fall-off of contrast along with increasing the spectral width of the radiation, especially for the more order fringes. In the case, the analysis of this fringe pattern is difficult to many popular methods, but even so the proposed method can still extract its center position and measure the radius values while those order fringes are clear to distinguish and fulfill the sampling theorem. We still believe that the technique provide a new way of image processing in precision measurement and fine interferometry, especially in the analysis of circular fringes pattern.

Acknowledgment
This project is supported by the National Natural Science Foundation of China (61108038); Natural Science Foundation of Inner Mongolia of China (2016MS0620, 2015MS0616); Science Foundation of Inner Mongolia University of Technology of China (X201210).

Authors' contributions
All authors read and approved the final manuscript.

Competing interests
The authors declare that they have no competing interests.

Author details
[1]College of Mechanical Engineering, Inner Mongolia University of Technology, Huhhot 010051, China. [2]College of Information Engineering, Inner Mongolia University of Technology, Hohhot 010080, China.

References
1. Rajshekhar, G., Rastogi, P.: Fringe analysis: premise and perspectives. Optics & Lasers in Engineering 50(8), iii–x (2012)
2. Winston, A.W., Baer, C.A., Allen, L.R.: A simple film thickness gauge utilizing Newton's rings. Vacuum 9(5), 302 (1959)
3. Wahl, KJ., Chromik, R.R., Lee, G.Y.: Quantitative in situ measurement of transfer film thickness by a Newton's rings method. Wear 264(7), 731–736 (2008)
4. Gentle, C.R., Halsall, M.: Measurement of Poisson's ratio using Newton's rings. Opt. Lasers Eng. 3(2), 111–118 (1982)
5. Abdelsalam, D.G., Shaalan, M.S., Eloker, M.M., Kim, D.: Radius of curvature measurement of spherical smooth surfaces by multiple-beam interferometry in reflection. Opt. Lasers Eng. 48(6), 643–649 (2010)
6. Yua, X.L., Yao, Y., Shi, W. J., Sun, Y.X., Chen, D.Y.: Study on an automatic processing technique of the circle interference fringe for fine interferometry. Optik 121(9), 826–830 (2010)
7. Cai, L.Z., Liu, Q., Yang, X.L.: A simple method of contrast enhancement and extremum extraction for interference fringes. Optics & Laser Technology 35(4), 295–302 (2003)
8. Dobroiu, A., Alexandrescu, A., Apostol, D., Nascov, V., Damian, V.: Centering and profiling algorithm for processing Newton's rings fringe patterns. Opt. Eng. 39(12), 3201–3206 (2000)
9. Nascov, V., Apostol, D., Garoi, F.: Statistical processing of Newton's rings using discrete Fourier analysis. Opt. Eng. 46(2), 28201 (2007)

Table 1 Calculated radius values of the experimental interferogram

Fringes number	1th	2th	3th	4th	5th	6th	7th	8th	9th
Radius/pixel	65.5	96.5	119.8	139.5	156.5	171.8	185.8	198.8	211.5

10. Kaufmann, G.H., Galizzi, G.E.: Evaluation of a method to determine interferometric phase derivatives. Opt. Lasers Eng. **27**(5), 451–465 (1997)

11. Nascov, V., Dobroiu, A., Apostol, D., Damian, V.: Statistical errors on Newton fringe pattern digital processing. Proc. SPIE **5581**, 788–796 (2004)

12. Sokkara, T.Z.N., Dessoukya, H.M.E., Shams-Eldinb, M.A., El-Morsy, M.A.: Automatic fringe analysis of two-beam interference patterns for measurement of refractive index and birefringence profiles of fibres. Opt. Lasers Eng. **45**(3), 431–441 (2007)

13. Okada, K., Yokoyama, E., Miike, H.: Interference fringe pattern analysis using inverse cosine function. Electronics & Communications in Japan **90**(1), 61–73 (2007)

14. Dias, P.A., Dunkel, T., Fajado, D.A.S., Gallegos, E.L., Denecke, M., Wiedemann, P., Schneider, F.K., Suhr, H.: Image processing for identification and quantification of filamentous bacteria in in situ acquired images. BioMedical Engineering OnLine **15**, 64 (2016)

15. Xia, M.L., Wang, L., Lan, Z.X., Chen, H.Z.: High-throughput screening of high Monascus pigment-producing strain based on digital image processing. J. Ind. Microbiol. Biotechnol. **43**(4), 451–461 (2016)

16. Li, Y.H., Chen, X.J., Liu, W.J., Yu, Z.H.: Center positioning of circular interference fringe patterns for fine measurement. Optik **125**(12), 2796–2799 (2014)

17. Hermann, E., Bleicken, S., Subburaj, Y., García-Sáez, A.J.: Automated analysis of giant unilamellar vesicles using circular Hough transformation. Oxford Journals **30**(12), 1747–1754 (2014)

18. Turker, M., Koc-San, D.: Building extraction from high-resolution optical spaceborne images using the integration of support vector machine (SVM) classification, Hough transformation and perceptual grouping. Int. J. Appl. Earth Obs. Geoinf. **34**, 58–69 (2015)

19. Lalitha, N.V., Srinivasa Rao, C.H., Jaya Sree, P.V.Y.: An efficient audio watermarking based on SVD and Cartesian-Polar transformation with synchronization. Lecture Notes in Electrical Engineering **372**, 365–375 (2015)

20. Zhou, S.B., Shen, A.Q., Li, G.F.: Concrete image segmentation based on multiscale mathematic morphology operators and Otsu method. Advances in Materials Science & Engineering **2015**, 1–11 (2015)

2

Focusing of THz waves with a microsize parabolic reflector made of graphene in the free space

Taner Oguzer[1], Ayhan Altintas[2] and Alexander I. Nosich[3*]

Abstract

Background: The scattering of H- and E-polarized plane waves by a two-dimensional (2-D) parabolic reflector made of graphene and placed in the free space is studied numerically.

Methods: To obtain accurate results we use the Method of Analytical Regularization.

Results: The total scattering cross-section and the absorption cross-section are computed, together with the field magnitude in the geometrical focus of reflector. The surface plasmon resonances are observed in the H-case. The focusing ability of the reflector is studied in dependence of graphene's chemical potential, frequency, and reflector's depth.

Conclusions: It is found that there exists an optimal range of frequencies where the focusing ability reaches maximum values. The reason is the quick degradation of graphene's surface conductivity with frequency.

Keywords: Graphene reflector, Focusing ability, Integral equation, Analytical regularization

Background

Graphene, which is a monolayer (1 nm) or a very thin (2-3 nm) stack of a few layers of graphite ([1–3], see Fig. 1-c in [3]), is a non-conventional material famous for being electrically conductive, mechanically strong and optically transparent. Due to the inductive nature of the associated complex-valued surface impedance, it can support the Surface Plasmon (SP) wave [1]. This wave can be strongly reflected back from the edges of patterned graphene so that natural SP modes (standing waves) can occur, in the Fabry-Perot type manner. This phenomenon has been observed at the frequencies varying from the infrared for the nano-size flat graphene samples [3] to the THz range for the micro-size ones [2]. It is already exploited in the nanosensor devices [3, 4].

Important feature of graphene is that its conductivity can be controlled by applying an external electrostatic biasing field which modifies graphene's chemical potential. Usually this requires a dielectric substrate although suspended graphene is also realizable [5]. Therefore in the modeling, one

can consider a curved graphene strip located in the free space, and assume that the d-c bias is still present. Note also that the edge effects become important only if a graphene strip width is smaller than 100 nm. For wider strips one can disregard the edge effects and use the electron conductivity model developed for infinite graphene layer. In the THz range this requirement is well satisfied for micro-size strips.

One of the interesting questions in this area is how well the THz wave can be focused with a curved reflector made of graphene. The goal of this paper is to answer this question for a 2-D parabolic reflector as depicted in Fig. 1. We perform such a study using the electromagnetic boundary value problem (BVP), which includes the resistive-sheet boundary condition originally derived for thinner-than-skindepth imperfect (partially transparent) metal layers [6, 7].

We consider both the H- and the E-polarization cases where electric field is in the plane of reflector's cross-section and in parallel to reflector, respectively. It should be noted that, similarly to the full-wave modeling of perfectly electrically conducting (PEC) reflectors, finite-difference time-domain method can be considered as one of possible computational instruments. However it leads to huge number of unknowns due to the discretization of large physical domain and also has a disadvantage in the

---

* Correspondence: anosich@yahoo.com
[3]Laboratory of Micro and Nano Optics, Institute of Radio-Physics and Electronics NASU, Kharkiv 61085, Ukraine
Full list of author information is available at the end of the article

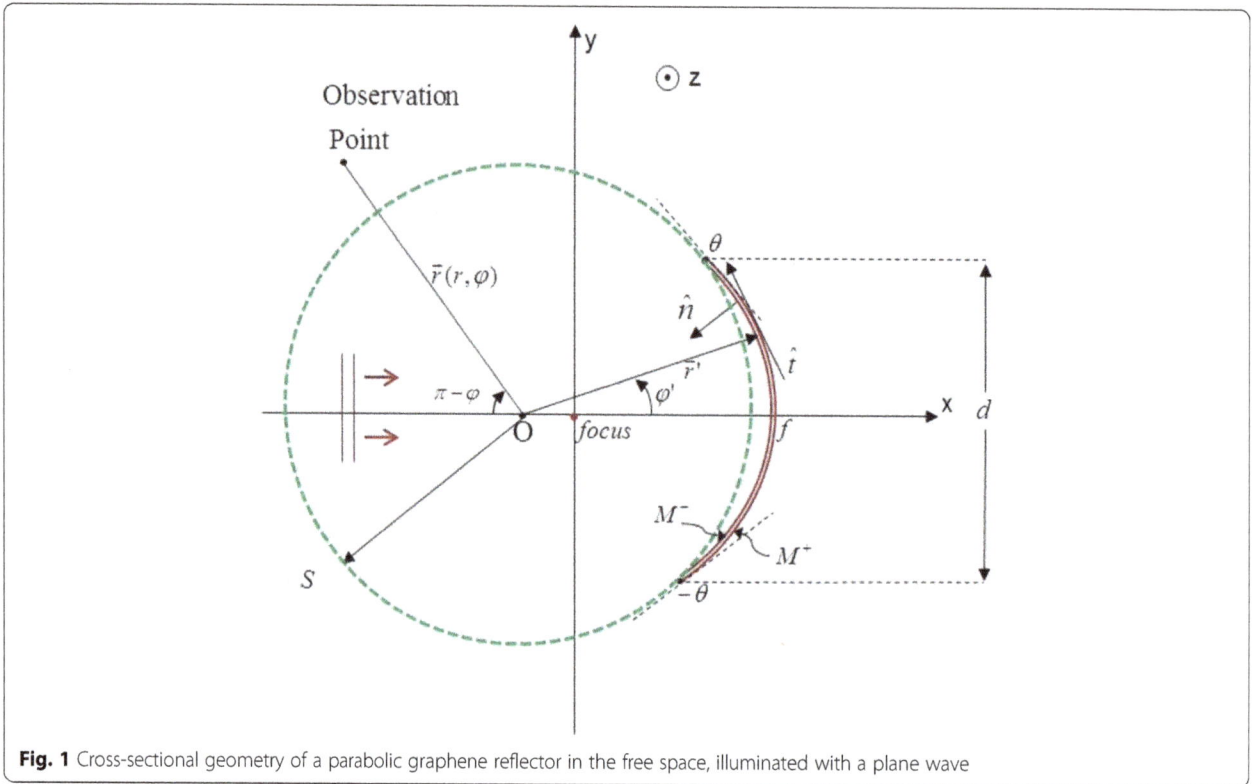

Fig. 1 Cross-sectional geometry of a parabolic graphene reflector in the free space, illuminated with a plane wave

inability to satisfy the far field radiation condition. The method of moments (MoM) procedure can be also applied to treat singular integral equations (SIE) derived for arbitrary reflectors. However conventional MoM with local basis and testing functions has overall accuracy at the level of 2-3 digits even when treating the medium size reflectors (10 wavelengths). If better accuracy is needed or larger reflectors are interested in, one hits non-realistic computation times or complete failure of the code because of the quick growth of the matrix condition number. Another alternative is the high frequency techniques like geometrical and physical optics, which work much faster however do not produce accurate full-wave results.

All mentioned above is especially important in the case of H-polarization where the associated SIE has hyper-type singularity. Attractive way out of that pitfall is offered by the method of analytical regularization (MAR) [8]. With MAR, the kernel of the associated SIE for the current on the reflector is presented as a sum of two parts, a more singular part (usually static) and a remainder. Then the more singular part is analytically inverted by using some special technique like the Riemann-Hilbert Problem (RHP) method [7, 9, 10]. The remainder leads to the Fredholm second-kind matrix equation that provides a convergent numerical solution. The same can be achieved by choosing the global expansion functions that are orthogonal eigenfunctions of the hyper-singular part of SIE operator and using them in a MoM-like Galerkin projection algorithm [11]. In either case

the SIE-MAR technique enables accurate and economic full-wave analysis of electromagnetic scattering problems for both PEC and imperfect reflectors. For instance, in [9], the H-wave scattering and the focusing were studied for the resistive 2-D reflectors having elliptical contours.

In the E-polarization case, the associated SIE has a logarithmically singular kernel [12] and hence is already a Fredholm second kind equation. This guarantees convergence of discretization schemes. Still projecting it on the set of entire-domain expansion functions brings additional advantages and makes the resulting numerical algorithm more economic. The E-polarized beam scattering and collimation by parabolic resistive reflectors was analyzed in this manner in [12].

Note that the scattering and absorption of THz waves by a single flat graphene strip and finite graphene-strip gratings was reduced to SIE and its Nystrom type solution was built in [13, 14]. Infinite graphene-strip grating in the free space was also studied by the MAR-RHP technique in [10]. In these works, the field characteristics were investigated as a function of graphene and grating parameters showing the presence of SP resonances. In more recent works [4, 15], the bulk refractive index sensitivities of the THz range SP resonances were studied for a micro-size graphene strip and a dielectric tube covered with graphene, respectively.

Following the mentioned and other works, we simulate graphene with the aid of the resistive-sheet boundary condition together with the Kubo formula for the graphene electron conductivity. We derive a corresponding SIE from the electromagnetic BVP and solve it using the RHP-based MAR solution. This type of algorithm provides us accurate data for the quantifications of the scattering, absorption and focusing characteristics. As a result, we obtain the frequency scans of the total scattering and absorption cross sections (TSCS and ACS) of a 2-D parabolic graphene reflector and in the H-case identify the SP resonances. Then we perform a study of the focusing ability of such a curved strip as a 2-D reflector for various graphene parameters.

Preliminary results of such analysis were reported at a conference [16]; here we present new and more complete numerical study and obtain better insight into the studied effects.

Methods

The problem geometry of a 2-D parabolic graphene reflector frontally illuminated by a plane wave is presented in Fig. 1. Reflector's contour M is defined as a finite parabolic profile. An auxiliary closed contour denoted as C is the contour M completed with the circular arc S, which must have the same curvature as the reflector at the latter's edge points. Such a smooth contour C is necessary for obtaining the regularized (i.e. Fredholm second kind) matrix equation - see [9, 12].

The rigorous formulation of the considered BVP involves the Helmholtz equation, the Sommerfeld radiation condition far from the reflector, the resistive boundary condition on M, and an edge condition such that the field power is limited in any finite domain around the reflector edge. Collectively, these conditions guarantee the uniqueness of the problem solution.

The resistive boundary condition on a zero-thickness sheet is a well-established model of a thin penetrable sheet, e.g. a metal thinner than skin depth or a very thin dielectric layer. In view of "atomic" thickness of graphene, the same boundary condition can also be used for a flat or curved graphene surface, avoiding introduction the thickness of graphene of 2-3 nm that generates meshing troubles in the use of purely numerical codes like COMSOL. It can be written as the following two equations valid at $\vec{r} \in M$:

$$\left(\vec{E}_{tan}^{+} + \vec{E}_{tan}^{-}\right)/2 = Z\,\vec{n} \times \left(\vec{H}_{tan}^{+} - \vec{H}_{tan}^{-}\right), \quad \vec{E}_{tan}^{+} = \vec{E}_{tan}^{-},$$

(1)

where the subscript "tan" indicates the tangential field, the superscripts "- " and "+" relate to the front and back faces of reflector, respectively, and $\vec{n}$ is understood as the unit vector normal to the concave side of reflector. The jump in the tangential magnetic field, $\vec{J} = \vec{H}_{tan}^{+} - \vec{H}_{tan}^{-}$, is unknown function of the electric surface-current density, and the co-efficient Z is graphene's surface impedance [1–5].

Note also that the surface impedance is related to the graphene surface electron conductivity σ as $Z = 1/\sigma$, and the conductivity can be found as the Kubo sum of intra-band and interband contributions [1–6]. As condition (1) was derived for infinite planar layer, in the modeling of the wave-scattering by finite surfaces it must be combined with the edge condition to provide the uniqueness of the BVP solution.

In the H-wave case, on using the boundary condition (1) we obtain a hyper-singular SIE for the surface current J_t on the reflector. On integrating by parts, it can be cast to the following form:

$$ZJ_t - \frac{iZ_0}{k} \frac{\partial}{\partial l} \int_M \left[\frac{\partial}{\partial l'} J_t\left(\vec{r}'\right)\right] G\left(\vec{r}, \vec{r}'\right) dl'$$

$$+ ikZ_0 \int_M J_t\left(\vec{r}'\right) \cos\left[\xi\left(\vec{r}\right) - \xi\left(\vec{r}'\right)\right] G\left(\vec{r}, \vec{r}'\right) dl'$$

$$= \frac{iZ_0}{k} \frac{\partial H_z^{in}}{\partial n},$$

(2)

where the 2-D Green's function G is a Hankel function of zero order and first kind satisfying the radiation condition, i.e. $G\left(\vec{r}, \vec{r}'\right) = (i/4)H_0^{(1)}(k_o R)$, $R = \left|\vec{r} - \vec{r}'\right|$, and the angle $\xi(\phi)$ is between the normal on M and the x-direction.

Now, we assume that the curve M can be characterized with the aid of the parametric equations $x = x(\phi), y = y(\phi)$, where $0 \leq |\phi| \leq \theta$, in terms of the polar angle, ϕ. Besides, we denote the differential length in the tangential direction at any point on M as $\partial l = a\beta(\phi)\partial\phi$. We introduce also a function $\beta(\phi) = r(\phi)/[a \cos \gamma(\phi)]$, where $\gamma(\phi)$ is the angle between the normal on M and the radial direction. Then we extend the surface-current density J_t with zero value to arc S and cast IE (2) to a dual equation on the arcs S and M [9].

To continue with the MAR, we add and subtract, from the integral kernels in (2), similar functions at a full circular contour of the same radius as S. The latter operators can be inverted analytically while the remaining ones have smooth kernels,

$$A(\phi, \phi') = H_0^{(1)}(kR) - H_0^{(1)}[2ka \sin(|\phi - \phi'|/2)], \quad (3)$$

$$B(\phi, \phi') = \cos[\xi(\phi) - \xi(\phi')]\beta(\phi)\beta(\phi')H_0^{(1)}\left[k\left|\vec{r}(\phi) - \vec{r}'(\phi')\right|\right]$$

$$- \beta^2(\phi)H_0^{(1)}[2ka \sin(|\phi - \phi'|/2)]$$

(4)

For the inversion of the singular operators, all functions including the incident field should be expanded in terms of the Fourier series in ϕ. Note that the functions

A and B are continuous and have also continuous first derivatives, while their second derivatives with respect to ϕ and ϕ' have only logarithmic singularities and hence belong to L_2. Therefore on the curve C their Fourier co-efficients in ϕ decay fast enough with larger indices and hence can be efficiently computed by the Fast Fourier Transform algorithm. Then the discretized version of the SIE and the zero current condition on the aperture S give us a dual series equation. Its semi-inversion, based on the MAR approach using the RHP technique [7, 10], finally produces an algebraic equation set [9]. This infin-ite matrix equation is of the Fredholm second kind hence the Fredholm theorems guarantee the existence of the unique solution and also the convergence of the approximate numerical solutions when truncating the set with progressively larger orders.

In the E-wave case, on using the boundary condition (1) we obtain the following log-singular IE for the surface current J_z on the reflector:

$$ZJ_z - ikZ_0 \int_M J_z\left(\overrightarrow{r}'\right) G\left(\overrightarrow{r}, \overrightarrow{r}'\right) dl' = E_z^{in}, \qquad (5)$$

As mentioned, convergence of usual discretizations of this equation is guaranteed by its Fredholm second-kind nature. Therefore we apply the projection to the set of entire-domain angular exponents [12]. In either polarization case we adapt the matrix truncation number to provide the 4-digit or better accuracy of computations.

The scattered electromagnetic field in the far zone of reflector is a cylindrical wave with functions H_z^{sc} or E_z^{sc} (depending on the polarization) reduced to $(2/i\pi kr)^{1/2} e^{ikr}$ $\phi(\phi)$, where $\phi(\phi)$ is the angular scattering pattern. Then TSCS can be obtained by using the following expression:

$$\sigma_{tsc} = \frac{2}{\pi k} \int_0^{2\pi} |\phi(\phi)|^2 d\phi, \qquad (6)$$

and ACS of a lossy graphene reflector can be found from the optical theorem,

$$\sigma_{abs} = -\frac{4}{k} \mathrm{Re}\, \phi(0) - \sigma_{tsc} \qquad (7)$$

Results and discussion

The numerical accuracy and convergence of the explained above in-house algorithms have already been verified in [9, 12]. In the current work, we apply it to the analysis of both the plane-wave scattering and absorp-tion and the effect of focusing by the graphene reflector.

Therefore, besides of TSCS and ACS defined above, we also calculate another parameter, which serves as a simple figure of merit of the focusing ability (FA), in the plane-wave focusing by a parabolic graphene reflector. In view of the unite-amplitude plane wave incidence, FA

can be reasonably defined as the total field magnitude at the geometrical-focus point of parabola.

In Fig. 2, the values of ACS and TSCS are plotted as a function of frequency for two graphene reflectors with the fixed size of $d = 200$ μm (small-size reflector) and $d = 1000$ μm (medium-size reflector), respectively, the both having the same fixed focal ratio f/d.

The oscillations observed on the plots are due to the SP resonances, especially well visible in ACS behavior. Note also that the absorption is by an order of magni-tude smaller than the scattering, and the both drop with frequency because of the growth of surface impedance.

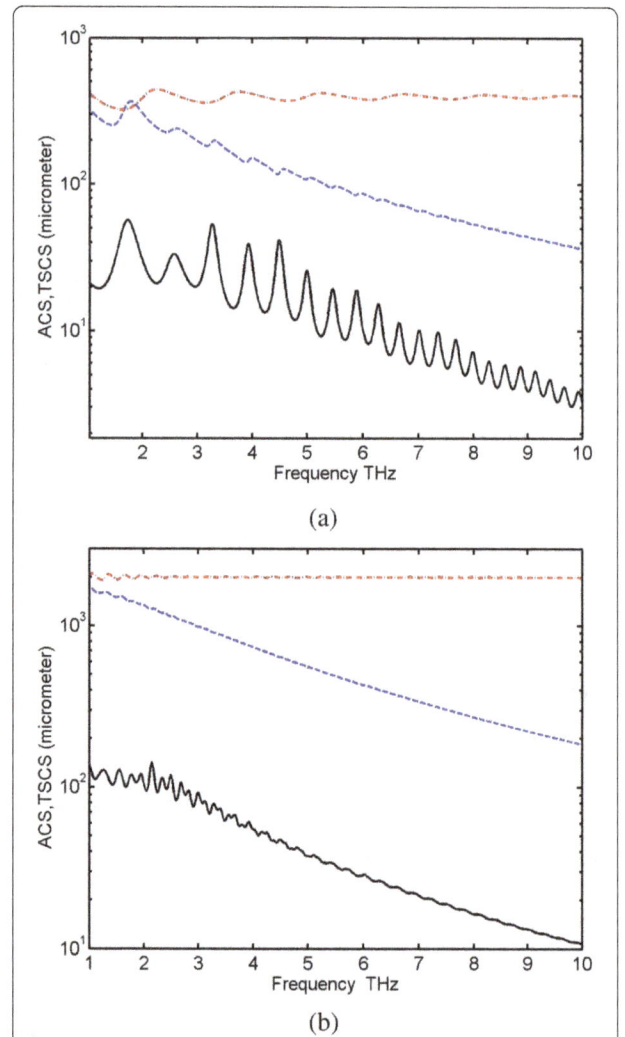

(a)

(b)

Fig. 2 H-case: Wave scattering and absorption by parabolic graphene reflectors versus the frequency in the THz range, for small-size reflector, $d = 200$ μm (**a**) and medium-size reflector, $d = 1000$ μm (**b**) Solid lines (black) and dashed lines (blue): ACS and TSCS for $\mu_c = 1$ eV. Dash-dotted lines (red): TSCS for the PEC reflector. The other parameters are the relative focal distance $f/d = 0.3$, the temperature $T = 300$ K, and the electron relaxation time $\tau = 1$ ps

The frequency scans of FA are plotted in Fig. 3 for the same two reflectors as in Fig. 2. It can be seen that the growth in μ_c increases FA at all frequencies. This happens because higher values of chemical potential μ_c lead to the lower values of the surface impedance of graphene that makes it less transparent. Then the curves get closer to the PEC case however still depart from it if the frequency becomes higher.

Periodic ripples on the plots of FA are explained by the free-space interference of the waves scattered by the edges of reflector: this explanation is becomes evident if one takes into account that their period is the same for the PEC and the graphene cases and is determined by reflector's size.

To obtain a fuller vision of the focusing ability of graphene reflector, we present a color map of this

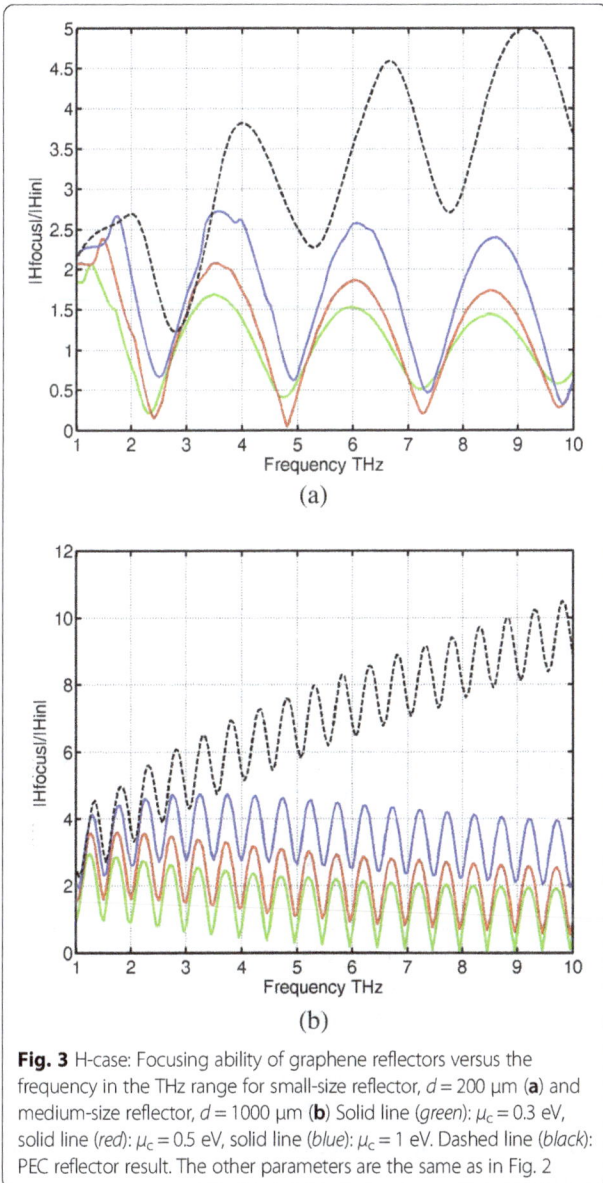

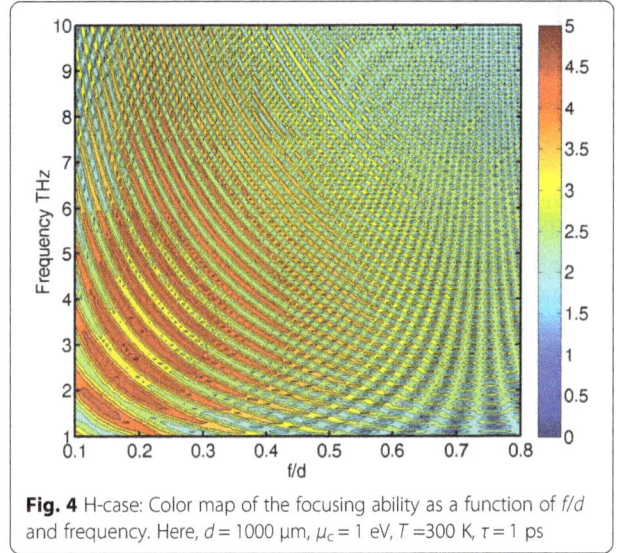

Fig. 4 H-case: Color map of the focusing ability as a function of f/d and frequency. Here, $d = 1000$ µm, $\mu_c = 1$ eV, $T = 300$ K, $\tau = 1$ ps

quantity as a function of two parameters: the focal ratio f/d and the frequency in the THz range – see Fig. 4. One can see that the optimal value of f/d, which provides maximum FA, is slightly below the value of 0.25 known to be optimal for PEC reflectors. New feature, as visible both from Figs. 3 and 4, is existence of an optimal frequency range where the focusing ability reaches maximum. This is apparently explained by the fact that, if the frequency grows, then the initial positive effect of increasing the electrical size of reflector becomes gradually overweighed by the negative effect of increasing the absolute value of graphene's impedance. Location and width of the optimal frequency band depends on the chemical potential, i.e. on graphene's doping.

Finally, in Fig. 5 we present the total near-field pattern for the graphene reflector with the aperture of $d = 450$ µm

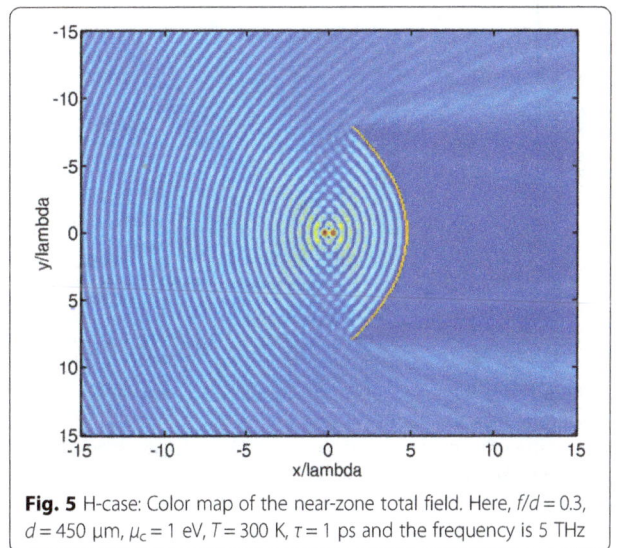

(a)

(b)

Fig. 3 H-case: Focusing ability of graphene reflectors versus the frequency in the THz range for small-size reflector, $d = 200$ µm (**a**) and medium-size reflector, $d = 1000$ µm (**b**) Solid line (green): $\mu_c = 0.3$ eV, solid line (red): $\mu_c = 0.5$ eV, solid line (blue): $\mu_c = 1$ eV. Dashed line (black): PEC reflector result. The other parameters are the same as in Fig. 2

Fig. 5 H-case: Color map of the near-zone total field. Here, $f/d = 0.3$, $d = 450$ µm, $\mu_c = 1$ eV, $T = 300$ K, $\tau = 1$ ps and the frequency is 5 THz (this is 7.5λ). Note the splitting of the focal domain to two bright spots along the axis of symmetry – this is a side effect, at the given frequency, of the finite size of reflector. Besides, one can see clearly observable interference of the waves scattered by the edges of the parabolic reflector in front of it and the presence of shadow behind it. Still this shadow is not very dark because the graphene reflector is partially transparent.

The further Figs. 6, 7, 8, and 9 present the numerical data analogous to in Figs. 2, 3, 4, and 5 however computed for the E-polarized wave incidence. Note the absence of the surface-plasmon resonances on the plots of ACS and FA as a function of frequency in Fig. 6a (compare to Fig. 2a) and Fig. 7a (compare to Fig. 3a), i.e. for a small-size reflector.

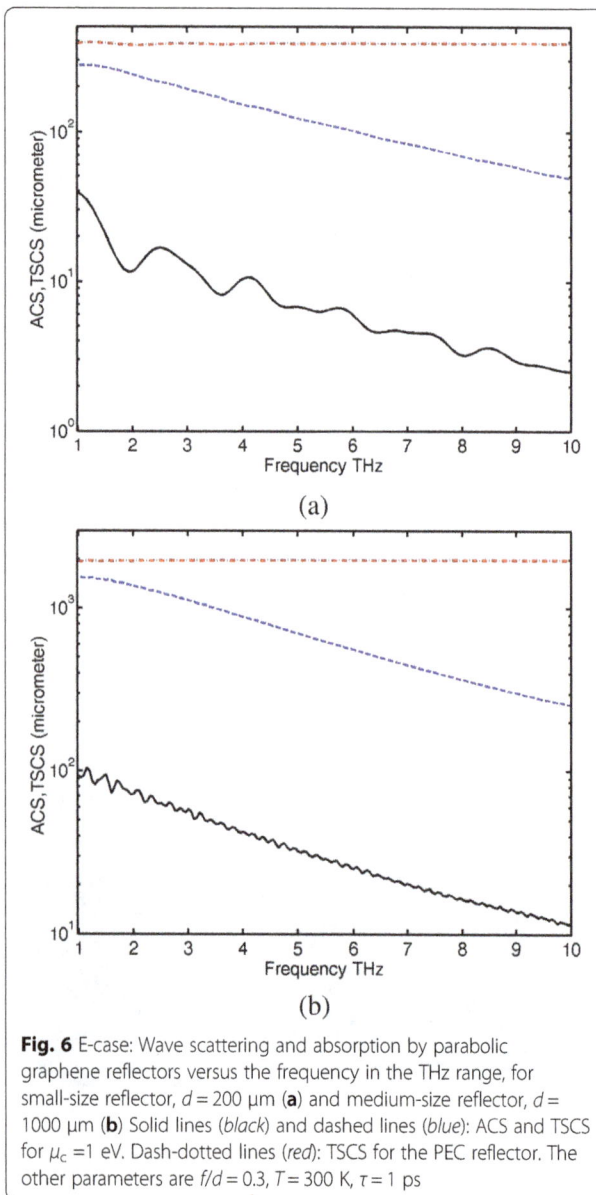

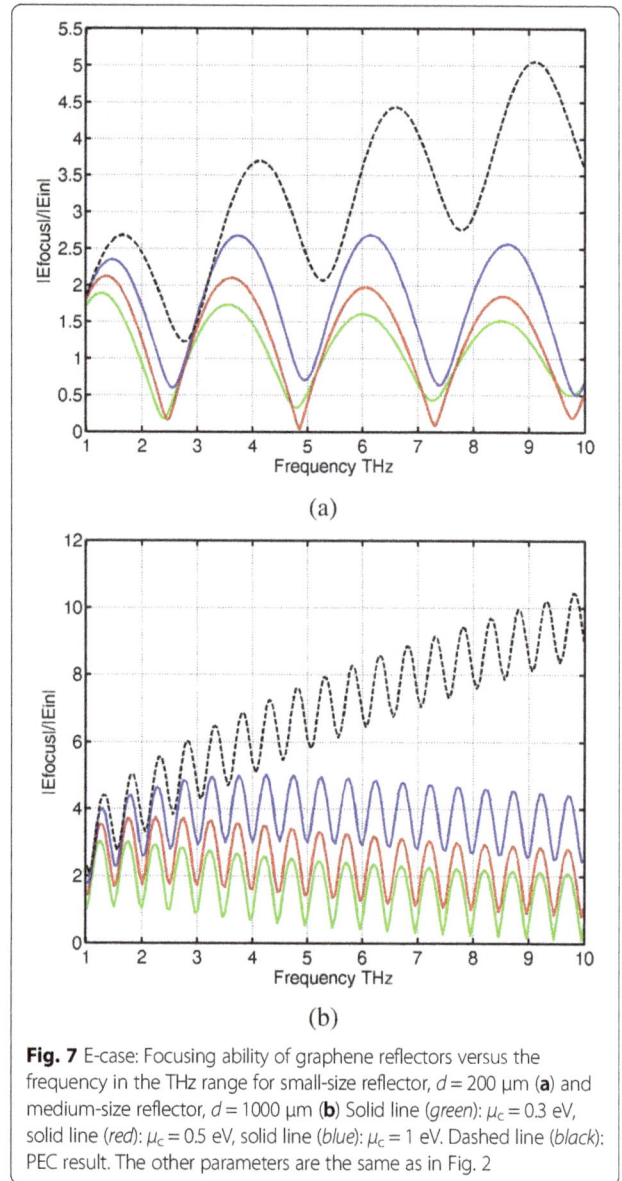

(a)

(b)

Fig. 6 E-case: Wave scattering and absorption by parabolic graphene reflectors versus the frequency in the THz range, for small-size reflector, $d = 200$ μm (**a**) and medium-size reflector, $d = 1000$ μm (**b**) Solid lines (*black*) and dashed lines (*blue*): ACS and TSCS for $\mu_c = 1$ eV. Dash-dotted lines (*red*): TSCS for the PEC reflector. The other parameters are $f/d = 0.3$, $T = 300$ K, $\tau = 1$ ps

(a)

(b)

Fig. 7 E-case: Focusing ability of graphene reflectors versus the frequency in the THz range for small-size reflector, $d = 200$ μm (**a**) and medium-size reflector, $d = 1000$ μm (**b**) Solid line (*green*): $\mu_c = 0.3$ eV, solid line (*red*): $\mu_c = 0.5$ eV, solid line (*blue*): $\mu_c = 1$ eV. Dashed line (*black*): PEC result. The other parameters are the same as in Fig. 2

One can notice obvious similarities between plots and patterns for the H-case and the E-case if a graphene reflector is at least medium-size and the frequency is above 3 THz. This is apparently because the focusing of waves by a finite parabolic reflector, even a semitransparent one, is essentially a high-frequency or quasi-optical effect. The main parameter in this case is just the electric size of reflector in terms of the free-space wavelength. The effect of the surface plasmon resonances is almost negligible at high frequencies, as well as dependence on the polarization in general. Note that in the E-polarization case the near-field portrait (Fig. 9) shows only one bright spot close to he geometrical focus of parabola.

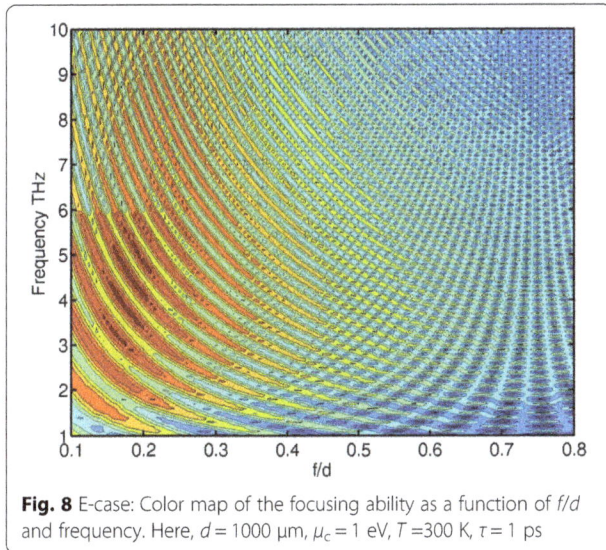

Fig. 8 E-case: Color map of the focusing ability as a function of f/d and frequency. Here, $d = 1000$ μm, $\mu_c = 1$ eV, $T = 300$ K, $\tau = 1$ ps

surface-plasmon resonances are present at lower THz frequencies in the H-wave case however their effect on the performance of micro-size graphene reflectors is small.

Abbreviations
2-D: Two-dimensional; ACS: Absorption cross-section; BVP: Boundary-value problem; FA: Focusing ability; MAR: Method of analytical regularization; MoM: Method of moments; PEC: Perfect electric conductor; RHP: Riemann-Hilbet problem; SIE: Singular integral equation; SP: Surface plasmon; THz: Terahertz; TSCS: Total scattering cross-section

Funding
This work was not supported by any specific funding.

Authors' contributions
TO carried out the computations and drafted the manuscript. AA participated in the coordination of the study and in the interpretation of computed results. AN conceived of the study and finalized the manuscript. All authors read and approved the final manuscript.

Conclusion

To summarize, a micro-size 2-D graphene reflector with parabolic profile, symmetrically illuminated by the H-polarized and E-polarized plane waves has been analyzed numerically using the MAR approach. The results show that the focusing ability of such a reflector is on par with a PEC reflector in the range of the frequency and the graphene parameters where the surface impedance of the latter is small. As follows from the Kubo formalism, this entails a necessity of working with higher values of chemical potential and electron relaxation time. This also means that for every fixed size of reflector there exists a band of optimal THz frequencies and the focusing ability is severely degraded at higher frequencies because of degradation of graphene's surface conductivity. The

Competing interests
The authors declare that they have no competing interests.

Author details
[1]Department Electrical and Electronics Engineering, Dokuz Eylul University, Buca, 35160 Izmir, Turkey. [2]Department Electrical and Electronics Engineering, Bilkent University, 06800 Ankara, Turkey. [3]Laboratory of Micro and Nano Optics, Institute of Radio-Physics and Electronics NASU, Kharkiv 61085, Ukraine.

References
1. Depine, R.A.: Graphene Optics: Electromagnetic Solution of Canonical Problems. IOP Concise Phys, Morgan and Claypool Publ, Bristol (2016)
2. Low, T., Avouris, P.: Graphene plasmonics for terahertz to mid-infrared applications. ACS Nano **8**, 1086–1101 (2014)
3. Rodrigo, D., Limaj, O., Janner, D., Etezadi, D., GarcíadeAbajo, F.J., Pruneri, V., Altug, H.: Mid-infrared plasmonic biosensing with graphene. Science **349**, 165–168 (2015)
4. Shapoval, O.V., Nosich, A.I.: Bulk refractive-index sensitivities of the THz-range plasmon resonances on a micro-size graphene strip. J Phys D Appl Phys **49**, 055105/8 (2016)
5. Du, X., Skachko, I., Barker, A., Andrei, E.Y.: Approaching ballistic transport in suspended graphene. Nat Immunol **3**, 491–495 (2008)
6. Orta, R., Savi, P., Tascone, R.: The effect of finite conductivity on frequency selective surface behavior. Electromagnetics **10**, 213–227 (1990)
7. Nosich, A.I., Okuno, Y., Shiraishi, T.: Scattering and absorption of E and H-polarized plane waves by a circularly curved resistive strip. Radio Sci **31**, 1733–1742 (1996)
8. Nosich, A.I.: Method of analytical regularization in computational photonics. Radio Sci **51**, 1421–1430 (2016)
9. Oğuzer, T., Altintas, A., Nosich, A.I.: Analysis of the elliptic profile cylindrical reflector with a non-uniform resistivity using the complex source and dual series approach: H-polarization case. Opt Quant Electron **45**, 797–812 (2013)
10. Zinenko, T.L.: Scattering and absorption of terahertz waves by a free-standing infinite grating of graphene strips: analytical regularization analysis. J Opt **17**, 055604/8 (2015)
11. Lucido, M.: A new high-efficient spectral-domain analysis of single and multiple coupled microstrip lines in planarly layered media. IEEE Trans Microw Theory Techn **60**, 2025–2034 (2012)
12. Oguzer, T., Altintas, A., Nosich, A.I.: Integral equation analysis of an arbitrary-profile and varying-resistivity cylindrical reflector illuminated by an E-polarized complex-source-point beam. J Opt Soc Am A **26**, 1525–1532 (2009)

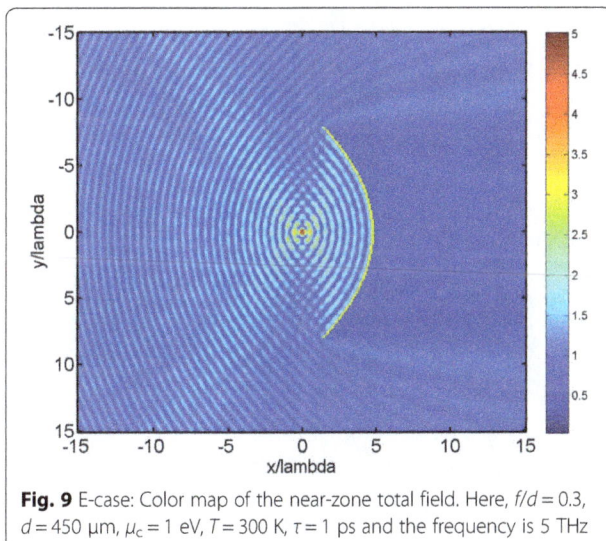

Fig. 9 E-case: Color map of the near-zone total field. Here, $f/d = 0.3$, $d = 450$ μm, $\mu_c = 1$ eV, $T = 300$ K, $\tau = 1$ ps and the frequency is 5 THz

13. Balaban, M.V., Shapoval, O.V., Nosich, A.I.: THz wave scattering by a graphene strip and a disk in the free space: integral equation analysis and surface plasmon resonances. J Opt **15**, 114007/9 (2013)

14. Shapoval, O.V., Gomez-Diaz, J.S., Perruisseau-Carrier, J., Mosig, J.R., Nosich, A.I.: Integral equation analysis of plane wave scattering by coplanar graphene-strip gratings in the THz range. IEEE Trans Terahertz Sci Technol **3**, 666–673 (2013)

15. Velichko, E.A.: Evaluation of a dielectric microtube with a graphene cover as a refractive-index sensor in the THz range,". J Opt **18**, 035008/11 (2016)

16. Oguzer, T., Altintas, A.: Focusing ability of a microsize graphene-based cylindrical reflector in the THz range illuminated by H-polarized electromagnetic plane wave. Proc Int Conf Math Methods Electromagn Theory (MMET-2016) **Lviv**, 232–235 (2016)

3

Reduced Graphene Oxide nano-composites layer on fiber optic tip sensor reflectance response for sensing of aqueous ethanol

M. A. A. Rosli[1*], P. T. Arasu[2], A. S. M. Noor[1,3], H. N. Lim[4] and N. M. Huang[5]

Abstract

In this study, the used of tapered optical fiber tip as sensors coated with reduced Graphene Oxide (rGO) is investigated. The resultant rGO nanocomposites coated on the tapered fiber sensor were characterized by X-ray Diffraction (XRD), Raman spectroscopy, and field emission scanning electron microscopy (FESEM). Optimization of the rGO layer and the tapering parameters are found and the sensing capability of the device is tested using different concentrations of ethanol in water. The nanocomposite layer improved the performance of the sensor by demonstrating high sensitivity to aqueous ethanol when interrogated in the visible region using a spectrometer in the optical wavelength range of 500–700 nm. The reflectance response of the rGO coated fiber tip reduced linearly, upon exposure to ethanol concentrations ranging between 20-80 %.

Keywords: Reduced graphene oxide, Optical fiber sensor, Reflectance

Background

The use of optical fiber as a sensor gained much interest in the last decade. Optical fiber sensors have several advantages over electrical based sensors in many chemical and biological applications [1]. The most intriguing advantages lies in the miniaturization and response time. The development of submicron-sized optical fiber sensors is the technology based on nanofabricated optical fiber tips [2, 3]. Optical fiber sensors have been demonstrated in measuring the pH of buffer solutions inside micron-size holes in polycarbonate membrane [4]. These submicron pH sensors have millisecond response times due extremely small sizes.

The basis of fiber optic based sensors lies in the geometry of the fiber itself. Fiber optic is made of a plastic or glass core surrounded by a layer of cladding material [5]. The difference in density or refractive indices between these two materials enables the light propagation in an optical fiber in accordance with the principle of total internal reflection [6]. Optical fiber are mainly used as a sensors for physical changes, multimode fiber(MMF) can also be used to sense

refractive index change but commonly involve tapered MMF and coating it with other materials to expose the core to the new surrounding area [7]. In this process the waist size of the MMF is reduced, to a point which all the core and cladding becomes a new core, so that the MMF area can be immersed in the sample and the sample can act as the new cladding to the fiber. Exclusively, the sensitivity of the sensor increase as cladding thickness decrease. This technique is particularly effective but it is more relevant to the change of the total internal reflection, TIR [6].

The optical and electronic properties of graphene attracts tremendous interest in the science of optical sensing. Graphene has high optical transparency and mobility [8]. In additiion, graphene characteristics are flexibility, robustness and environmental stability [9]. In particular, graphene oxide(GO) and reduced graphene oxide(rGO) have been used as an composite layer in energy storage denses [10], biomedical applications [11] and electronic components [12].

Al-Qazwini et al. [13] shares that the performance of an surface plasmon resonance (SPR) based optical fiber sensor using finite-difference time domain. The results show that the performance of the fiber sensor can be optimized by choosing a proper combination of metal layer thickness of 40–60 nm and residual

* Correspondence: anwarulspeaker@gmail.com
[1]Department of Computer and Communication Systems Engineering, Faculty of Engineering, Universiti Putra Malaysia, 43400UPM Serdang, Selangor, Malaysia
Full list of author information is available at the end of the article

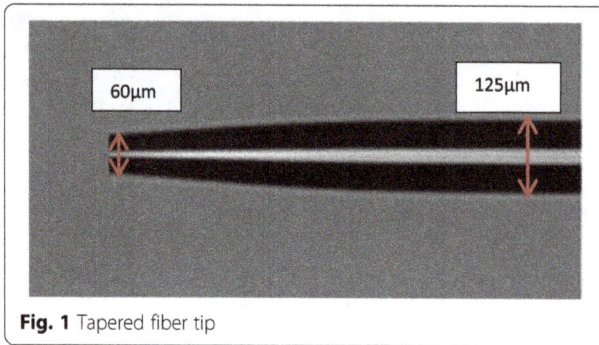
Fig. 1 Tapered fiber tip

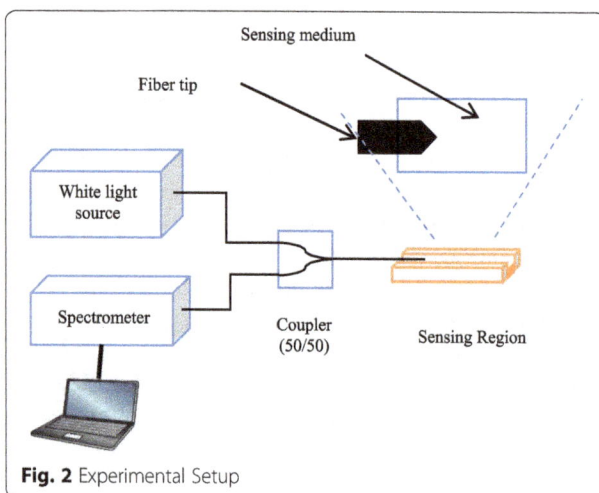
Fig. 3 Raman spectra of rGO thin film on glass substrate showing the D, G, S3 and 2D peaks

cladding thickness of 400–500 nm. In additon, they investigated an SPR-based optical fiber sensor by modeling a simple planar waveguide structure composed of four superimposed layers substituting the gold-coated polished single-mode optical fiber.

In this paper, we report a tapered optical fiber tip coated with rGO for sensing ethanol in water. Reduced graphene oxide is prepared from reduction of graphene oxide by thermal, chemical or electrical treatments [8]. Graphene film fabrication from solutions of GO have attracted considerable attention because these procedures are suitable for mass production. However, GO is an insulator, and therefore a reduction process is required to make the GO film conductive. rGO is used for the sensing nanocomposite layer due to its high ccnductivity [14] and the solubility in ethanol is higher [15] than GO compared to the work by Shabaneh et al. [16], rGO have much more simple fabrication process compared to GO. By varying the tapering profile, and improvisation of the rGO thickness, we enhanced the sensitivity of the sensor.

Methods
Multimode fiber (MMF) with a core and cladding diameter of 62.5 μm and 125 μm respectively was used in this

research [17, 18]. Vytran glass processing workstation (GPX 3000 series) was used to taper the multimode fiber to obtain a waist diameter of 60 μm with waist length of taper is 5 mm, and for both downtaper and uptaper is 3 mm. After that, the tapered fiber will be cleaved in the middle as shown in Fig. 1, obtained from the camera of the Vytran workstation. It shows that the tapered tip has a diameter of 60 μm and has a smooth taper with a clean cleaved tip.

The fiber tips are then coated with different concentrations of RGO using drop casting method covered the tapered area. To do this, the fiber tips are placed in a 70 °C oven for 20 min to completely dry the tips and prepare the tips for the annealing process. Then 1 ml RGO of 0.2, 0.5, 0.75 and 1.0 mg/mol concentrations are dropped on each tip respectively. These fiber tips are then returned to the oven for annealing at 70 °C for 1 h.

The experimental setup of the project is shown in Fig. 2. Ocean optic whitelight source (HL 2000 ocean optics) and spectrophotometer (USB 4000 ocean optics) were used as the input and detector respectively. The fiber tip is placed in a flow cell and the reflection of the light from the tip is captured through a coupler. The spectraSuite software captures and presents the data in a graph.

Results and discussion
Figure 3 shows the Raman spectrum of the rGO thin film on glass substrates. The measurements are carried

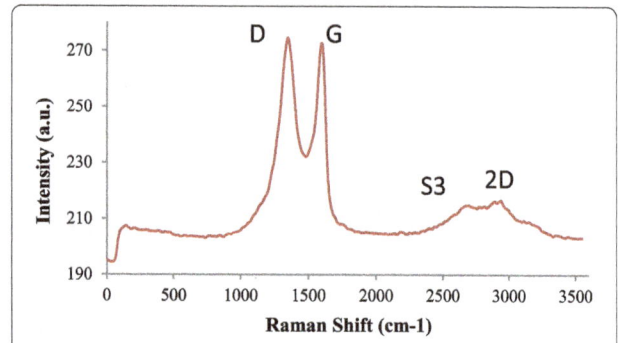
Fig. 2 Experimental Setup

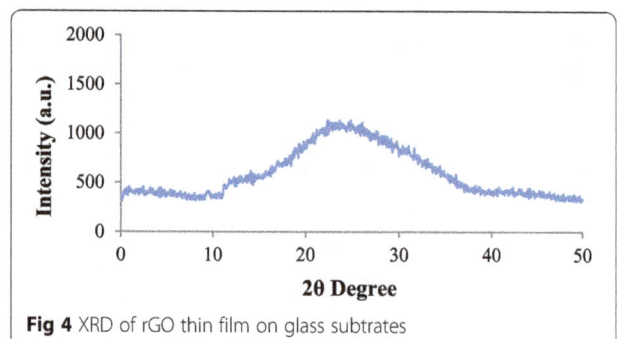
Fig 4 XRD of rGO thin film on glass subtrates

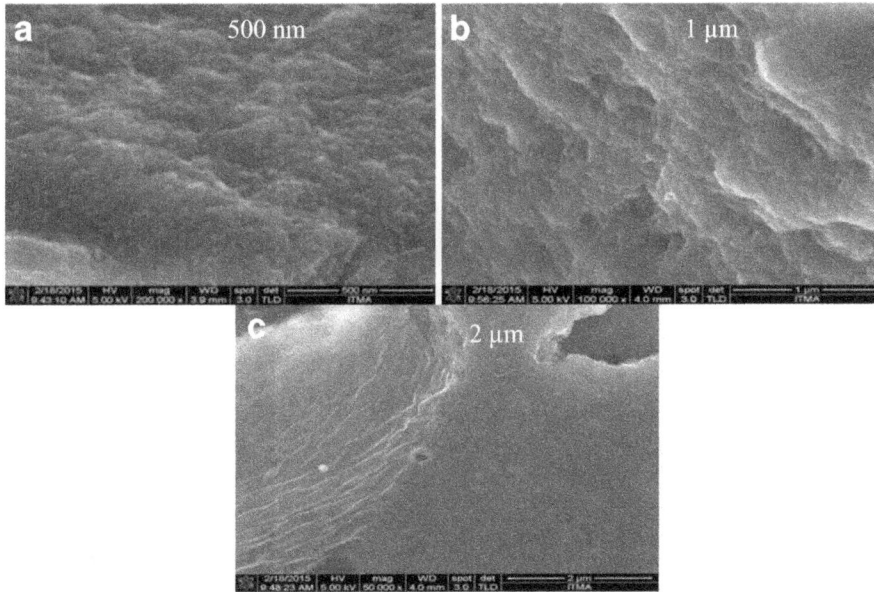

Fig. 5 FESEM image of reduced graphene oxide sheets on glass substrate at (**a**) 500 nm (**b**) 1 micron and (**c**) 2 microns scale

out by Raman spectrometer (Renishaw) using laser source with $\lambda = 514$ nm. The spectrum reveals the four characteristic D, G, 2D and S3 peaks of RGO. The D peak at about 1336 cm-1 generates from the breathing modes of six- membered rings that are activated via structural imperfections caused by the attachment of hydroxyl and epoxide groups on the carbon basal plane. The G peak at 1585 cm-1 duly corresponds to the first-order scattering of the E2g phonon mode at the Brillouin zone center. The 2D peak at 2831 cm-1 is the second order of the D peak and the S3 peak at 2585 cm-1 is due to the imperfect activated grouping of phonons.

XRD analysis of the rGO nanocomposites on the substrates is shown in Fig. 4. In this research, a broad peak is observed from 10 to 40° with the highest intensity at

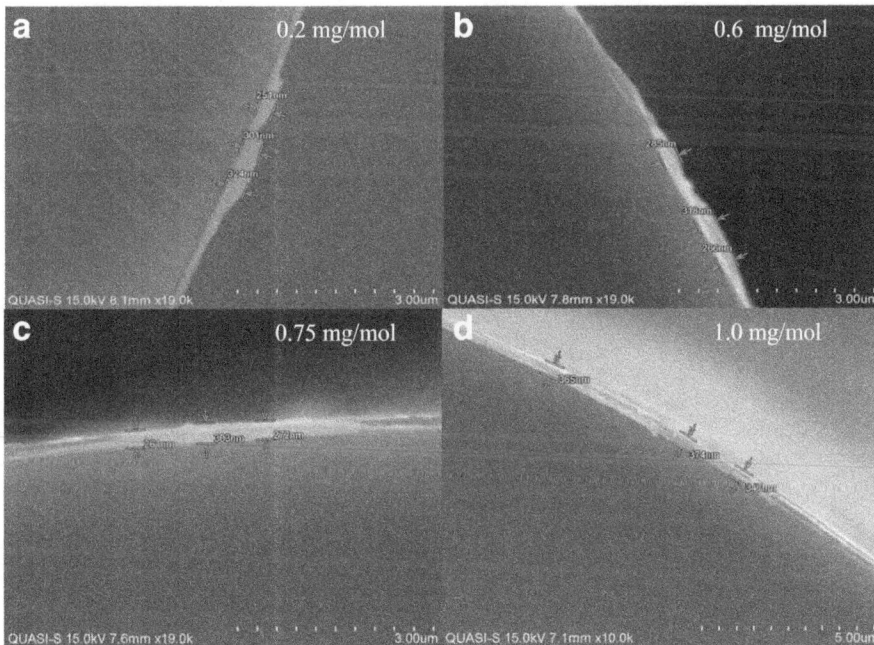

Fig. 6 Cross-sectional SEM image of reduced graphene oxide layers deposited on optical fiber tip for concentrations (**a**) 0.2 mg/mol (**b**) 0.6 mg/mol (**c**) 0.75 mg/mol (**d**) 1.0 mg/mol

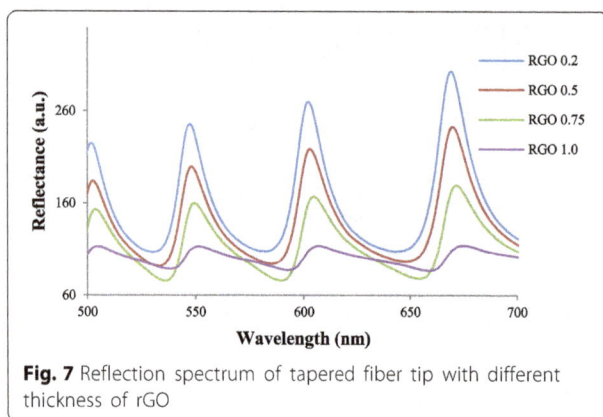

Fig. 7 Reflection spectrum of tapered fiber tip with different thickness of rGO

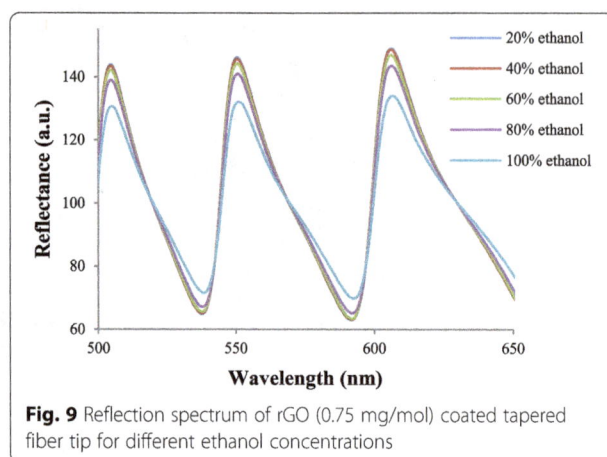

Fig. 9 Reflection spectrum of rGO (0.75 mg/mol) coated tapered fiber tip for different ethanol concentrations

23.65° which may be attributed to partial restacking of exfoliated graphene layers [19–21].

The morphology of the rGO nanocomposites is observed by field emission scanning electron microscopy (FESEM) and scanning electron microscopy(SEM). FESEM image is obtained using FEI Nova Nano SEM400 with 5.0 kV source. Figure 5a to c shows the FESEM image of the rGO nanocomposites. It reveals that the rGO nanocomposites are in the form of nano sheets.

Figure 6a to d show the SEM images of the cross section of the fiber tip showing the thickness of the rGO layers with concentrations of 0.2, 0.5, 0.75, 1.0 mg/mol The average thickness of the rGO layer with concentrations 0.2, 0.5, 0.75, 1.0 mg/mol is 292 nm, 290 nm, 279 nm and 362 nm respectively.

The folding of the rGO sheets contributes to a darker shade on the SEM image. This implies that the graphene oxide is single to a few layers thick. Furthermore, the rGO sheets adhered well to the substrate, promoting reflectance on the rGO nanocomposites. The thickness of the nano structured rGO nanocomposites is estimated to be approximately 20–30 mm, due to the overlapping of the nanosheets. This SEM is performed to verify the uniformity of the coating of rGO films on the substrates. These micro characterization results are significant to

verify the morphology of the rGO nanostructured thin films.

The reflectance spectrums of the fiber optic tips with different concentrations of rGO are investigated. As shown in Fig. 7, the reflectance decreases as the concentration of the rGO is increased. This is due to higher absorbance of the light signal as the thickness of the rGO layer increases. The interaction between the different concentrations of ethanol molecules and rGO on the tapered fiber tip transforms the optical characteristic of the rGO films, resulting in the proportional response of the developed sensors towards ethanol [6].

In order to find the optimum thickness for rGo nanocomposites the fiber optic tips are tested with 5 concentrations of ethanol, from 20 to 100 %. The results are summarized in Fig. 8. For all the rGO coated tips, the intensity drops as the ethanol concentration increases. The fiber optic tip with the 0.75 mg/mol tip gave the stable response for all wavelengths. Interference are specially required the changes of parameters such as taper waist diameter, length, and transition form [7, 22, 23].

The reflectance spectrum and dynamic response of the sensor is investigated for the 0.2 mg/mol, 0.5 mg/mol, 0.75 mg/mol and 1.0 mg/mol rGO coated tip at a wavelength range of 500 to 650 nm. The results show that at 0.75 mg/mol rGO, that intensity decreases as the concentration of the ethanol is increased as shown in Fig. 9. Table 1 shows the percentage different between difference ethanol concentration in water. From Table 1, it

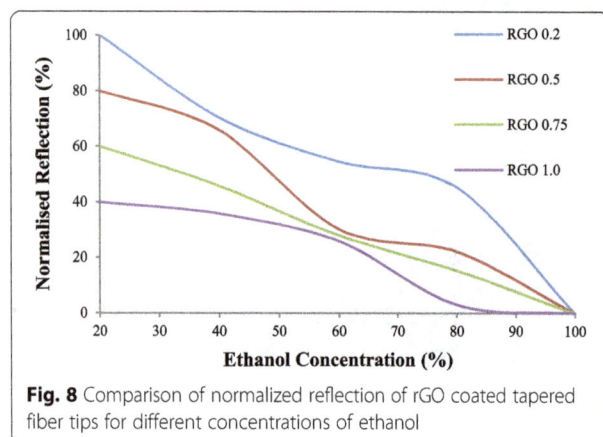

Fig. 8 Comparison of normalized reflection of rGO coated tapered fiber tips for different concentrations of ethanol

Table 1 Sensitivity for different concentration of ethanol

Ethanol	Δ_{ref}	$\Delta_{r1} - \Delta_{r2}$
20 %	a_1	–
40 %	a_2	0.349
60 %	a_3	1.77
80 %	a_4	3.336
100 %	a_5	9.645

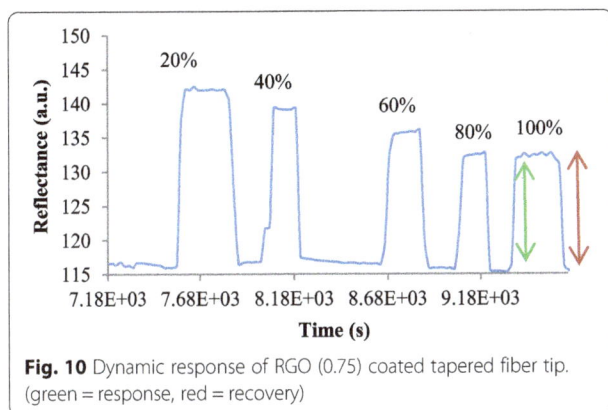

Fig. 10 Dynamic response of RGO (0.75) coated tapered fiber tip. (green = response, red = recovery)

can be conclude that sensitivity for the developed sensor is approximately 3.0196.

For the fiber tip with 0.75 mg/mol rGO concentration, the dynamic response is presented in Fig. 10. It shows that the higher the concentration of ethanol, the higher the reflectance response of the sensor. The sensor shows high sensitivity that can be observed from the dynamic response of the reflectance plotted. For every ethanol concentration, the sensor responds to a stable level and when the ethanol is removed, the sensor recovers and returns to a stable baseline. The response and recovery time was obtained as 40 and 70 s respectively.

Conclusion

The performance of an ethanol sensor using a tapered fiber tip coated with a new material, rGO was investigated in this research. We have successfully designed and fabricated a tapered fiber tip optic sensor coated with reduced graphene oxide (rGO) as new sensing layer to detect different concentrations of ethanol in water.

The results show that this enhancement of the sensing surface is able to deliver high sensitivity for the detection of various concentration of aqueous ethanol. The tapered fiber optic tip also gives a stable repeatable response towards ethanol concentrations as well as fast response and recovery time of 40 and 70 s respectively. Moreover the introduction of the layer of rGO nano-particle also helps increase its structural strength. The experimental data also determines that the thickness of the rGO layer to produce the optimal results is approximately 280 nm.

It show that fiber tip optic with a higher concentration rGO which mean of higher refractive index coating are more sensitive compared to the lower concentration rGO-coated fiber tip.

Acknowledgments

The work reported in this paper has partly supported by the Universiti Putra Malaysia's Research University Grant Schemes (Ref: 05-01-12-1626RU and 05-02-12-2015RU) and Ministry of Higher Education, Malaysia's Fundamental Research Grant Scheme (Ref: 03-04-10-795FR).

Authors' contributions

MAAR: Sensing experiments. PTA: Material characterization. HNL & NMH: GO synthesis. ASMN: Data analysis. All authors read and approved the final manuscript.

Competing interests

The authors declare that they have no competing interests.

Author details

[1]Department of Computer and Communication Systems Engineering, Faculty of Engineering, Universiti Putra Malaysia, 43400UPM Serdang, Selangor, Malaysia. [2]Communication Technology Section, Universiti Kuala Lumpur-British Malaysia Institute, 53100 GOMBAK, Kuala Lumpur, Malaysia. [3]Research Centre of Excellence for Wireless and Photonic Network, Faculty of Engineering, Universiti Putra Malaysia, 43400UPM, Serdang, Selangor, Malaysia. [4]Department of Chemistry, Faculty of Science, Universiti Putra Malaysia, 43400UPM, Serdang, Selangor, Malaysia. [5]Physics Department, Low Dimensional Materials Research Centre, University of Malaya, 50603 Kuala Lumpur, Malaysia.

References

1. Banerjee, A., Mukherjee, S., Verma, R.K., Jana, B., Khan, T.K., Chakraborty, M., Das, R., Biswas, S., Saxena, A., Singh, V.: Fiber optic sensing of liquid refractive index. Sensors Actuators B Chem. 123(1), 594–605 (2007)
2. Rickelt, L.F., Ottosen, L.D.M., Kühl, M.: Etching of multimode optical glass fibers: A new method for shaping the measuring tip and immobilization of indicator dyes in recessed fiber-optic microprobes. Sensors Actuators B Chem. 211, 462–468 (2015)
3. Shabaneh, A.A., Girei, S.H., Arasu, P.T., Rahman, W.B.W.A., Bakar, A.A.A., Sadek, A.Z., Lim, H.N., Huang, N.M., Yaacob, M.H.: Reflectance response of tapered optical fiber coated with graphene oxide nanostructured thin film for aqueous ethanol sensing. Opt. Commun. 331, 320–324 (2014)
4. Lin, J.: Recent development and applications of optical and fiber-optic pH sensors. Trends Anal. Chem. 19(9), 541–552 (2000)
5. Fidanboylu and Efendioglu, HS: Fiber optic sensors and their applications. Symp. A Q. J. Mod. Foreign Lit. 1–6, 2009.
6. Mukherjee, A., Munsi, D., Saxena, V., Rajput, R., Tewari, P., Singh, V., Ghosh, A. K., John, J., Wanare, H., Gupta-Bhaya, P.: Characterization of a fiber optic liquid refractive index sensor. Sensors Actuators B Chem. 145(1), 265–271 (2010)
7. Tian, Y., Wang, W., Wu, N., Zou, X., Wang, X.: Tapered optical fiber sensor for label-free detection of biomolecules. Sensors 11(4), 3780–3790 (2011)
8. Pei, S., Cheng, H.M.: The reduction of graphene oxide. Carbon N. Y. 50(9), 3210–3228 (2012)
9. Gadipelli, S., Guo, Z.X.: Graphene-based materials: Synthesis and gas sorption, storage and separation. Prog. Mater. Sci. 69, 1–60 (2015)
10. Wang, Y.-P., Tian, B., Sun, W.-R., Liu, D.-Y.: Analytic study on the mixed-type solitons for a (2 + 1)-dimensional N-coupled nonlinear Schrödinger system in nonlinear optical-fiber communication. Commun. Nonlinear Sci. Numer. Simul. 22(1–3), 1305–1312 (2015)
11. Lee, B.H., Min, E.J., Kim, Y.H.: Fiber-based optical coherence tomography for biomedical imaging, sensing, and precision measurements. Opt. Fiber Technol. 19(6), 729–740 (2013)
12. Carenco, A., Scavennec, A.: Active opto-electronic components. Comptes Rendus Phys. 4(1), 85–93 (2003)
13. Al-Qazwini, Y., Noor, A.S.M., Arasu, P.T., Sadrolhosseini, A.R.: Investigation of the performance of an SPR-based optical fiber sensor using finite-difference time domain. Curr. Appl. Phys. 13(7), 1354–1358 (2013)
14. Chan Lee, S., Some, S., Wook Kim, S., Jun Kim, S., Seo, J., Lee, J., Lee, T., Ahn, J.-H., Choi, H.-J., Chan Jun, S.: Efficient Direct Reduction of Graphene Oxide by Silicon Substrate. Sci. Rep. 5, 12306 (2015)
15. Konios, D., Stylianakis, M.M., Stratakis, E., Kymakis, E.: Dispersion behaviour of graphene oxide and reduced graphene oxide. J. Colloid Interface Sci. 430, 108–112 (2014)
16. Arasu, P., Noor, A., Shabaneh, A.: Absorbance properties of gold coated fiber Bragg grating sensor for aqueous ethanol. J. Eur. Opt. Soc. 9, 14018 (2014)

17. Aguilar-Soto, J.G., Antonio-Lopez, J.E., Sanchez-Mondragon, J.J., May-Arrioja, D.A.: Fiber Optic Temperature Sensor Based on Multimode Interference Effects. J. Phys. Conf. Ser. **274**, 012011 (2011)
18. Wang, P., Brambilla, G., Ding, M., Semenova, Y., Wu, Q., Farrell, G.: High-sensitivity, evanescent field refractometric sensor based on a tapered, multimode fiber interference. Opt. Lett. **36**(12), 2233–2235 (2011)
19. Loryuenyong, V, Totepvimarn, K, Eimburanapravat, P, Boonchompoo, W, Buasri, A: Preparation and Characterization of Reduced Graphene Oxide Sheets via Water-Based Exfoliation and Reduction Methods. Advances in Material Science and Engineering. pp. 1–5 (2013)
20. Cao, N, Zhang, Y: Study of Reduced Graphene Oxide Preparation by Hummers' Method and Related Characterization. J. Nanomater. pp. 1–5 (2015)
21. N. T. Sl, Nanoinnova Technologies SL C/Faraday 7, 28049 Madrid http://www.nanoinnova.com. pp. 7–9.
22. Shabaneh, A., Girei, S., Arasu, P., Mahdi, M., Rashid, S., Paiman, S., Yaacob, M.: Dynamic response of tapered optical multimode fiber coated with carbon nanotubes for ethanol sensing application. Sensors (Switzerland) **15**(5), 10452–10464 (2015)
23. Yadav, T.K., Mustapa, M.A., Abu Bakar, M.H., Mahdi, M.A.: Study of single mode tapered fiber-optic interferometer of different waist diameters and its application as a temperature sensor. J. Eur. Opt. Soc. **9**, 8–12 (2014)

4

Bidirectional MM-Wave Radio over Fiber transmission through frequency dual 16-tupling of RF local oscillator

K. Esakki Muthu[1*] and A. Sivanantha Raja[2]

Abstract

In this paper for the first time, a 60 GHz bidirectional Millimeter Wave (MM-Wave) Radio over Fiber (RoF) transmission through a new frequency dual 16-tupling of 3.75 GHz local oscillator (LO) is demonstrated. The proposed system is constructed with parallel combination of two cascaded stages of MZMs. The upper cascaded stage and the Lower cascaded stages are biased at the Maximum Transmission Point (MATP). By suitable adjustments of LO phase and amplitude, optical sidebands with spacing of 8 times the input LO frequency is generated. These sidebands are then separated using filters to achieve dual 16-tupling. A good agreement between numerical derivations and the simulation results are achieved. Further, a simulation is performed to access the dual bidirectional transmission performance for the double and single tone modulation with 2.5 Gbps data transmission. The transmission distance is limited to 25 km for the double tone modulation due to bit walk of effect. A 60 km link distance is achieved with single tone modulation. The dispersion induced power penalties less than 0. 5 dB at 10^{-9} BER is observed for both up and down streams.

Keywords: Millimeter Wave, 16-tupling, Radio over Fiber, Mach-Zehnder Modulator

Background

From the past two decades demands for the high data rate wireless services are ever increasing. To support such a growing demand, a higher bandwidth carrier frequency is required. As the lower frequency spectrum up to 30 GHz is congested, the MM-Wave band (30–300GHz) is now considered to be the promising candidate for the support of the emerging data traffic since it offers 270 GHz bandwidth. Though, a major bottle neck of the MM-Wave communications are the MM-Wave generation and transmission since the electrical generation suffers from the limited frequency response of the available conventional electronics components and, the MM-Wave transmission suffers from huge free space/cable losses [1]. A viable solution for this issue enables us to use the hybrid system which combines both wireless and optical communications, where the MM-Waves are generated by optical methods at the central station (CS) and distributed over an optical fiber to the base

station (BS). The optical generation and distribution of MM-Wave enjoys the low loss and huge bandwidth of the optical fibers. Several optical MM-Wave generation methods have been proposed including direct modulation of the laser diode, optical heterodyning of the highly correlated laser sources and external modulation. Among these, external modulation based frequency multiplication techniques have become popular due to higher modulation bandwidth, tunability and high stability [2, 3]. Several multiplication techniques have been proposed with different multiplication factors such as doubling [3], tripling [4], quadrupling [5–7], sextupling [8], octupling [9], 12-tupling [10], 16- tupling [11, 12] and so on. A MM-Wave generation technique with highest frequency multiplication factor reduces the need of high frequency RF local oscillator at the CS. However, cost effective design of CS as well as the BS is a challenge. A full duplex RoF system which shares a single laser source is a one of the cost effective techniques. In addition to this, reduction of high frequency RF LO at the CS will bring down the cost of the entire system. Hence, a full duplex RoF transmission with frequency dual quadrupling was proposed in [13], dual sextupling [14] and dual octupling presented in [15] these methods were using

* Correspondence: esaionly@gmail.com
[1]University VOC College of Engineering, University VOC, Thoothukudi, Tamilnadu 630003, India
Full list of author information is available at the end of the article

a single laser source for both upstream and downstream by supporting two BSs simultaneously. However, as the frequency multiplication factor is too low, there is a demand for a high frequency RF local oscillator at the CS. To achieve a higher frequency multiplication factor, two or more MZMs were employed either in series or in parallel configuration. A cascaded combination of two MZMs was used to generate frequency quadrupling [16], sextupling [8, 17] and octupling [18]. Three arm MZM was used for the generation of sextupling in [19] and three parallel MZMs were used to generate sextupling, 12-tupling and 18 tupling in [20]. A frequency octupling was generated using 4 MZMs in [21, 22].

In this paper, we propose a novel frequency dual 16-tupling of the given RF local oscillator which will considerably eliminate the need for a high frequency RF local oscillators at the CS compared to the other techniques proposed in [13, 15] and also as it supports full duplex transmission by wavelength reuse, cost of both CS and BS can be greatly minimized. A mathematical proof of the proposed scheme is presented and for the proof of concept a simulation is conducted with full bidirectional MM-W RoF transmission.

The structure of this paper is as follows, the mathematical principle behind the proposed technique is discussed in the section II. The simulation work and results of proposed scheme is presented in section III and finally, conclusion is presented in the chapter IV.

Principle

The principle behind the proposed scheme is depicted in Fig. 1. The proposed configuration is composed of parallel combination of two cascaded MZMs. A CW DFB laser source $E_o(t) = E_o cos\omega_c t$ drives the both upper and the lower cascaded stages of the MZMs. The upper cascaded stage consists of MZM1 and MZM2 and the

lower cascaded stage consists of MZM3 and MZM4, they are all biased at its Maximum Transmission Point (MATP).

The both the electrodes of the MZM1 is driven by $v_1(t) = V_{rf} sin \omega_m t$, and the transfer function of the MZM1 can be expressed as,

$$E_{MZM1}(t) = \frac{E_o}{4} cos\omega_c t \left[\begin{array}{c} e^{j\frac{\pi V_{rf}}{V_\pi} sin(\omega_m t) + j\frac{\pi V_{b2}}{V_\pi}} \\ +e^{-j\frac{\pi V_{rf}}{V_\pi} sin(\omega_m t) + j\frac{\pi V_{b1}}{V_\pi}} \end{array} \right]$$

(1)

where V_{rf} is the RF signal amplitude, V_π is the switching bias voltage of the MZM, V_{b1} and V_{b2} are the bias voltage of the electrodes, by setting $V_{b1} = V_{b2} = V_\pi/2$, the Eq. (1) can be rewritten as,

$$E_{MZM1}(t) = \frac{E_o}{4} cos\omega_c t \left[j\left(e^{jm sin(\omega_m t)} + e^{-m sin(\omega_m t)} \right) \right]$$ (2)

where $m = \pi V_{rf}/V_\pi$ is the modulation index. Using the Bessel function of first kind the Eq. (2) can be written as,

$$E_{MZM1}(t) = j\frac{E_o}{2} cos\omega_c t \left[\left(J_0(m) + 2\sum_{n=1}^{\infty} J_{2n}(m) cos2n\omega_m t \right) \right]$$

(3)

where J(.) is the Bessel function of order n. The Eq. (2) shows that the output of MZM1 contains even order sidebands along with central carrier. And the electrodes of the MZM2 are driven by $v_2(t) = V_{rf} sin\left(\omega_m t + \frac{\pi}{2}\right)$, the transfer function of the MZM2 can be expressed as

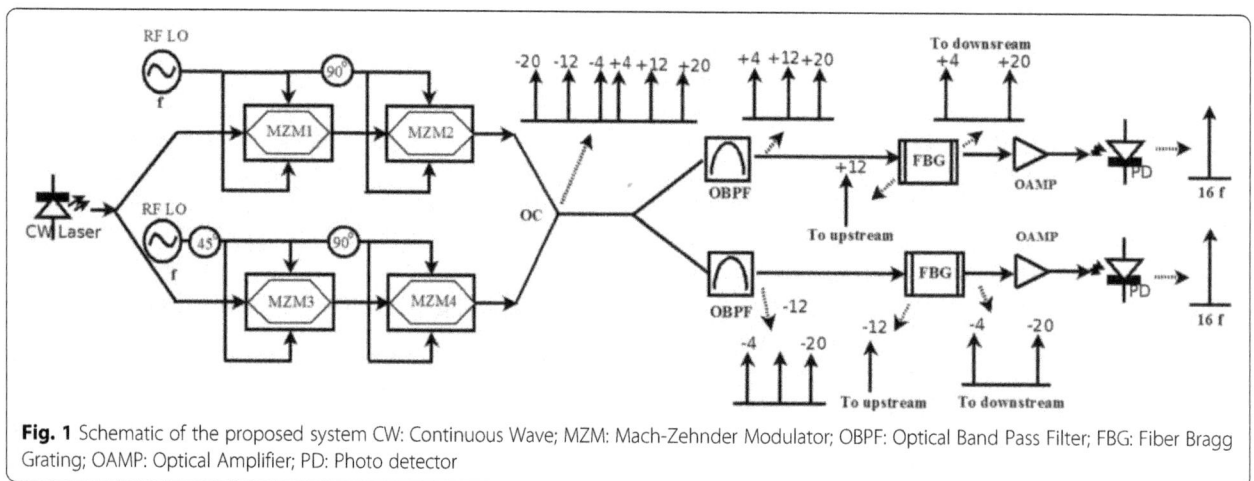

Fig. 1 Schematic of the proposed system CW: Continuous Wave; MZM: Mach-Zehnder Modulator; OBPF: Optical Band Pass Filter; FBG: Fiber Bragg Grating; OAMP: Optical Amplifier; PD: Photo detector

$$E_{MZM2}(t) = j\frac{E_o}{4}cos\omega_c t\left[\left(e^{jm\sin\left(\omega_m t+\frac{\pi}{2}\right)} + e^{-m\sin\left(\omega_m t+\frac{\pi}{2}\right)}\right)\right] \tag{4}$$

Using Bessel function,

$$E_{MZM2}(t) = j\left[\left(J_0(m) + \sum_{n=1}^{\infty}(-1)^n J_{2n}(m)cos2n\omega_m t\right)\right] \tag{5}$$

Then the output of the upper cascaded stage can be expressed as,

$$E_U(t) = E_{MZM1}(t) * E_{MZM2}(t) \tag{6}$$

$$E_U(t) = -\frac{E_o}{2}cos\omega_c t\left\{\left[J_0(m) + 2\sum_{n=1}^{\infty}J_{2n}(m)\cos(2n\omega_m t)\right] * \left[J_0(m) + 2\sum_{n=1}^{\infty}(-1)^n J_{2n}(m)^2\cos(2n\omega_m t)\right]\right\} \tag{7}$$

By eliminating common terms and higher order terms whose magnitudes are very low, the Eq. (7) can be reduced to,

$$E_U(t) = -\frac{E_o}{2}cos\omega_c t\{J_0{}^2(m) + 2J_0(m)\sum_{n=1}^{\infty}(-1)^n J_{2n}(m)\cos(2n\omega_m t)) + 2J_0(m)\sum_{n=1}^{\infty}J_{2n}(m)\cos(2n\omega_m t) + 2J_0(m)\left[\sum_{n=1}^{\infty}(-1)^n J_{2n}{}^2(m)\{1+\cos(4n\omega_m t)\}\right]\} \tag{8}$$

Simplification of the above eqn. results,

$$\begin{aligned} E_U(t) = \frac{E_o}{2}cos\omega_c t\{&-J_0{}^2(m) + J_2{}^2(m) + J_2{}^2(m)cos4\omega_m t \\ &+ 4J_0J_4(m)cos4\omega_m t - 2J_4{}^2(m) - 2J_4{}^2(m)cos8\omega_m t \\ &+ 2J_6{}^2(m) + 2J_6{}^2(m)cos12\omega_m t - 4J_0J_8(m)cos8\omega_m t \\ &- 2J_8{}^2(m) - 2J_8{}^2(m)\,cos8\omega_m t + 2J_{10}{}^2(m) \\ &+ J_{10}{}^2(m) + 2J_{10}{}^2(m)cos20\omega_m t - 4J_0J_{12}(m)cos12\omega_m t \\ &- 2J_{12}{}^2(m) - 2J_{12}{}^2(m)cos24\omega_m t + J_{14}{}^2(m) \\ &+ 2J_{14}{}^2(m)cos28\omega_m t - 4J_0J_{16}(m)cos16\omega_m t - 2J_{16}{}^2(m) \\ &- 2J_{16}{}^2(m)cos32\omega_m t - 2J_{18}{}^2(m) \\ &- 2J_{18}{}^2(m)cos36\omega_m t - 4J_0J_{20}(m)cos20\omega_m t \\ &- 2J_{20}{}^2(m) - 2J_{20}{}^2(m)cos40\omega_m t - \cdots\} \end{aligned} \tag{9}$$

From Eq. (9) one can observe that output of the upper cascaded stage consists sidebands which are integer multiples of four. In the lower cascaded stage can output of the MZM3 is driven by $v_3(t) = V_{rf}\sin\left(\omega_m t + \frac{\pi}{4}\right)$, and setting $V_{b1} = V_{b2} = 0$, the transfer function of the MZM3 can be written as be written as,

$$E_{MZM3}(t) = \frac{E_o}{4}cos\omega_c t\left[\begin{array}{c} e^{jm\sin\left(\omega_m t+\frac{\pi}{4}\right)} \\ +e^{-m\sin\left(\omega_m t+\frac{\pi}{4}\right)} \end{array}\right] \tag{10}$$

Using Bessel function of first kind,

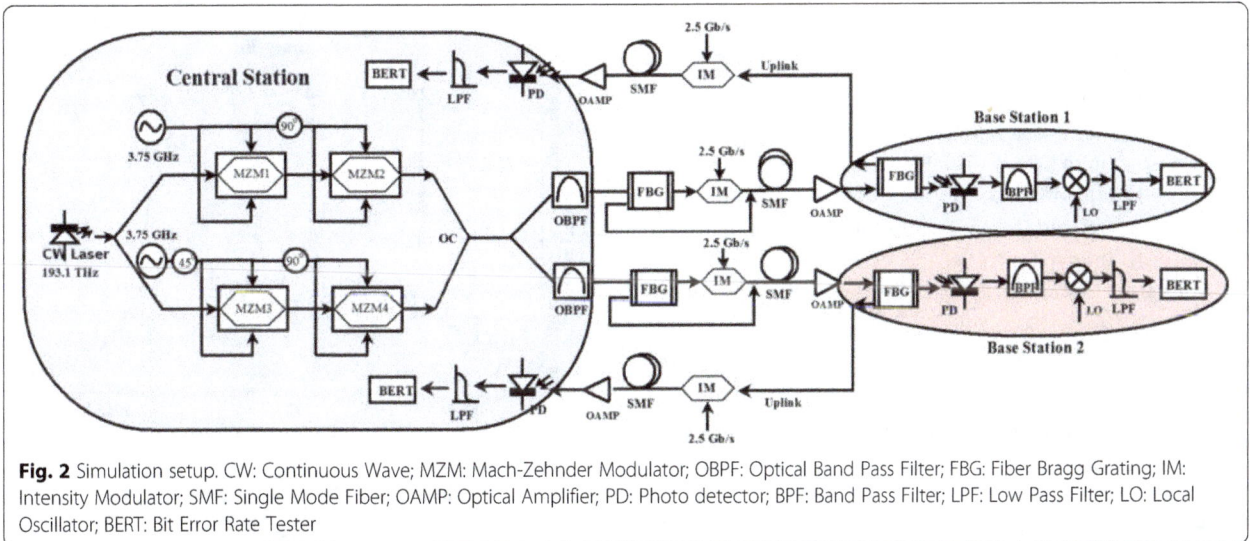

Fig. 2 Simulation setup. CW: Continuous Wave; MZM: Mach-Zehnder Modulator; OBPF: Optical Band Pass Filter; FBG: Fiber Bragg Grating; IM: Intensity Modulator; SMF: Single Mode Fiber; OAMP: Optical Amplifier; PD: Photo detector; BPF: Band Pass Filter; LPF: Low Pass Filter; LO: Local Oscillator; BERT: Bit Error Rate Tester

$$E_{MZM3}(t) = \frac{E_o}{2}cos\omega_c t \left[2\sum_{n=1}^{\infty} \frac{J_0(m)+}{J_{2n}(m)cos2n(\omega_m t + \pi/4)} \right]$$

(11)

The output of the MZM4 driven by the $v_4(t) = V_{rf}$ $sin\left(\omega_m t + \frac{3\pi}{4}\right)$ and setting $V_{b1} = V_{b2} = 0$ can be written as,

$$E_{MZM4}(t) = \frac{E_o}{4}cos\omega_c t \left[\begin{array}{c} e^{jm\sin\left(\omega_m t+\frac{3\pi}{4}\right)} \\ +e^{-jm\sin\left(\omega_m t+\frac{3\pi}{4}\right)} \end{array} \right]$$

(12)

Using Bessel function,

$$E_{MZM4}(t) = \left[J_0(m) + 2\sum_{n=1}^{\infty}(-1)^n J_{2n}(m)cos2n\left(\omega_m t + \frac{\pi}{4}\right) \right]$$

(13)

Hence the output of the lower cascaded stage can be expressed as,

$$E_L(t) = E_{MZM3}(t) * E_{MZM4}(t)$$

(14)

$$E_U(t) = \frac{E_o}{2}cos\omega_c t \left\{ J_0^2(m) + 2J_0(m)\sum_{n=1}^{\infty}(-1)^n J_{2n}(m)\cos2n\left(\omega_m t + \frac{\pi}{4}\right) \right.$$
$$+2J_0(m)\sum_{n=1}^{\infty}J_{2n}(m)\cos2n\left(\omega_m t + \frac{\pi}{4}\right)$$
$$\left. +2J_0(m)\left[\sum_{n=1}^{\infty}(-1)^n J_{2n}^2(m)\left\{1+\cos4n\left(\omega_m t + \frac{\pi}{4}\right)\right\}\right] \right\}$$

(15)

Simplification of the above eqn. results,

$$E_L(t) = \frac{E_o}{2}cos\omega_c t \{ J_0^2(m) - J_2^2(m) - J_2^2(m)cos(4\omega_m t + \pi)$$
$$+4J_0J_4(m)cos(4\omega_m t + \pi) + 2J_4^2(m) + 2J_4^2(m)cos(8\omega_m t + 2\pi)$$
$$-2J_6^2(m) - 2J_6^2(m)cos(12\omega_m t + 3\pi) - 4J_0J_8(m)(cos8\omega_m t + 2\pi)$$
$$+2J_8^2(m) + 2J_8^2(m)\cos(16\omega_m t + 4\pi) - 2J_{10}^2(m)$$
$$-2J_{10}^2(m)cos(20\omega_m t + 5\pi) - 4J_0J_{12}(m)cos12\omega_m t + 3\pi)$$
$$+2J_{12}^2(m) + 2J_{12}^2(m)cos(24\omega_m t + 6\pi) - J_{14}^2(m)$$
$$-2J_{14}^2(m)cos(28\omega_m t + 7\pi)$$
$$-4J_0J_{16}(m)cos(16\omega_m t + 4\pi) + 2J_{16}^2(m)$$
$$+2J_{16}^2(m)\cos(32\omega_m t + 8\pi) - 2J_{18}^2(m)$$
$$-2J_{18}^2(m)cos(36\omega_m t + 9\pi)$$
$$-4J_0J_{20}(m)cos(20\omega_m t + 5\pi) + 2J_{20}^2(m)$$
$$+2J_{20}^2(m)cos(40\omega_m t + 10\pi) - \cdots \}$$

(16)

The resultant optical output after the optical coupler can be expressed as,

$$E_O(t) = E_U(t) + E_L(t)$$

(17)

Cancelling the terms which are integer multiples of eight and eliminating the higher order terms due to its low power, the Eq. (18) can be expressed as,

$$E_O(t) = \frac{E_o}{2}cos\omega_c t \left\{ \left[4J_2^2(m) - 8J_0J_4(m)\right]cos4\omega_m(t) \right.$$
$$+ \left[4J_6^2(m) - 8J_0J_{12}(m)\right]cos12\omega_m t$$
$$\left. + \left[4J_{10}^2(m) - 8J_0J_{20}(m)\right]cos20\omega_m t \right\}$$

(18)

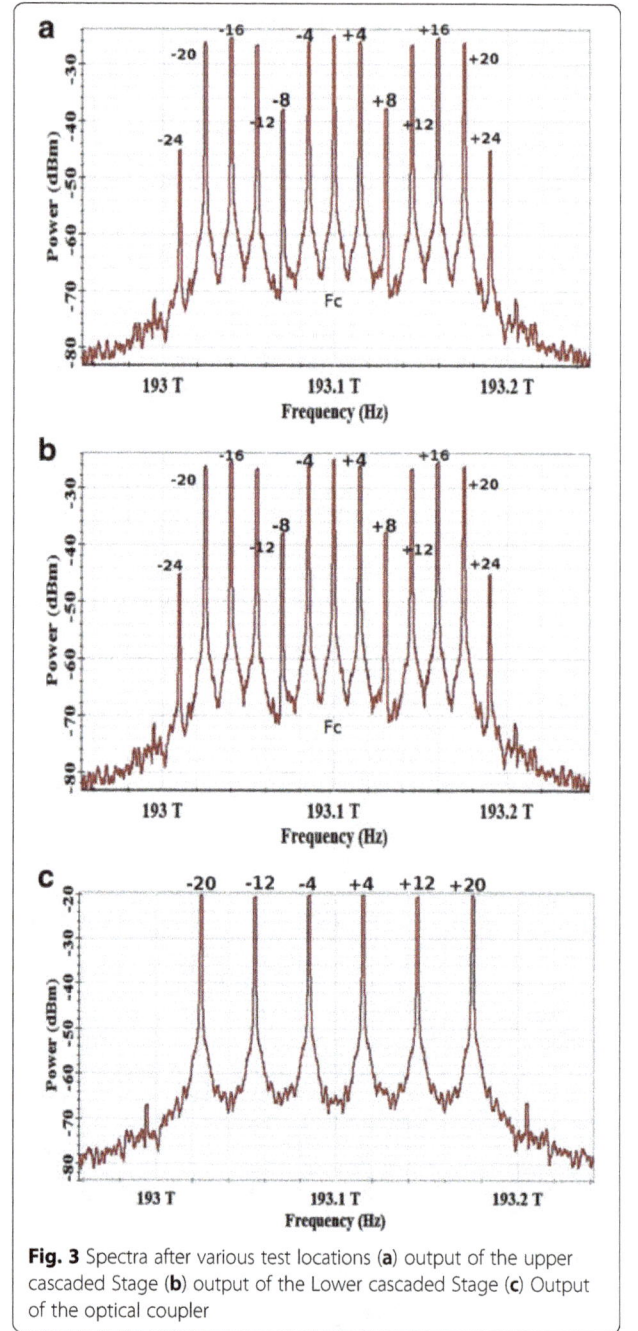

Fig. 3 Spectra after various test locations (a) output of the upper cascaded Stage (b) output of the Lower cascaded Stage (c) Output of the optical coupler

The Eq. (18) clearly shows that there are only $(\omega_c \pm 4\omega_c)$, $(\omega_c \pm 8\omega_c)$, $(\omega_c + 12\omega_c)$, and $(\omega_c + 20\omega_c)$ order sidebands. For want of dual 16-tupling the upper sidebands $(\omega_c + 4\omega_c)$, $(\omega_c + 12\omega_c)$ and $(\omega_c + 20\omega_c)$ should be separated and then only $(\omega_c + 4\omega_c)$ and $(\omega_c + 20\omega_c)$ sidebands will be allowed to beat at the photo detector to result a MM-Wave frequency output which is 16 time the input local oscillator frequency. At the same time, the sideband corresponding to the $(\omega_c + 12\omega_c)$ will be reused by the BS1 for the uplink data transmission. Similarly, the lower sidebands $(\omega_c - 4\omega_c)$ and $(\omega_c - 20\omega_c)$ sidebands will be beating at the photo detector to generate 16 tupled MM-Wave and the sideband $(\omega_c - 12\omega_c)$ will be reused at the BS2 for the uplink transmission. Hence, along with dual 16 tupled MM-Wave generation, a bidirectional MM-Wave RoF communication can also be established between the CS and BS.

Simulation, Results and Discussions

A simulation setup used for the verification of the proposed system is shown in Fig. 2. A 10 MHz spectral width continuous wave (CW) laser source with 193.1 THz central frequency is split equally and launched in to the upper and lower cascaded stages of this configuration. Both the MZMs in the upper cascaded stage and the lower stages are biased at its MATP with switching bias voltage of 4 V. The upper stage has two cascaded MZMs (MZM1 and MZM2). In this, the MZM1 is driven by 3.75 GHz Radio Frequency Local Oscillator (RF LO) and the MZM2 is driven by the same LO with 90° phase shift. Bias voltage V_{b1} and V_{b2} are set to 2 V for the upper cascaded stage. The lower stage has two cascaded MZMs (MZM3 and MZM4). The MZM3 is driven by a LO with 45° phase shift and the MZM4 is driven by LO with 135° phase shift. Amplitude of all the RF LOs are set to 4 V. The bias voltage of the lower cascaded stage is set to zero and the extinction ratios of all the MZMs are set to 100 dB. As both the MZMs in the upper cascaded stage is biased at MATP, the output of the upper cascaded stage has sidebands which are integer multiples of four along with the carrier as shown in Fig. 3(a). The output of the lower cascaded stage is shown in Fig. 3(b). It contains similar sideband, but the carrier (ω_c) and sidebands which are integer multiples of 4 such as $(\omega_c \pm 8\omega_c)$, $(\omega_c \pm 16\omega_c)$ and $(\omega_c \pm 24\omega_c)$ are exactly out of phase with the upper cascaded stage. Then the output of the upper and lower cascaded stages are combined using an optical coupler. The coupler output shown in Fig. 3(c), it shows only the in-phase components such as, $(\omega_c \pm 4\omega_c)$, $(\omega_c \pm 12\omega_c)$ and $(\omega_c \pm 20\omega_c)$. Further, the coupler output is equally split and then, the upper sidebands $(\omega_c + 4\omega_c)$, i.e., 193.115 THz, $(\omega_c + 12\omega_c)$ i.e., 193.145 THz and $(\omega_c + 20\omega_c)$ i.e., 193.175 THz are separated utilizing an optical band pass filter (OBPF) with 193.145 THz central frequency and 100 GHz bandwidth, similarly the lower sidebands $(\omega_c - 4\omega_c)$ i.e., 193.085 THz,

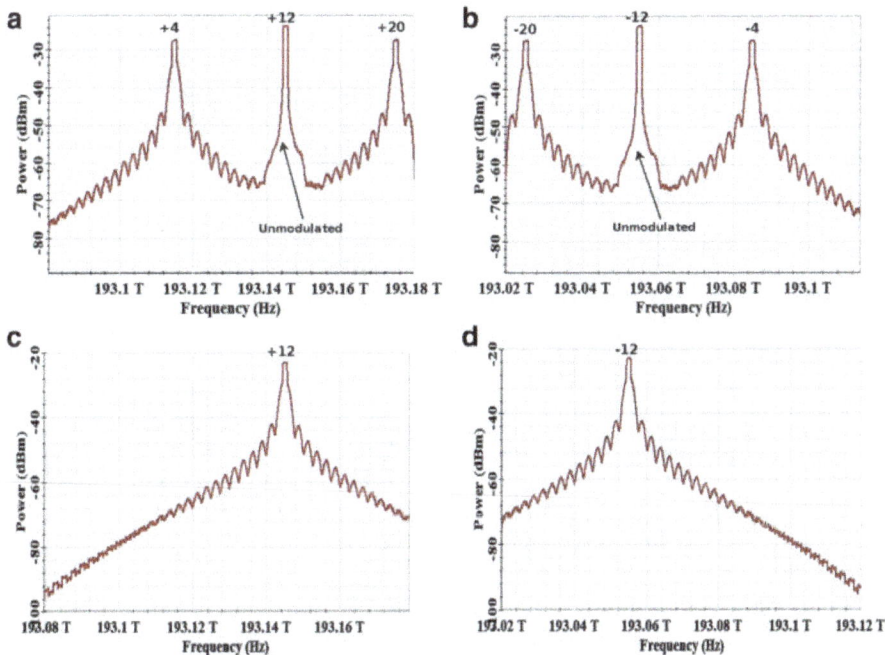

Fig. 4 Spectra after various test locations (**a**) Downstream data modulation on +4th and +20th order sidebands from CS to BS1 (**b**) Downstream data modulation on −4th and −20th order sidebands from CS at BS2 (**c**) Upstream data modulation on +12th order sideband from BS1 to CS (**d**)) Upstream data modulation on −12th order sideband from BS1to CS

$(\omega_c - 12\omega_c)$ i.e., 193.055 THz and $(\omega_c - 20\omega_c)$ i.e., 193.025 THz are separated using an OBPF with a central frequency of 193.055 THz and bandwidth of 100 GHz bandwidth. In the upper sidebands, +12th order sideband

(195.145 THz) is reflected using a Fiber Bragg Grating (FBG) with the bandwidth of 10 GHz and now we can see that the remaining +4th (193.115 THz) and +20th (193.175 THz) order sidebands are differ by the frequency

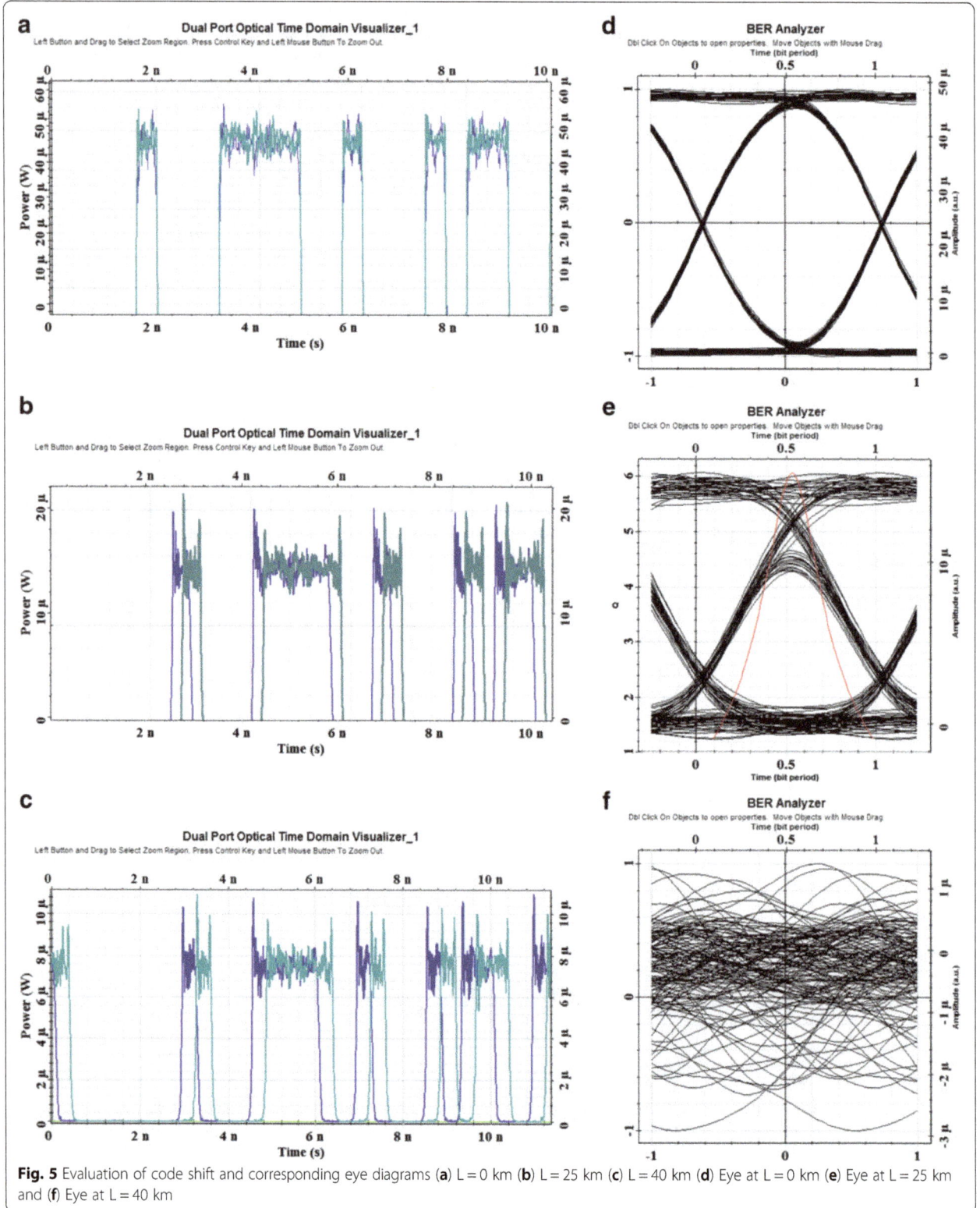

Fig. 5 Evaluation of code shift and corresponding eye diagrams (**a**) L = 0 km (**b**) L = 25 km (**c**) L = 40 km (**d**) Eye at L = 0 km (**e**) Eye at L = 25 km and (**f**) Eye at L = 40 km

16 times the input RF LO. These sidebands are modulated with the 2.5 Gb/s NRZ data of length 2^{7-1} using an intensity modulator and then combined with the unmodulated +12th (195.145 THz) order sideband as shown in Fig. 4(a). This field is transported from CS to the BS1 over a 25 km SMF with 0.2 dB/km attenuation coefficient and 16.75 ps/nm-km dispersion coefficient. The dispersion slope and PMD coefficient of the fiber is 0.075 ps/nm^2.km and 0.05 ps/√km. At the BS1, before the photo detection process another FBG with 10 GHz bandwidth is tuned to reflect the unmodulated tone located at 195.145 THz which is reused for carrying the 2.5 Gb/s NRZ upstream data over 25 km SMF from BS1 to CS. The upstream data modulated spectrum from the BS1 is shown in Fig. 4(c). The losses of both the links are compensated using an optical amplifier with a noise Figure of 4. Similarly at the lower sideband spectra, a 2.5 Gb/s NRZ data modulated over −4th (195.085 THz) and −20th (195.025 THz) order sidebands are combined with the unmodulated −12th (195.055 THz) order sideband and then transmitted from CS to BS2 over 25 km SMF. The data modulated spectrum for the BS2 is shown in Fig. 4(b). At the BS2 prior to the photo detection, the tone corresponding to 195.055 THz is separated using the FBG and reused for the 2.5 Gb/s upstream data transmission to the CS. The upstream data modulated spectrum from the BS2 is

shown in Fig. 4(d). The downstream signal received at the both the BSs are detected using a PIN photo-detector with 0.7 A/W responsivity, 100e^{-24} W/Hz thermal power density and 10 nA dark current and then passed through a band pass filter having a centre frequency of 60 GHz with a bandwidth 1.5 times the bitrate so as to select the data modulated 60 MM-Wave signal. This signal is then demodulated and then passed through a low pass filter with a bandwidth of 0.75 times the bitrate. The upstream data received at the CS is detected using the PIN detector and demodulated. BER performance of both downstream and the upstream are measured by varying the received optical power using an optical attenuator.

In the downstream, compared to the back-to-back case, the BER performance is degraded due to the dispersion induced RF power fading and the bit-walk off effect on the dual tone modulation. Since the dispersion causes the data on each sideband to undergo different time and phase shift, the data bits walk off from each other as the function of fiber length L [23]. The code shift between the sidebands is measured for various length of the fiber and it is shown in Fig. 4. From the Fig. 5(a)-(c), it is clear that the codes on the sidebands are exactly coinciding when the fiber length L = 0, the shift is acceptable when L = 25 km and the data completely walks off when L = 40 km and beyond. The Eye diagrams of the corresponding lengths are

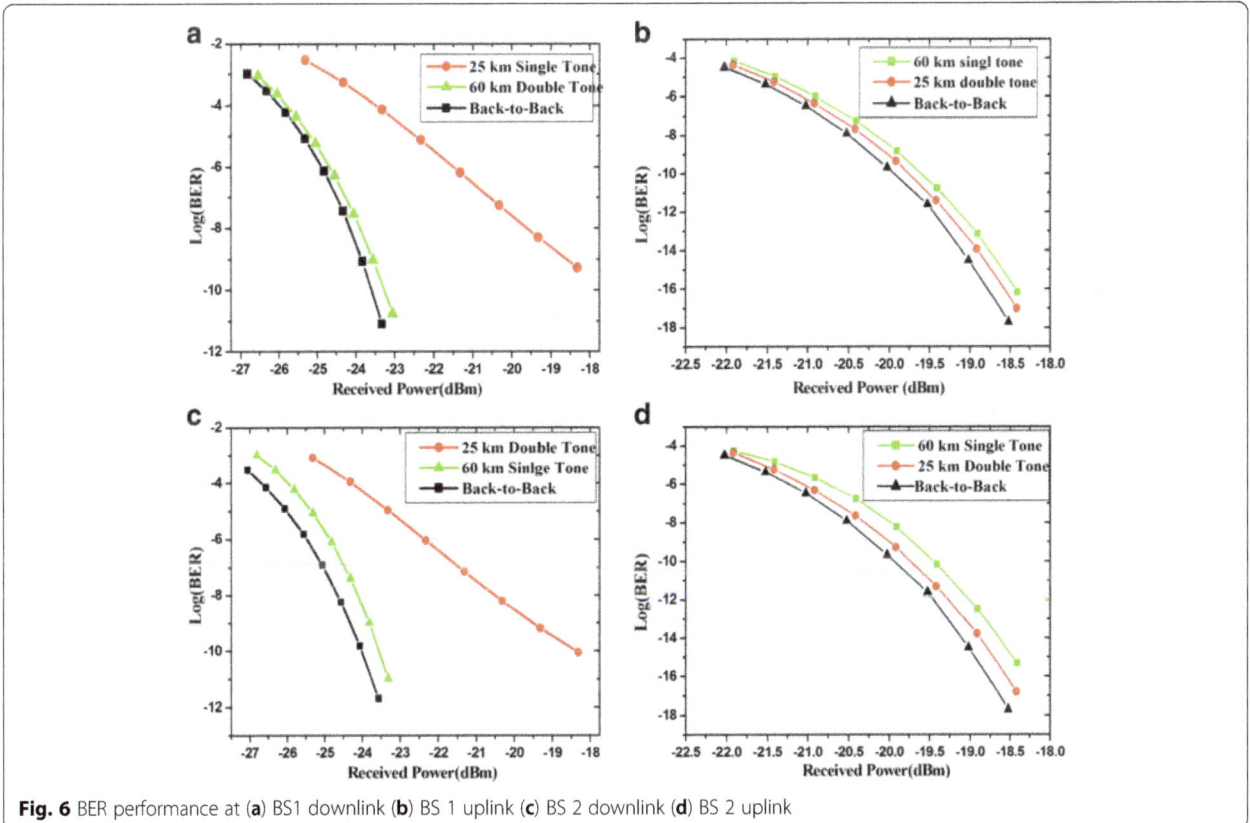

Fig. 6 BER performance at (a) BS1 downlink (b) BS 1 uplink (c) BS 2 downlink (d) BS 2 uplink

also shown in Fig. 5(d)-(f). As the code shift is more the eye completely closes. Hence the transmission distance is limited to 25 km in case of the double sideband data modulation with a power penalty of 5 dB at 10^{-9} BER for the both BSs downlink. The BER performance against the received power for BS1 and BS2 are shown Fig. 6(a) and 5(c) respectively. But the BER performance of the upstream transmission is very close to the back-to-back (B-t-B) BER due to the fact that there is no modulated RF carrier and hence no power fading and bit walk off effect. Therefore, the dispersion induced power penalty at the 10^{-9} BER for the upstream transmission is less than 0.3 dB. The transmission distance can be increased by modulating the data over only one of the sidebands. As the data modulated over only one sideband, the effect of dispersion induced power fading and the bit walk off effects are eliminated. Hence the transmission distance of the scheme is extended to 60 km. Further increase in the transmission distance is attributed to the pulse broadening due to the dispersion. The BER performance of the single tone modulation is compared with B-t-B case and double tone modulation and is shown in the Fig. 6(a)-(d). The dispersion induced power penalty for the single tone modulation is less than 0.5 dB for the down and upstream at 10^{-9} BER.

Conclusion

A new approach for generating the frequency dual 16-tupling is proposed and demonstrated using parallel configuration of two stage cascaded MZMs. This system simultaneously supported two base stations with bidirectional data transmission between BS and CS by wavelength reuse without additional requirements; hence costs of the BSs are also reduced significantly. Transmission Performance is evaluated for both double tone and single tone modulation formats. The simulation result showed 5 dB dispersion induced power penalty in the downstream data transmission and less than 0.3 dB for the upstream transmission over 25 km SMF at the BER of 10^{-9} for the double tone data modulation. With the single tone modulation transmission, the link distance is extended to 60 km with dispersion induced power penalty less than 0.5 dB for both upstream and downstream.

Acknowledgements
The authors thankfully acknowledge the Department of Science and Technology (DST), New Delhi for their Fund for Improvement of S&T Infrastructure in Universities and Higher Educational Institutions – (FIST) grant through the order No.SR/FST/College-061/2011(C) to procure the Optiwave suite Simulation tools.

Authors' contribution
KEM has worked out the mathematical derivations of the proposed system and carried out all the simulation works. ASR has developed the concept and involved in the manuscript preparation. Both authors read and approved the final manuscript.

Competing interests
The authors declare that they have no competing interests.

Author details
[1]University VOC College of Engineering, University VOC, Thoothukudi, Tamilnadu 630003, India. [2]A.C.College of Engineering and Technology, Karaikudi, Tamilnadu 630004, India.

References
1. Capmany, J., Novak, D.: Microwave photonics combines twoworlds. Nat. Photonics **1**, 319–330 (2007)
2. Qi, G., Yao, J.P., Seregelyi, J., Paquet, S., Belisle, C.: Generation and distribution of wide-band continuously tunable millimeter wave signal with an optical external modulation technique. IEEE Trans. Microwave Theory Tech. **53**, 3090–3097 (2005)
3. O'Rcilly, J.J., Lane, P.M., Heidemann, R., Hofstetter, R.: Opical generation of verynarrow linewidth Millimeter wave signals. Electron. Lett **28**(25), 2309–2311 (1992)
4. Wang, Q., Rideout, H., Zeng, F., Yao, J.P.: Millimeter-Wave frequency tripling based on four-wave mixing in a semiconductor optical amplifier. IEEE Photon. Technol. Lett. **18**(23), 2460–2462 (2006)
5. Lin, C.T., Shih, P.T., Chen, J., Xue, W.Q., Peng, P.C., Chi, S.: Optical millimeter wave signal generation using frequency quadrupling technique and no optical filtering. IEEE Photon. Technol. Lett. **20**(12), 209–211 (2008)
6. Mohamed, M., Zhang, X., Hraimel, B., Wu, K.: Analysis of frequency quadrupling using a singleMach-Zehnder modulator for millimeter-wavegeneration and distribution over fiber systems. Opt Express **16**(14), 10786–10802 (2008)
7. Yu, S., Gu, W., Yang, A., Jiang, T., Wnag, C.: A Frequency quadrupling optical mm-Wave generation for hybrid fiber-wireless systems. IEEE J. Sel. Areas Commun/Supplement- Part 2 **31**(12), 797–803 (2013)
8. Mohamed, M., Zhang, X., Hraimel, B., Wu, K.: Frequency sixupler for millimeter-wave over fiber systems. Opt Express **15**, 10141–10151 (2008)
9. Ma, J., Xin, X., Yu, J., Yu, C., Wnag, K., Huang, H., Rao, L.: Optical millimeter wave generated by octupling the frequency of the local oscillator. J. Opt. Netw. **7**, 837–845 (2008)
10. Zhu, Z., Zhao, S., Li, Y., Chu, X., Wnag, X., Zhao, G.: A radio over fiber system with frequency 12-Tupling optical millimeter wave generation to overcome chromatic dispersion. Quantum Electron. Lett. **49**, 919–922 (2013)
11. Chen, H., Ning, T., Jian, W., Pei, L., Li, J.: D-band millimeter-wave generator based on a frequency 16-tupling feedforward modulation technique. Opt. Eng. **52**, 0761041–0761044 (2013)
12. Zhu, Z., Zhao, S., Chu, X., Dong, Y.: Optical generation of millimeter – wave signals via frequency 16-tupling without an optical filter. Opt. Commun. **354**, 40–47 (2015)
13. Hu, J.H., Huang, X.G., Xie, J.L.: A full-duplex radio-over-fiber systems based on dual quadrupling-frequency. Opt. Commun. **284**, 729–734 (2011)
14. Yang, K., Huang, X.G., Zhu, J.H., Fang, W.J.: Transmission of 60 GHz wired/wireless based on full-duplex radio-over-fiber using dual-sextupling frequency. IET Commun. **6**, 2900–2906 (2012)
15. Cheng, G., Guo, B., Liu, S., Fang, W.: A novel full-duplex radio-over-fiber systems based on dual octupling-frequency for 82 GHz W-band radio frequency and wavelength reuse for uplink connection. Optik **125**, 4072–4076 (2014)
16. Zhao, Z., Wen, Y., Zhang, H.: Simplified optical millimeter wave generation conFigureureuration by frequency quadrupling using two cascaded Mach-Zehnder modulators. Opt Lett **34**, 3250–3252 (2009)
17. Qin, Y., Sun, J.: Frequency sextupling technique using two cascaded dual-electrode Mach-Zehnder modulators interleaved with Gaussian band pass filter. Opt. Commun. **285**, 2911–2916 (2012)
18. Chen, Y., Wen, A., Shang, L.: Analysis of an optical mm-wave generation scheme with frequency octupling using two cascaded Mach-zehnder modulators. Opt. Commun. **283**, 4933–4941 (2010)
19. Chen, Y., Wen, A., Yin, X., Shang, L.: A photonic mm-wave frequency sextupler using an Inegrated Mach-Zehnder modulator with three arms. Fiber Integr. Opt. **31**, 196–207 (2012)
20. Chen, Y., Wen, A., Guo, J., Shang, L., Wang, Y.: A novel optical mm-wave generation scheme based on three parallel Mach-Zehnder modulators. Opt. Commun. **284**, 1159–1169 (2011)

21. Shang, L., Wen, A., Li, B., Wang, T., Chen, Y., Li, M.: A filterless optical millimeter wave generation based on frequency octupling. Optik **123**, 1183–1186 (2012)
22. Hasan, M., Hall, T.J.: A Photonic frequency octo-tupler with reduced RF drive power and extended spurious sideband suppression. Opt. Laser Technol. **81**, 115–121 (2016)
23. Zhoua, M., Ma, J.: Influence of fiber dispersion on the transmission performance of a quadruple frequency optical millimeter wave with two signal modulation format. Opt Swtiching Netw **9**, 343–350 (2012)

5

Numerical analysis of temperature-controlled terahertz power splitter

Li Yang and Li Jiu-Sheng[*]

Abstract

Background: As a significant terahertz functional device, terahertz beam splitters with high performance are highly required to meet the need for terahertz communication, terahertz image, and terahertz sensor systems.

Results: The proposed 1×6 power splitter becomes 1×4 power splitter with the aid of the localized temperature change at the frequency of 1.0THz. The total output power is equivalent to 97.8% of the input power for the six-channel splitter and 95.4% for the four-channel splitter. The dimension of the device is of $35a \times 27a$.

Conclusions: A temperature-controlled terahertz power divider based on photonic crystal multimode interference structure and Y-junction photonic crystal waveguides is an efficient mechanism for the power divider of terahertz waves. The proposed device paves a promising way for the realization of terahertz wave integrated device.

Keywords: Terahertz wave, Tunable power splitter, Photonic crystal

Background

With the rapid development of terahertz technology, it becomes very critical and urgent to control the terahertz wave transmission efficiently. Nowadays, to manipulate the terahertz wave propagation has been one of the intensively hot research topics in both science and engineering fields. The devices for manipulation the terahertz wave include such as filters, power dividers, switches, de-multiplexers, absorbers, modulators, and so forth [1–5]. Power splitter is one key component in terahertz wave communication system, therefore, it becomes particularly important to study a terahertz power divider with high performance. In recent years, there have some research reports on the power divider in the literatures [6–12]. For example, in [8], C. Berry et. al. proposed a terahertz beam splitter using sub-wavelength silver grating fabricated on a high-density polymer substrate. In [9], C. Homes et. al. employed a thick silicon wafer as beam splitter for far-infrared and terahertz spectroscopy. In [10], B. Ung et. al. fabricated a terahertz beam-splitter based on low-density polyethylene plastic sheeting coated with a conducting silver layer. In [11] J.Li et. al. designed a terahertz wave polarization beam splitter using a cascaded multimode interference structure. In

[12] T. Niu et. al. demonstrated a terahertz beam splitter based on periodic sub-array. For current beam splitters, there still exist some problems, such as large dimensions, the need for multilayer structures, and with non-tunable, etc. As a significant terahertz functional device, terahertz beam splitters with high performance are highly required to meet the need for terahertz communication, terahertz image, and terahertz sensor systems.

In this work, we propose a temperature-controlled multi-channel terahertz wave power splitter based on Y-junction photonic crystal waveguides and multi-mode interference photonic crystal structure (see Fig. 1). Using the localized high-temperature in photonic crystal, we can shift the frequency position of the guide mode in photonic crystal waveguide and efficiently manipulate the output port power of the device. Both the plane wave expansion method and the finite-difference time-domain method are performed to investigate the mode properties and the transmission characteristics of the temperature-controlled terahertz power splitter. Interestingly, our results may pave the way for designing a tunable multi-channel terahertz power splitter using localized temperature-controlled photonic crystal for the thermo-optic effect. Using such a terahertz beam splitter, the whole length of the device is reduced to about 1/10 that of a conventional design [11].

* Correspondence: lijsh2008@126.com
Centre for THz Research, China Jiliang University, Hangzhou 310018, China

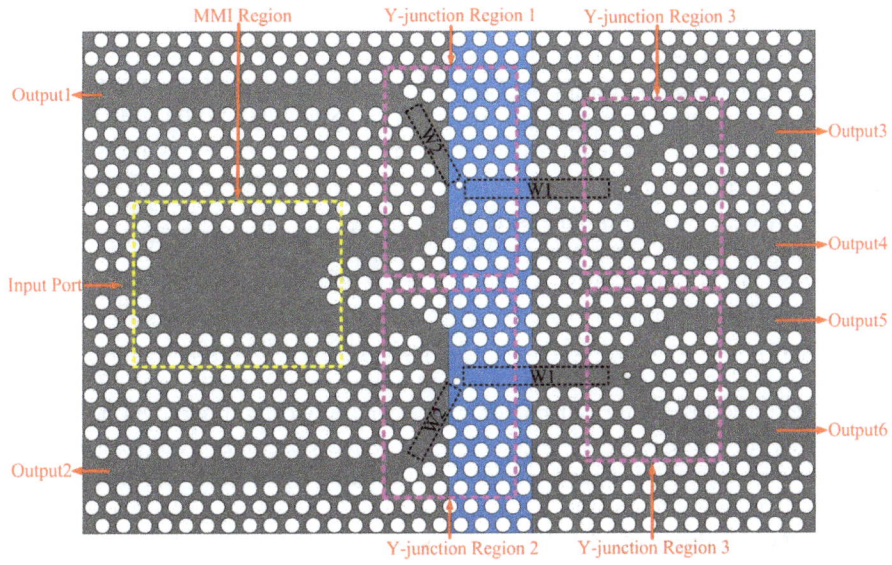

Fig. 1 Configuration of the proposed tunable multi-channel terahertz wave power splitter

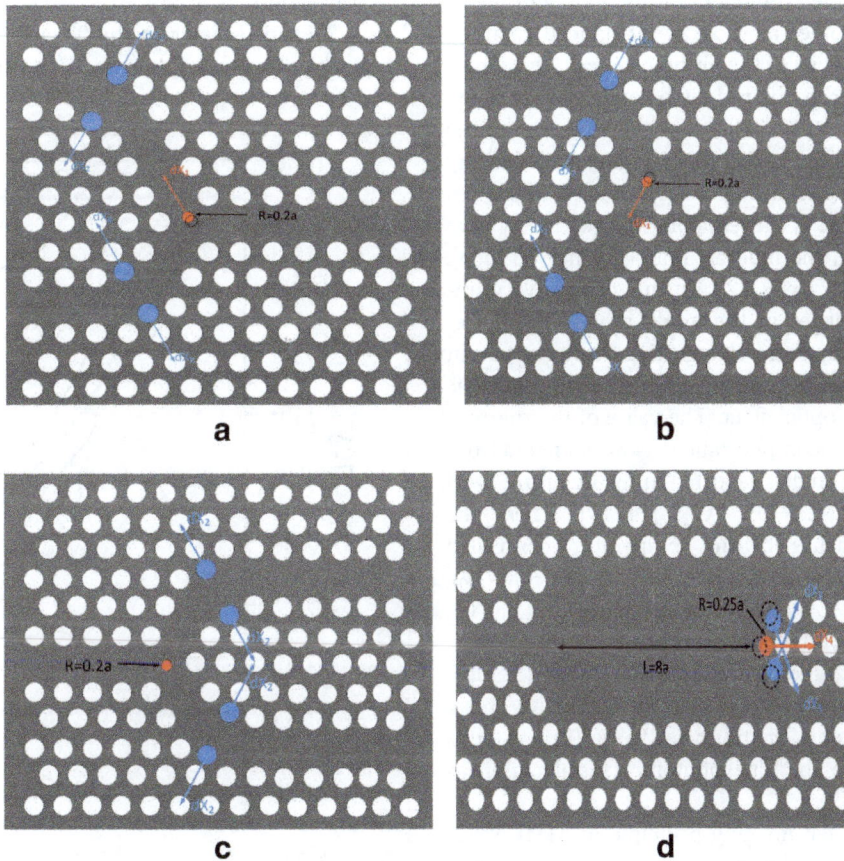

Fig. 2 Configuration of Y-junction and MMI region, **a** Y-junction Region 1, **b** Y-junction Region 2, **c** Y-junction Region 3, **d** MMI Region

Device design

The schematic view of the proposed temperature-controlled multi-channel terahertz power splitter is shown in Fig. 1. In this figure, the device consists of four photonic crystals Y-junction waveguides (marked with Y-junction Region 1, Y-junction Region 2, and two Y-junction Region 3), and a multi-mode interference structure (marked with MMI Region). A two-dimensional photonic crystal consists of a triangular lattice array of air holes with the radius of $r = 0.32a$. The localized high-temperature area in the photonic crystal is marked by using light blue. The background material is high resistivity silicon with refractive index of 3.45 (a is lattice constant of the structure). Here, the loss of the silicon with high resistivity is very weak. In the Y-junction Region 1 and 2, there have two photonic crystal waveguides named with W1 and W2. Both W1 and W2 regions are surrounded by dotted lines. Figure 2 shows the details of the three kinds of photonic crystal Y-junctions region and a MMI region. The photonic crystal Y-junction region has two 60° waveguide bends. In this study, we adjust two air hole's position shown in light blue air holes (see Fig. 2a, b and c). The air holes with the radius of $r = 0.32a$ are moved oppositely along the symmetric axis of the bend with $dx_2 = 0.26a$ [13]. Furthermore, to overcome the mode-mismatch of the Y-junction, in the middle of the Y-junction region, another air hole is introduced (marked with red), and the radius of the air hole is $r = 0.2a$. In additional, in Fig. 2a and b, the red air hole are moved along the arrows with $dx_1 = 0.3a$. To reduce the reflection of the MMI region, we optimize the MMI region by adjusting some air holes as shown in Fig. 2d. The radius of the red air hole with $r = 0.25a$ is moved along the red arrow with $dx_4 = 0.3a$. Similarly, the blue air holes are moved along the blue arrows with $dx_3 = 0.2a$. In our design, very importantly, small changes of the guide mode cut-off frequency can cause dramatic change the terahertz output intensity by the partially controlled temperature photonic crystal with thermo-optic effect. The value of the thermo-optic coefficient of silicon photonic crystal estimated from a variety of different studies reported in the literature averaged approximately $2.4 \times 10^{-4} °C^{-1}$ [14–19]. Then, the refractive index change of the temperature-controlled photonic crystal material is calculated by $\Delta n = (n - n_0) = \Delta T \times 2.4 \times 10^{-4} °C^{-1}$. At room temperature ($T = 25$ °C), the refractive index of the photonic crystal material is $n_0 = 3.45$. In this letter, to realize the function of temperature-controlled terahertz power splitter, the refractive index change of the photonic crystal is set to be 0.15 (At this time, $T = 650$ °C.) in the blue photonic crystal region as shown in Fig. 1.

According to the self-imaging principle in MMI waveguides, a self-image will be reproduced at the position of L, which can be calculated by [20]

$$L = m(3L_\pi) \tag{1}$$

where m denotes the periodic number of imaging along the multi-mode photonic crystal waveguide ($m = 0, 1, 2, ...$), L_π is defined as the beat length of the two lowest-order modes which is the position of the two-fold image first emerges and can be expressed by

$$L_\pi = \pi (\beta_0 - \beta_1) \tag{2}$$

where β_0 and β_1 are the propagation constants of the fundamental and the lowest-order even modes, respectively. In the MMI region, the exciting field entering the multimode waveguide is symmetric. Therefore, in our design, only even modes will be excited. To demonstrate the validity of our proposed method, the theoretically calculated L has been checked by numerical simulation according to the two-fold image mechanism.

Figure 3 shows the dispersion curves of MMI region in our presented devices by use of the plane-wave expansion method at room temperature of 25 °C. It can be noted that five guide modes (marked with 0th, 1st, 2nd, 3rd and 4th) can propagate in the MMI region. Note that there is an absolute photonic band-gap within the normalized frequency range from 0.213 (a/λ) to 0.294 (a/λ) (see the blue solid lines). Here, a is lattice constant, λ is the wavelength in free space. According to Eq. (2) and Fig. 3, β_0 and β_1 refer to the 0th and 2nd mode, respectively. Thus, we can obtain the $\beta_0 = 0.48 \times 2\pi/a$ and $\beta_1 =$

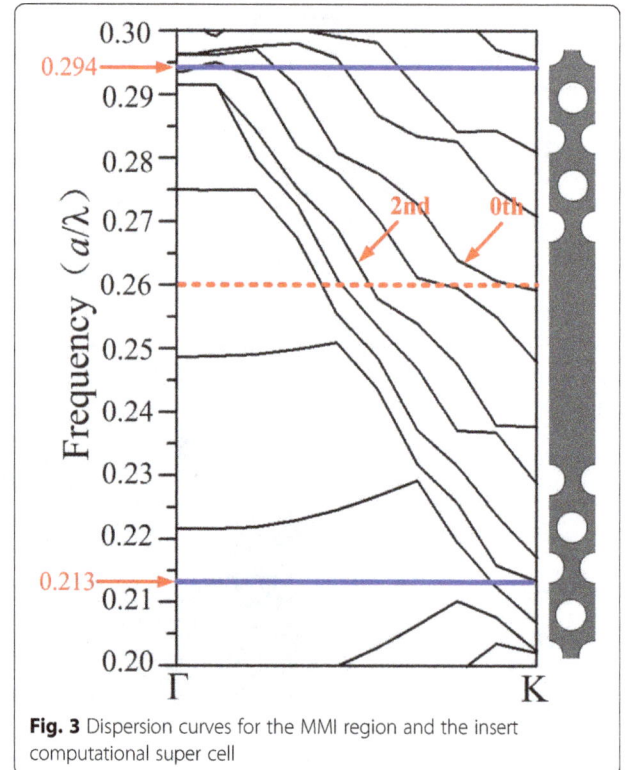

Fig. 3 Dispersion curves for the MMI region and the insert computational super cell

$0.27 \times 2\pi/a$. Substituting $\beta_0 = 0.48 \times 2\pi/a$ and $\beta_1 = 0.27 \times 2\pi/a$ into the Eq. (1) with $m = 1$ (The length of MMI region is set to be the same length of the first self-image), and then we can get $L = 3L_\pi = 7.14a$. One can see that multiple guided modes exist for almost all the frequencies in the complete photonic band gap, and all of them may operate in the photonic crystal waveguide. Therefore, in this letter, the working frequency point is set to be 0.26 (a/λ) (i.e. $f = 1\,\text{THz}$), and the lattice constant is $a = 78\,\mu\text{m}$.

Figure 4 depicts the dispersive relation of the W1 and W2 waveguides for the TM-polarization. The region between the two blue solid lines indicates the complete photonic band-gap, which is for the frequency range of 0.213 $(a/\lambda) \sim 0.294$ (a/λ). According to Fig. 4a, for W1 waveguide, one sees that there have a photonic band-gap in the frequency region from 0.263 (a/λ) to 0.267 (a/λ) for TM-polarization mode (see the red shaded region in the Fig. 4a) at 25 °C. When the localized photonic crystal temperature becomes 650 °C, the photonic band-gap in the ranges from 0.253 (a/λ) to 0.256 (a/λ) for TM-polarization mode (see the green shaded region in the Fig. 4a). Since the working frequency range is set to be 0.26 (a/λ) as mentioned above (see the red dash line in the Fig. 4a), the guide mode can propagate in W1 waveguide at the localized photonic crystal temperature with 25 °C and 650 °C. From Fig. 4b, for W2 waveguide, it can be observed that there have a photonic band-gap in the region of 0.263 $(a/\lambda) \sim 0.267$ (a/λ) for TM polarization mode at 25 °C. Similarly, when the localized photonic crystal temperature becomes 650 °C, a photonic band-gap changes the frequency region of 0.258 $(a/\lambda) \sim 0.262$ (a/λ) for TM-polarization mode (see the green shaded region in the Fig. 4b). In this case, for the working frequency of 0.26 (a/λ) (see the red dash line in the Fig. 4b), the TM polarization mode can propagate through photonic crystal waveguide W2 at the localized temperature of 25 °C while it can not pass through W2 at the localized photonic crystal temperature of 650 °C. Note that the terahertz wave in the presented terahertz beam splitter is controlled by changing of the localized photonic crystal temperature.

Methods

In order to be more intuitive understanding, a commercial available software module Rsoft FullWave with the finite-difference time-domain method, is employed to simulate the terahertz wave propagation in the proposed device (see Fig. 1). The perfectly matched layers are located around the designed structure as the absorbing boundary condition. The fineness of the finite-difference time-domain cells (i.e. Δx and Δy) are set as 0.05. The Δt coefficient is 0.95 and the total calculation time is 50,000. In order to excite photonic crystal waveguide mode in the device, a continuous wave source is lunched at the input port of the photonic crystal waveguide.

Results and discussions

As the temperature of the blue photonic crystal region is 25 °C, Fig. 5a shows the steady state field distribution. According to the Figure, the terahertz wave transmits through the four Y-junction photonic crystal waveguides and a multi-mode interference photonic crystal structure, finally distributes its energy into six output ports with the equal power. In Fig. 5b, the input terahertz wave is equally distributed in the four output ports, i.e., from output 3 to output 6, when the temperature of the blue photonic crystal region becomes 650 °C. In order to investigate the structure quantitatively, we have also calculated the output power as a function of the working frequency, as shown in Fig. 6. In Fig. 6a, for the six-channel power splitter, the output power of each output port can reach a maximum of 16.5% at 1.0THz, and its corresponding total output power is of 97.8%. At this time,

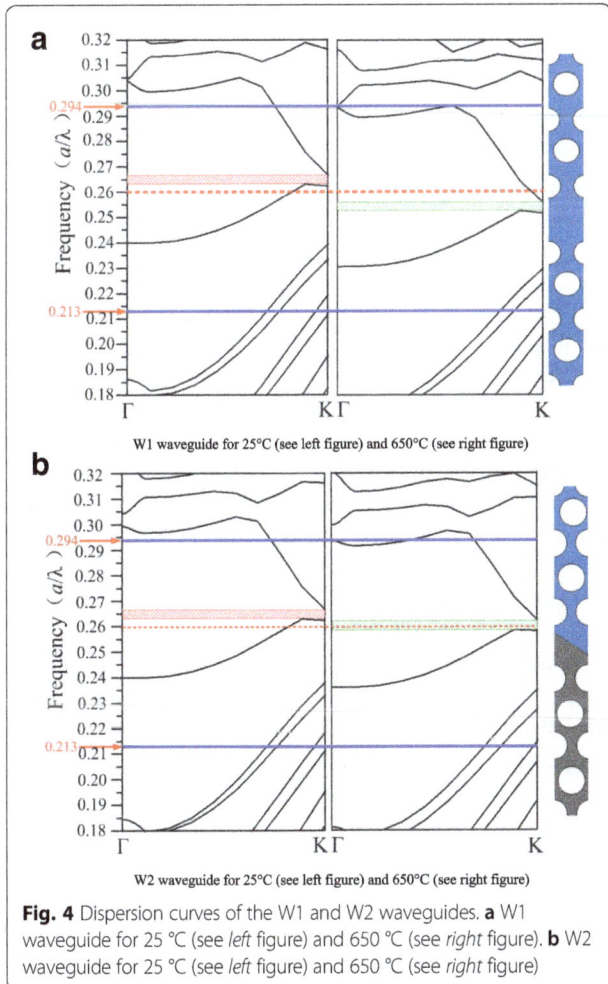

Fig. 4 Dispersion curves of the W1 and W2 waveguides. **a** W1 waveguide for 25 °C (see *left* figure) and 650 °C (see *right* figure). **b** W2 waveguide for 25 °C (see *left* figure) and 650 °C (see *right* figure)

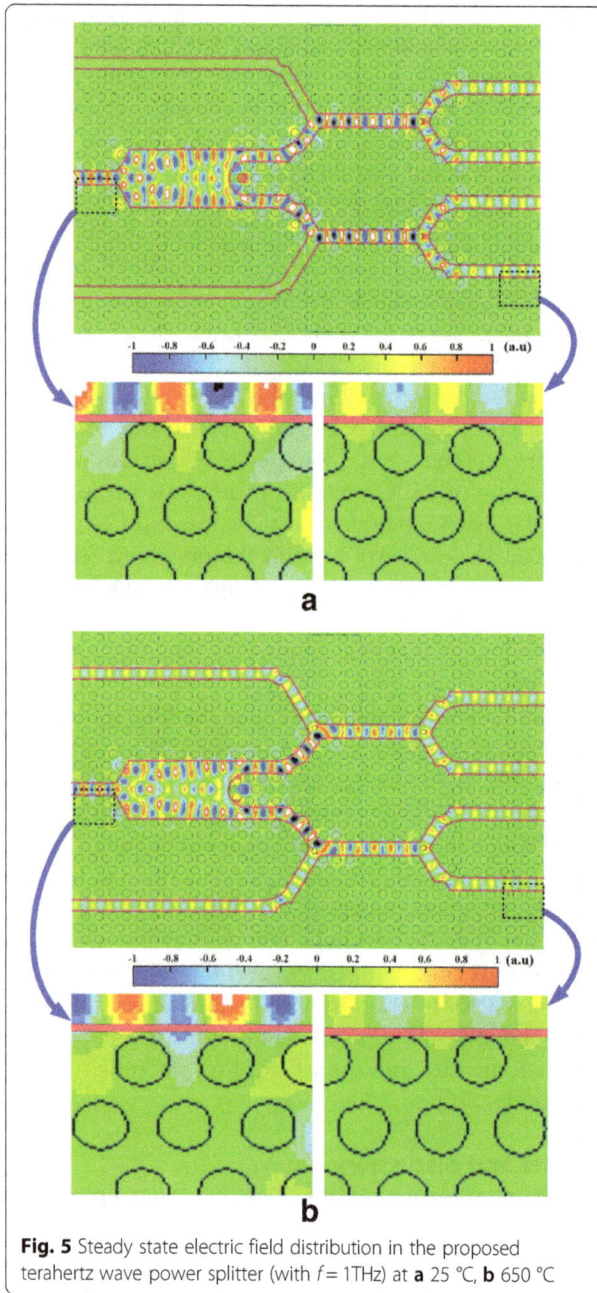

Fig. 5 Steady state electric field distribution in the proposed terahertz wave power splitter (with $f=1$THz) at **a** 25 °C, **b** 650 °C

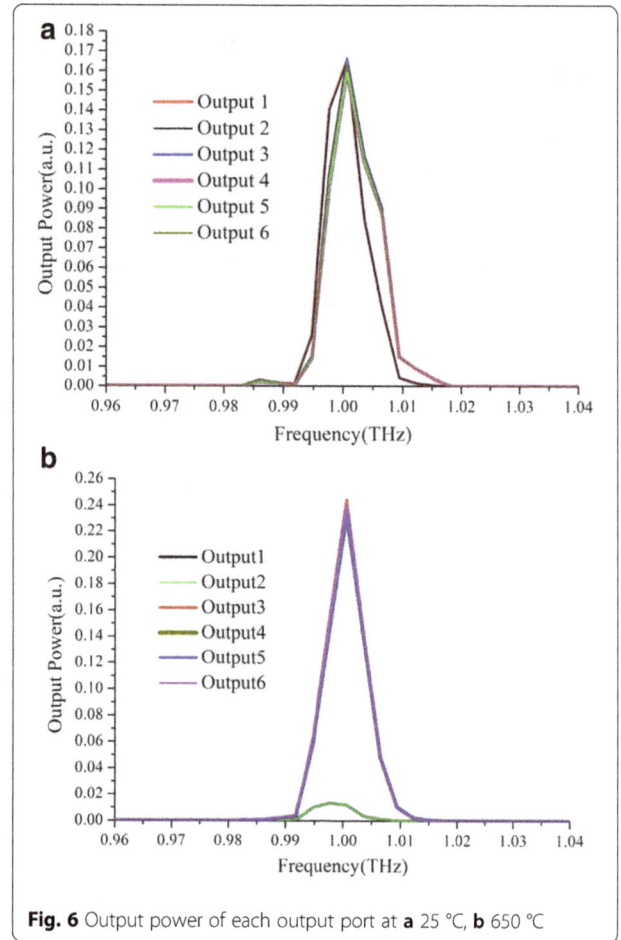

Fig. 6 Output power of each output port at **a** 25 °C, **b** 650 °C

1/10 that of a conventional design [11]. The wave-guiding loss could be significantly reduced by using crystalline or polymer material featuring low absorption in terahertz region. For instance, the proposed terahertz power splitter could be manufacture from sapphire or high-resistivity silicon using the advanced methods of shaped crystal growth, such as edge-defined film-fed growth (EFG) technique (or Stepanov technique) [21]. Moreover, it could be made of polymers (such as COC, HDPE, or TPX) by implementing drawing [22] or additive manufacturing (2D and 3D printing) principles [23]. Obviously, the geometry of proposed structure should be adjusted for the refractive index of the particular material, and for its temperature-induced changes.

the localized photonic crystal temperature is 25 °C. For the four-channel power splitter, the output power of each output port can reach a maximum of 24.4% at 1.0THz, and its corresponding total output power is of 95.4%, indicated by Fig. 6b. At this time, the localized photonic crystal temperature is 650 °C. The detail output powers of each output port are collected in Table 1. The simulated propagation pattern of electric field agrees well with the theoretical calculated prediction results. The total size of the whole device is of $35a \times 27a$ (i.e. 2.73×2.11 mm^2), which is about

Table 1 Output power of each output port

Temperature	Output power					
	Output 1	Output 2	Output 3	Output 4	Output 5	Output 6
25 °C	16.5%	16.5%	16.5%	15.9%	15.9%	16.5%
650 °C	1.3%	1.3%	23.3%	24.4%	23.3%	24.4%

Conclusions

To sum up, we have designed a temperature-controlled power splitter based on photonic crystal operating in the terahertz regime. The terahertz wave power splitter was evaluated using plane wave expansion and finite-difference time-domain method. The calculated and simulated results shows that the proposed power splitter constituted by four Y-junctions and a multi-mode interference structure with embedded localized temperature-controlled photonic crystal, which can be tuned by changing the external applied temperature. The numerical simulation results are quite consistent with the analytical predictions. The design proposed here is more amenable to fabrication and also offers greater flexibility in applications of terahertz manipulation devices.

Acknowledgments

The authors would like to thank anonymous reviewers for their valuable comments to make the paper suitable for publication.

Funding

This work was supported by the National Natural Science Foundation of China Grant No. 61379024.

Authors' contributions

YL designed and performed simulations, and analyzed data. JL performed simulations, prepared the finally drafted the manuscript and the revised manuscript. Both authors read and approved the final manuscript.

Competing interests

The authors declare that they have no competing interest.

References

1. Lee, S., Choi, M., Kim, T., Lee, S., Liu, M., Yin, X., Choi, H., Lee, S., Choi, C., Choi, S., Zhang, X., Min, B.: Switching terahertz waves with gate-controlled active graphene metamaterials. Nat. Mater. **9**, 3433–3439 (2012)
2. Zhang, H., Guo, P., Chen, P., Chang, S.: Liquid-crystal-filled photonic crystal for terahertz switch and filter. J. Opt. Soc. Am. B **26**, 101–107 (2009)
3. Tao, H., Bingham, C., Strikwerda, A., Pilon, D., Shrekenhamer, D., Landy, N., Fan, K., Zhang, X., Padilla, W., Averitt, R.: Highly flexible wide angle of incidence terahertz metamaterial absorber: Design, fabrication, and characterization. Phys. Rev. B **78**, 241103(R) (2008)
4. Robinson, S., Nakkeeran, R.: Investigation on two dimensional photonic crystal resonant cavity based bandpass filter. Optik **123**, 451–457 (2012)
5. Rodriguez, B., Yan, R., Kelly, M., Fang, T., Tahy, K., Hwang, W., Jena, D., Liu, L., Xing, H.: Broadband graphene terahertz modulators enabled by intraband transitions. Nat. Commun. **3**, 780 (2012)
6. Xiao, S., Qiu, M.: Surface-mode microcavity. Appl. Phys. Lett. **87**, 111102 (2005)
7. Park, I., Lee, H., Kim, H., Moon, K., Lee, S., Hoan, B., Park, S., Lee, E.: Photonic crystal power-splitter based on directional coupling. Opt. Express **12**, 3599 (2004)
8. Berry, C., Jarrahi, M.: Broadband terahertz polarizing beam splitter on a polymer substrate. J. Infrared Millim. Terahz. Waves **33**, 127–130 (2012)
9. Homes, C., Carr, G., Lobo, R., LaVeigne, J., Tanner, D.: Silicon beam splitter for far-infrared and terahertz spectroscopy. Appl. Opt. **46**, 7884 (2007)
10. Ung, B., Fumeaux, C., Lin, H., Fischer, B., Ng, B., Abbott, D.: Low-cost ultra-thin broadband terahertz beam-splitter. Opt. Express **20**, 4968 (2012)
11. Li, J., Liu, H., Zhang, L.: Terahertz wave polarization beam splitter using a cascaded multimode interference structure. Appl. Opt. **53**, 5024 (2014)
12. Niu, T., Withayachumnankul, W., Upadhyay, A., Gutruf, P., Abbott, D., Bhaskaran, M., Sriram, S., Fumeaux, C.: Terahertz reflectarray as a polarizing beam splitter. Opt. Express **22**, 16148 (2014)
13. Ren, G., Zheng, W., Zhang, Y., Wang, K., Du, X., Xing, M., Chen, L.: Mode analysis and design of a low-loss photonic crystal 60° waveguide bend. J. Lightwave Technol. **26**, 2215 (2008)
14. Jellison, G., Burke, H.: The temperature dependence of the refractive index of silicon at elevated temperatures at several laser wavelengths. J. Appl. Phys. **60**, 841–843 (1986)
15. Corte, F., Montefusco, M., Moretti, L., Rendina, I., Cocorullo, G.: Temperature dependence analysis of the thermo-optic effect in silicon by single and double oscillator models. J. Appl. Phys. **88**, 7115–7119 (2000)
16. Cocorullo, G., Corte, F., Rendina, I.: Temperature dependence of the thermo-optic coefficient in crystalline silicon between room temperature and 550 K at the wavelength of 1523 nm. Appl. Phys. Lett. **74**, 3338 (1999)
17. Ghosh, G.: Temperature dispersion of refractive indices in crystalline and amorphous silicon. Appl. Phys. Lett. **66**, 3570 (1995)
18. Tinker, M., Lee, J.: Thermal and optical simulation of a photonic crystal light modulator based on the thermo-optic shift of the cut-off frequency. Opt. Express **13**, 7176–7188 (2005)
19. Tinker, M., Lee, J.: Thermo-optic photonic crystal light modulator. Appl. Phys. Lett. **86**, 221111 (2005)
20. Soldano, L., Pennings, E.: Optical multi-mode interference devices based on self-imaging principles and applications. J. Lightwave Technol. **13**, 615–627 (1995)
21. Zaytsev, K., Katyba, G., Kurlov, V., et al.: Terahertz photonic crystal waveguides based on sapphire shaped crystals. IEEE Trans. Terahertz Sci. Technol. **6**(4), 576–582 (2016)
22. Nielsen, K., Rasmussen, H., Adam, A., et al.: Bendable, low-loss topas fibers for the terahertz frequency range. Opt. Express **17**(10), 8592 (2009)
23. Ma, T., Guerboukha, H., Girard, M., et al.: 3D printed hollow-core terahertz optical waveguides with Hyper uniform disordered dielectric reflectors. Adv. Opt. Mater. **4**(12), 2085–2094 (2016)

6

LED-based Vis-NIR spectrally tunable light source - the optimization algorithm

M. Lukovic[1]* ⓘ, V. Lukovic[1], I. Belca[2], B. Kasalica[2], I. Stanimirovic[3] and M. Vicic[2]

Abstract

Background: A novel numerical method for calculating the contributions of individual diodes in a set of light emitting diodes (LEDs), aimed at simulating a blackbody radiation source, is examined. The intended purpose of the light source is to enable calibration of various types of optical sensors, particularly optical radiation pyrometers in the spectral range from 700 nm to 1070 nm.

Results: This numerical method is used to determine and optimize the intensity coefficients of individual LEDs that contribute to the overall spectral distribution. The method was proven for known spectral distributions: "flat" spectrum, International Commission on Illumination (CIE) standard daylight illuminant D65 spectrum, Hydrargyrum Medium-arc Iodide (HMI) High Intensity Discharge (HID) lamp, and finally blackbody radiation spectra at various temperatures.

Conclusions: The method enables achieving a broad range of continuous spectral distributions and compares favorably with other methods proposed in the literature.

Keywords: Algorithmic solution, LEDs, Calibration source, Blackbody, Optical pyrometers

Background

Numerous variants of spectral light sources based on combined radiation of individual LEDs have been reported in the past 15 years [1–4]. Each LED has its own spectral characteristic and contributes to the overall output spectrum in a relatively narrow range. As the number of newly-developed semiconductor light sources increases, covering wider and wider spectral range, this kind of construction becomes increasingly popular [5–7]. This approach allows for the generating a broad range of different output spectral distributions of almost arbitrary shape. This in turn, enables various applications such as calibration of light-measuring instruments, ambient lighting, applications in forensic science, fluorescence applications [8–10], etc.

Special case of light sources based on combined emission from a set of LEDs (where each individual LED has its own spectral distribution) are calibration sources [11–17]. Such instruments often allow for the generation of arbitrary shaped output spectrum. For example, reference [9] describes a LED-based calibration source for an ultra-sensitive spectrometry system used for electro- and photo-luminescent measurements. However, extensive search through available literature produced only a few readily implementable prescriptions for synthesizing the output spectrum of a LED-based tunable source [15, 17]. The objective of this paper is to explore the possibilities for improving spectrum synthesis methods.

We explore the possibilities of generating various spectral shapes in the very near infrared region (VNIR) using a relatively large number of individual LEDs. It is our belief that the results of this work might be useful to other researchers in this field.

Special emphasis will be placed on generating a simulated "flat" spectrum and blackbody radiation spectra in the 700 - 1070 nm range for the 800 - 1300 °C temperature interval. However, the proposed methodology should be readily applicable to other spectral ranges and/or temperatures [18].

The basic problem of synthesizing the shape of a given spectral profile is determining the intensity of each individual LED that contributes to the overall spectrum. Each LED has a relatively narrow spectral distribution as illustrated in Fig. 1 [19]. In the first approximation, the spectral distribution of a single LED will be assumed to be Gaussian. The synthesized output spectrum is a sum of the contributions from each individual LED's

* Correspondence: milentije.lukovic@ftn.kg.ac.rs
[1]University of Kragujevac, Faculty of Technical Sciences, Cacak 32000, Serbia
Full list of author information is available at the end of the article

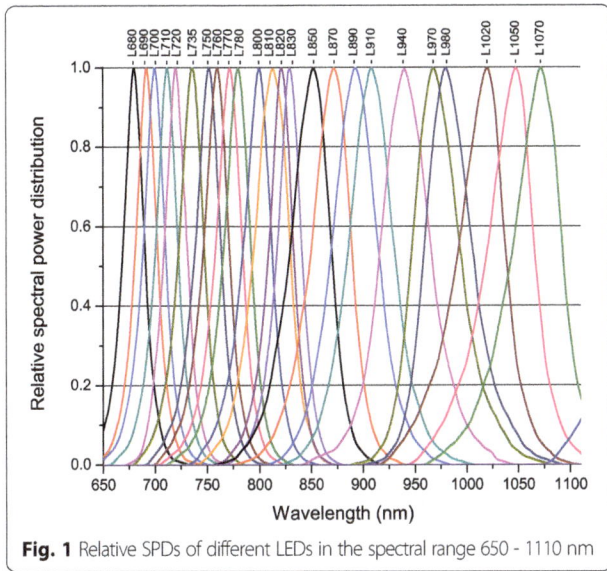

Fig. 1 Relative SPDs of different LEDs in the spectral range 650 - 1110 nm

normalized spectral power distribution (SPD) weighted by a certain factor. This factor in fact corresponds to the current that drives the particular LED, in order to get unity intensity at Gaussian center.

For a given target output spectral profile, it is therefore necessary to mathematically find the combination of values of the coefficients (driving current intensities) that best iterate the target spectrum. An algorithmic solution was developed to achieve this.

The algorithm was initially applied for the synthesis of a flat spectrum, i.e. for producing an output spectrum that has a constant intensity with respect to the wavelength in the given wavelength interval. The "flat" spectrum can be an excellent tool for direct measurements and evaluation of the responsivity function of optical sensors and systems like low-signal intensity measuring spectrometers, photo-multipliers, etc., where standard lamps and black bodies introduce large relative errors due to a great intensity variation (almost two orders of magnitude).

A further refinement of the algorithm was used to synthesize arbitrarily shaped spectrum profiles and in particular to simulate blackbody radiation. Deviation of the synthesized blackbody spectra from the theoretical curve was analyzed in detail. This was done in order to estimate the temperature reading errors when the LED-based source is used as a calibration source for optical pyrometers. A LED-based system that simulates black-body radiation would be a handy and practical solution for calibrating pyrometers in industrial installations, opposite to large blackbody furnaces.

Methods

As already mentioned, the main purpose of the proposed algorithm for a given number of LEDs is to find the coefficients (weight factors) that multiply the driving

currents of a LED in such way that the summary output spectral profile represents the best possible approximation of the desired output spectral profile.

In order to simplify calculations, we assumed that the SPDs of individual LEDs are normalized. In other words, the output intensity for the emission peak of each LED has a unity value. One has to bear in mind that in practical implementations LEDs feature different emission efficiencies. Therefore, the coefficients obtained through simulation need to be multiplied by an appropriate factor that accounts for different LED efficiencies.

Table 1 shows the relative light intensity emitted by individual LEDs per unit current (i.e. emission gain) in the wavelength interval 632-1548 nm. The LEDs used were: L680, L690, L700, L710, L720, L735, L750, L760, L770, L780, L800, L810, L820, L830, L850, L870, L890, L910, L940, L970, L980, L1020, L1050, L1070, L1200, L1300, L1450 and L1550. The labels and the intensity data were adopted from [19]. Figure 2 illustrates the typical SPD of a diode from the set (L910 in this example).

Determining the values of the sought coefficients in linear algebra comes down to solving m equations with n unknown coefficients a_j $(j = 1,...n)$ according to (1),

Table 1 Matrix M of relative LED SPDs in the spectral range from 632 nm to 1548 nm with an increment of 4 nm. LEDs: L680, L690, ..., L1550

Wavelength (nm)	LED					
	L680	L690	L700	...	L1450	L1550
632	0.006757	0	0	...	0	0
636	0.013514	0	0	...	0	0
640	0.027027	0	0	...	0	0
644	0.041892	0.006757	0	...	0	0
648	0.067568	0.013514	0.006757	...	0	0
652	0.108108	0.027027	0.02027	...	0	0
656	0.151351	0.041892	0.02973	...	0	0
660	0.22973	0.067568	0.040541	...	0	0
664	0.337838	0.108108	0.060811	...	0	0
668	0.486486	0.151351	0.090541	...	0	0
672	0.689189	0.22973	0.135135	...	0	0
676	0.891892	0.337838	0.189189	...	0	0
680	1	0.486486	0.27027	...	0	0
684	0.891892	0.689189	0.364865	...	0	0
688	0.689189	0.891892	0.5	...	0	0
692	0.445946	1	0.702703	...	0	0
696	0.27027	0.891892	0.905405	...	0	0
700	0.175676	0.689189	1	...	0	0
...	...	...	...	...	...	...
1544	0	0	0	...	0.040541	0.986486
1548	0	0	0	...	0.033784	1

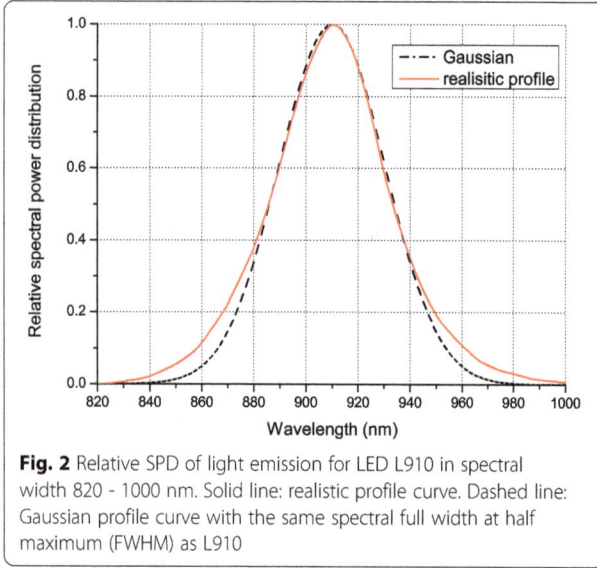

Fig. 2 Relative SPD of light emission for LED L910 in spectral width 820 - 1000 nm. Solid line: realistic profile curve. Dashed line: Gaussian profile curve with the same spectral full width at half maximum (FWHM) as L910

where **M** represents the matrix of relative SPDs derived from Table 1, depending on the number of chosen diodes n and the selected spectral bandwidth m.

$$a_1 M_{11} + a_2 M_{12} + \dots + a_n M_{1n} = I_1$$
$$a_1 M_{21} + a_2 M_{22} + \dots + a_n M_{2n} = I_2$$
$$\dots$$
$$a_1 M_{m1} + a_2 M_{m2} + \dots + a_n M_{mn} = I_m$$

$$(1)$$

The matrix form of (1) is given in (2) and (3), where **A** represents a matrix of unknown LED coefficients and **I** represent a matrix of targeted intensities. The element M_{ij} of the matrix **M** represents the spectral contribution on the i-th wavelength of the j-th LED. The matrix **M** has the dimensions $m \times n$ with non-zero elements mainly concentrated along its diagonal. Owing to the fact that $m > n$, mathematical problem (1) is actually an overdetermined system.

$$\begin{bmatrix} M_{11} M_{12} \dots M_{1n} \\ M_{21} M_{22} \dots M_{2n} \\ \dots \\ M_{m1} M_{m2} \dots M_{mn} \end{bmatrix} \cdot \begin{bmatrix} a_1 \\ a_2 \\ \dots \\ a_n \end{bmatrix} = \begin{bmatrix} I_1 \\ I_2 \\ \dots \\ I_m \end{bmatrix} \quad (2)$$

$$\mathbf{M} = \begin{bmatrix} M_{11} M_{12} \dots M_{1n} \\ M_{21} M_{22} \dots M_{2n} \\ \dots \\ M_{m1} M_{m2} \dots M_{mn} \end{bmatrix}; \quad \mathbf{A} = \begin{bmatrix} a_1 \\ a_2 \\ \dots \\ a_n \end{bmatrix}; \quad \mathbf{I} = \begin{bmatrix} I_1 \\ I_2 \\ \dots \\ I_m \end{bmatrix}$$

$$(3)$$

The optimal solution to this kind of problem can be sought by several different methods (e.g. see [20–24]).

We proposed yet another innovative approach, starting from the following assumptions:

- The target spectral profile is well-defined.
- The interval of the simulated wavelengths is covered by n LED sources.
- Each LED's SPD can be initially approximated by a Gaussian, to accelerate calculations. However, final calculations are performed using real SPDs.
- The contribution of each LED to the summary spectrum is determined by a coefficient a_j ($j = 1,\dots n$), which is proportional to the driving current of that LED.
- All coefficients a_j must be in the interval $lb_a < a_j < ub_a$. The values of the lower lb_a and the upper ub_a interval boundary depend on the shape of the target.

The system of Eq. (1) in the new approach is solved by allowing for the summed intensities I_1, I_2, $\dots, I_m$ to slightly deviate from the prescribed values of the target intensities I_{T1}, I_{T2}, $\dots, I_{Tm}$. The coefficients a_1, $a_2,\dots$, a_n are varied within their expected range and for each variation a standard deviation from the target intensities I_{T1}, I_{T2}, $\dots, I_{Tm}$ is calculated and stored. This procedure enables finding of the variation of coefficients a_1, a_2, $\dots$, a_n that yields the minimum deviation of the combined LEDs spectral distribution from the target SPD. The outline of the proposed optimization algorithm is presented in the flowchart shown in Fig. 3.

Each coefficient a_j is determined with resolution res, which gives a number of possible values for each a_j as:

$$N = (ub_a - lb_a)/res \quad (4)$$

Under these assumptions, it is possible to generate Vr different spectra (variations with repetition):

$$Vr = N^n \quad (5)$$

The criterion for selecting the best variation is minimal standard deviation from the target SPD. Taking into account the spectral range in which optical pyrometers would operate and the purpose for which it would be used, the spectral interval of interest in our research was 700 - 1070 nm. Choosing of the best values for the coefficients a_1, $a_2,\dots$, a_n by finding the variation that produces the minimum deviation of the synthesized spectrum in mentioned spectral region, was limited by the availability of LEDs on the market. Due to this constraint, we covered the interval by $n = 24$ LED models: L680, L690, L700, L710, L720, L735, L750, L760, L770, L780, L800, L810, L820, L830, L850, L870, L890, L910, L940, L970, L980, L1020, L1050 and L1070. To broaden the dynamic range, a group of four identical devices were used for each diode model. Since we wished to determine the

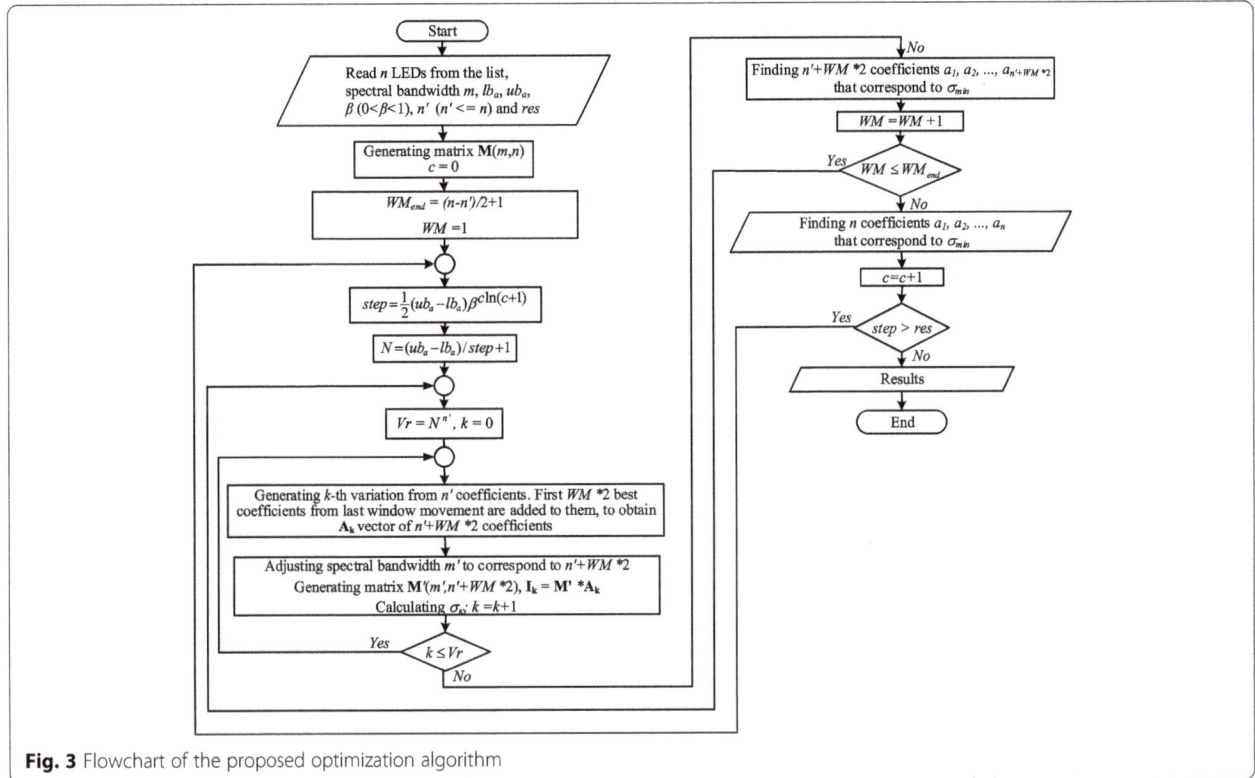

Fig. 3 Flowchart of the proposed optimization algorithm

coefficients accurately to the third decimal place ($res = 0.001$), according to (4) it followed that $N = 4000$. Based on (5), for 24 LED models the overall number of variations was $Vr = 4000^{24}$. The number of variations represented a formidable computing challenge and it was necessary to further reduce it. This was achieved by: (i) reducing the number of possible coefficient values and (ii) reducing the number of diodes that were simultaneously active during the optimization run.

The number of possible coefficient values was reduced by an iterative procedure, where $N = 3$ was kept fixed, while the interval and the resolution were simultaneously decremented.

The reduction in the number of active diodes during optimization was effectively achieved by shortening the wavelength interval for the optimization search. This procedure started by taking the first n' diodes (arranged by increasing wavelength). The number n' was chosen so that computer optimization over the shortened wavelength interval could be carried out in a reasonable time. The next step involved shifting of the "optimization window" by two spaces to the right. Thus, a new set on n' diodes underwent an optimization run. The coefficients left of the current window remained as calculated in the previous run. The procedure was repeated until the optimization window reached the rightmost diode (the diode with the longest wavelength). Figure 4 is a graphical representation of the procedure.

The search for the coefficients in the optimization window was performed using the following method: three values ($N = 3$) of the coefficients were evaluated for each iteration step and each diode. In the first

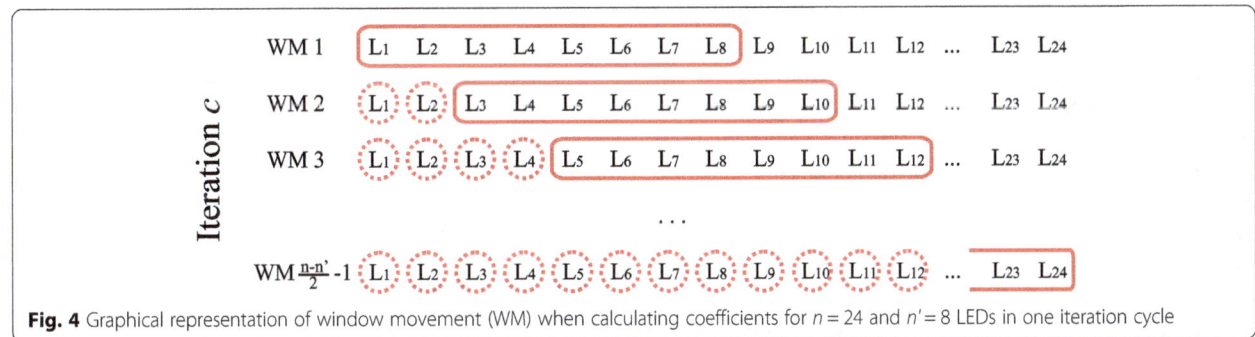

Fig. 4 Graphical representation of window movement (WM) when calculating coefficients for $n = 24$ and $n' = 8$ LEDs in one iteration cycle

iteration ($c = 0$), these values were given by (6) (brackets denote a set of elements) and represented the lower boundary (lb_a), the upper boundary (ub_a), and their arithmetic average.

$$a_{j,c=0} \in \left\{ lb_a \; ; \; \frac{ub_a - lb_a}{2} \; ; \; ub_a \right\} \tag{6}$$

The computer program evaluated all $3^{n'}$ variations of the coefficients and chose one that yielded a minimal deviation of intensities I_1, I_2, ...,I_m from the target intensities I_{T1}, I_{T2}, ...,I_{Tm}. For each evaluated coefficient variation, the criterion for minimal deviation was taken to be standard deviation σ_{min} according to (7):

$$\sigma = \sqrt{\frac{1}{m} \sum_{i=0}^{m} (I_i - I_{Ti})^2} \tag{7}$$

In the next iteration cycle, the value chosen from the previous cycle was surrounded by shorter interval boundaries (step) according to (8). The function $f(c)$ depends of the iteration number c. Selection of the step is of utmost importance for convergence. If the step is too narrow, the possibility exists that the real value of the coefficient will be missed. Conversely, if the step is too broad, the number of iterations increases and convergence is too slow.

$$step = \frac{1}{2}(ub_a - lb_a) \cdot f(c) \tag{8}$$

Numerous functions $f(c)$ were tested to find a way to systematically decrease the interval step. The simplest of these functions was given by (9).

$$f(c) = \beta^c \tag{9}$$

where β is the fitting parameter, $\beta \in (0, 1)$. However, the best results were achieved with the function given by (10):

$$f(c) = \beta^{c \cdot \ln(c+1)} \tag{10}$$

Consequently, (8) became:

$$step = \frac{1}{2}(ub_a - lb_a) \cdot \beta^{c \cdot \ln(c+1)} \tag{11}$$

After the initial iteration ($c = 0$) and determination of the best variation of the coefficients, the following iterations ($c \geq 1$) were performed in an identical manner, bearing in mind that the three possible values of each coefficient $a_{j,c}$ were chosen from the set given by (12) or, alternatively written, (13). The only condition that needed to be met was that the coefficients from the previous (c-1)-th iteration satisfy $lb_a < a_{j,c-1} < ub_a$. If a coefficient from the (c-1)-th iteration satisfied $a_{j,c-1} \leq lb_a$, than $a_{j,c}$ in the c-th iteration, it assumed values given by (14).

Finally, if a coefficient from the (c-1)-th iteration satisfied $a_{j,c-1} \geq ub_a$, then the three possible values for the c-th iteration were given by (15).

$$a_{j,c} \in \left\{ \begin{array}{c} a_{j,c-1} - step; \\ a_{j,c-1}; \\ a_{j,c-1} + step \end{array} \right\} \tag{12}$$

$$a_{j,c} \in \left\{ \begin{array}{c} a_{j,c-1} - \frac{1}{2}(ub_a - lb_a)\beta^{c \cdot \ln(c+1)} \; ; \\ a_{j,c-1} \; ; \\ a_{j,c-1} + \frac{1}{2}(ub_a - lb_a)\beta^{c \cdot \ln(c+1)} \end{array} \right\} \tag{13}$$

$$a_{j,c} \in \left\{ \begin{array}{c} lb_a \; ; \\ lb_a\left(1 - \frac{1}{2}\beta^{c \cdot \ln(c+1)}\right) + \frac{ub_a}{2}\beta^{c \cdot \ln(c+1)} \; ; \\ lb_a\left(1 - \beta^{c \cdot \ln(c+1)}\right) + ub_a\beta^{c \cdot \ln(c+1)} \end{array} \right\} \tag{14}$$

$$a_{j,c} \in \left\{ \begin{array}{c} ub_a \; ; \\ ub_a\left(1 - \frac{1}{2}\beta^{c \cdot \ln(c+1)}\right) + \frac{lb_a}{2}\beta^{c \cdot \ln(c+1)} \; ; \\ ub_a\left(1 - \beta^{c \cdot \ln(c+1)}\right) + lb_a\beta^{c \cdot \ln(c+1)} \end{array} \right\} \tag{15}$$

With each subsequent iteration, the intervals from which each coefficient was sampled, decreased. The procedure was repeated until all the intervals fall below the sought resolution res. The alternative criterion for the end of simulation was when the best variation of the coefficients' c-th iteration conceded with the best variation from the previous (c-1)-th variation.

Among various monotonically decreasing functions that we investigated, $\beta^{c \cdot \ln(c+1)}$ (Fig. 5) proved to be one of the simplest and most efficient for the determining the decreasing step during the iterations. This function

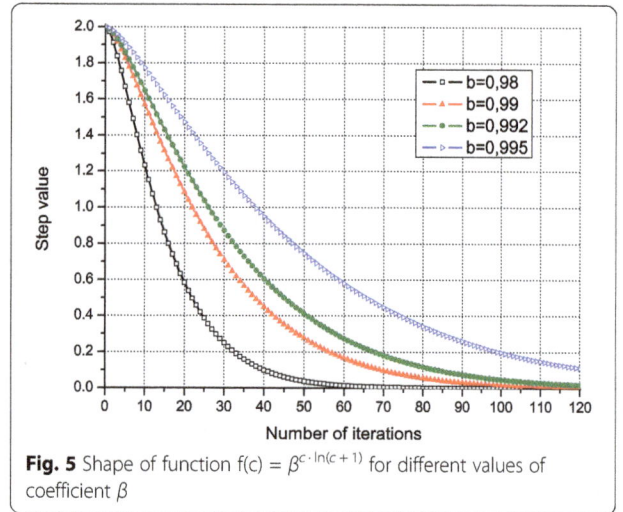

Fig. 5 Shape of function f(c) = $\beta^{c \cdot \ln(c+1)}$ for different values of coefficient β

had a single parameter β that needs to be defined prior to the simulation. The value of β determined the number of iterations. If β was small, the simulation executed quickly but was likely to miss the optimum set of coefficients. A higher value of β yielded better results at the expense of an increased number of iterations. Above certain values of β, the computing time increased with no noticeable improvement in accuracy. For most target spectral profiles, this point of diminishing returns was found to be at $\beta = 0.99$.

The logarithmic part of function (10) prevented the program from executing an unnecessarily large number of steps after a certain resolution was achieved. For example, if the currently achieved resolution is 0.01 ($lb_a = 0$, $ub_a = 4$, $\beta = 0.9$) for function (9), which has no logarithm, it takes 11 more iterations to get to the target resolution of 0.005 (Table 2). By contrast, function (10) achieves a resolution of 0.005 in only three additional steps for the same values of the coefficients (Table 3). We believe that the use of function (10) for this particular optimization problem significantly enhances the efficiency of the simulation and represents one of the most important improvements in comparison to previous algorithms.

One of the common approaches (e.g. see [15, 17]) for estimating how good the overlap is between the LED source and the target SPD is to introduce parameter p:

$$p = \frac{\sum_{380}^{780}\left|\sum_{i=1}^{n}k_i^{j-1}S_{LED_i}(\lambda) - S_{TARGET}(\lambda)\right|}{\sum_{380}^{780}S_{TARGET}(\lambda)} \qquad (16)$$

In our research we also used parameter p along with standard deviation σ as a criterion for choosing the best variation of coefficients a_j.

Table 2 Number of iteration cycles for $res = 0.005$ and function (9)

Iteration number c	Function: $\frac{1}{2}(ub_a - lb_a) \cdot \beta^c$
40	0.014781
41	0.013303
42	0.011973
43	0.010775
44	0.009698
45	0.008728
46	0.007855
47	0.007070
48	0.006363
49	0.005726
50	0.005154

Table 3 Number of iteration cycles for $res = 0.005$ and function (10)

Iteration number c	Function: $\frac{1}{2}(ub_a - lb_a) \cdot \beta^{dn(c+1)}$
15	0.012503
16	0.008428
17	0.005645
18	0.003757

Results and discussion

Simulation results

The proposed algorithm was encoded in the C programming language. The program outputs the best coefficients, the minimum achieved standard deviation, the average value of the intensities, and the maximum deviation of intensities from target values. The values of parameter p for different search steps (*step*) (Table 4) are also contained in the program's output. The results from Table 4 suggest that standard deviation of the maximum intensity deviation rapidly falls as the values of the *step* decrease.

A separate program written in C# used the text output of the main simulation and generated textual and graphical reports. Running of the actual program for different values of n' showed that it produced the best results for $n' = n$. However, with increasing n', the simulation time sharply increased and the simulation quickly became infeasible. With the computer currently at our disposal (PC, CPU 3 GHz, 4GB RAM), the limit for simulations of reasonable length was set at $n' = 13$.

Figure 6 illustrates typical simulation results for the "flat" target spectrum. It also indicates the position of the maximum of intensity deviation with respect to the intensity of the targeted spectrum. This particular example of spectrum synthesis serves a dual purpose: (i) the simplicity of the spectrum shape allows for easy analysis of the errors introduced by the algorithm and (ii) a physical device with a flat output spectrum can serve as a very useful tool for calibration and characterization measurements of various types of spectrophotometers.

Table 4 Representative values of minimum standard deviation σ_{min} (7) and parameter p (16) as a function of *step* for spectral range 700 - 1070 nm, $lb_a = 0$, $ub_a = 4$, $n' = 13$, $\beta = 0.99$

Step	σ_{min}	I_{av}	maxDev (%)	Parameter p
2.000	0.562685	1.166304	112.16	0.470257
1.520	0.369954	1.130186	67.96	0.316449
1.000	0.177625	0.997913	34.04	0.155436
0.517	0.087310	1.026825	26.60	0.068435
0.168	0.040357	1.003645	9.49	0.033639
0.015	0.021147	1.000960	5.95	0.017140
0.001	0.020861	0.999567	5.69	0.016990

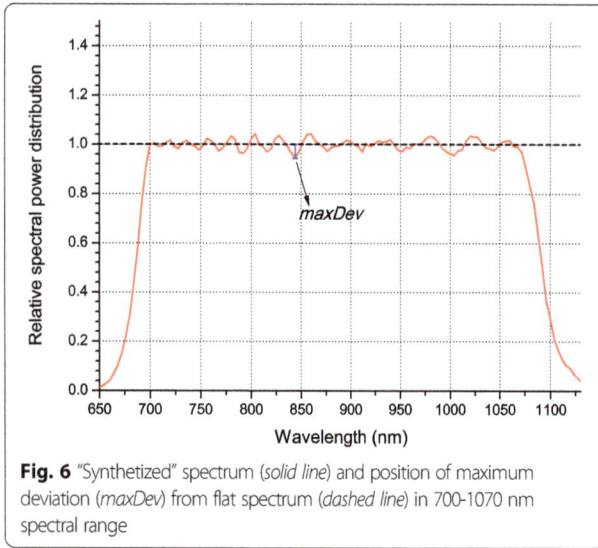

Fig. 6 "Synthetized" spectrum (*solid line*) and position of maximum deviation (*maxDev*) from flat spectrum (*dashed line*) in 700-1070 nm spectral range

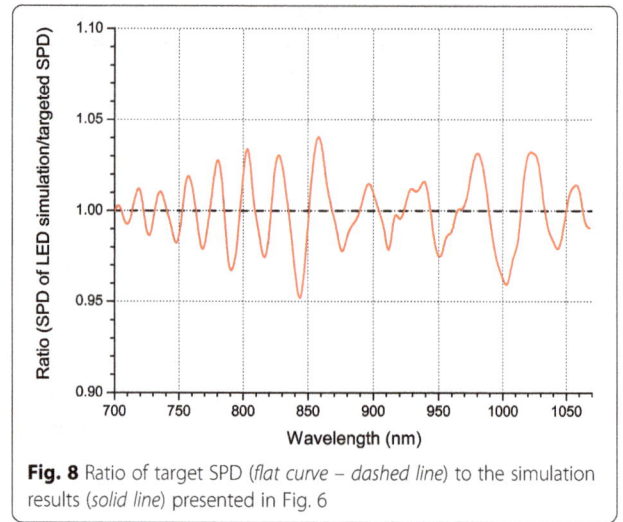

Fig. 8 Ratio of target SPD (*flat curve – dashed line*) to the simulation results (*solid line*) presented in Fig. 6

Figure 7 illustrates the typical convergence pattern of a single LED coefficient (L850 in this example), as a function of the iteration number c. Convergence of coefficients of the remaining LEDs in the array follows a similar pattern. It is readily apparent that the coefficient oscillates around the optimal value, with the amplitude of oscillations diminishing as c rises, according to the given logarithmic function (11). The iteration process is repeated until the amplitude of oscillations falls below the sought resolution.

Finally, Fig. 8 is a graphical representation of deviations from the targeted flat curve in the spectral range 700 - 1070 nm of the obtained spectrum. Due to the SPD characteristics of the LEDs used in the simulation (Figs. 1 and 2), deviations were mostly pronounced in the 840 - 860 nm spectral range (~5 %). In the rest of the spectrum they did not exceed 4 %.

Verification of the algorithm using programming package mathematica

The algorithmic solution presented in this paper was also verified with the computer algebra system MATHE-MATICA, which is highly applicable to problems that involve symbolic computations. Here, similar to what we did in our algorithmic solution, we adopted that the desired intensities I_1, I_2, ..., I_m have the same constant value, i.e. $I_1 = I_2 = ... = I_m = 1$. Also, we added ε_i, $i = \overline{1, m}$, to the right sides of the equations from (1), where ε_i is the difference between the i-th obtained intensity I_i and the desired light-emitting intensity value I equal to 1 (17, 18). We also considered the limits (19) for the unknowns a_j, $j = \overline{1, n}$.

$$\sum_{j=1}^{n} a_j M_{ij} = 1 + \varepsilon_i, \qquad i = \overline{1, m}, \qquad (17)$$

$$-1 \leq \varepsilon_i \leq 1, \qquad\qquad i = \overline{1, m}, \qquad (18)$$

$$0 \leq a_j \leq 4, \qquad\qquad j = \overline{1, n} \qquad (19)$$

The standard deviation of the dispersion of intensities (2) obtained in this way, compared to the desired intensity values equal to 1, needed to be minimized (21):

$$\sigma = \sqrt{\frac{1}{m} \sum_{i=0}^{m} (I_i - 1)^2} = \sqrt{\frac{1}{m} \sum_{i=0}^{m} \varepsilon_i^2} \qquad (20)$$

Therefore, we obtained the following single-objective optimization problem in (2), which was subjected to (17, 18, and 19).

$$\min \sqrt{\frac{1}{m} \sum_{i=0}^{m} \varepsilon_i^2} \qquad (21)$$

The above-mentioned optimization problem is a variation of the standard linear programming problem; it contains $n + m$ unknowns and is subject to the constraints

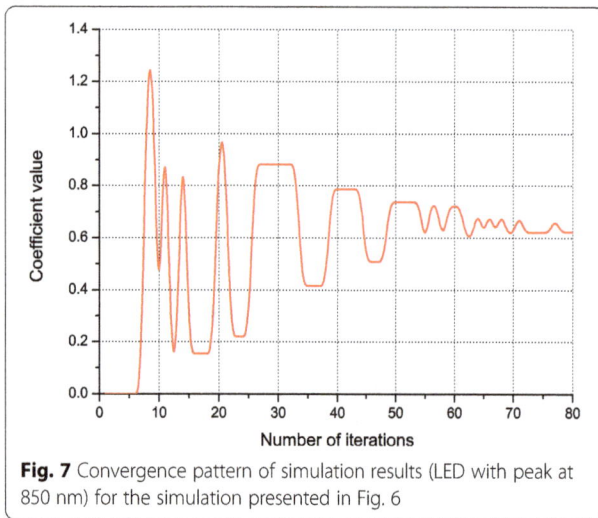

Fig. 7 Convergence pattern of simulation results (LED with peak at 850 nm) for the simulation presented in Fig. 6

determined by m equations and $n + m$ inequalities. There are several methods that can be applied to solve this problem (e.g. see Tikhonov's method [25–28]). The classical approach is to introduce so-called free variables, in order to transform the primal problem to its standard form, containing only equations. After that, the Simplex method is certainly the approach of choice for solving the obtained linear programming problem [29, 30].

A built-in MATHEMATICA function NMinimize[{f,-cons}, vars] minimizes the objective function f numerically, subject to the constraints provided by the list cons, and variables given by the list vars. Therefore, the following implementation was considered:

MinDeviation[M_List] := Module[$\{m, n, cons, f, vars\}$,

$\{m, n\}$ = Dimensions[$\mathbf{M}$];

$cons = $ Union$\left[\text{Table}\left[\sum_{j=1}^{24} \mathbf{M}[[i, j]] * a[j] == 1 + \text{eps}[i], \{i, 1, m\}\right],\right.$

$\quad$ Table$[-1 \le \text{eps}[i] \le 1, \{i, 1, m\}]$,

$\quad$ Table$\left[0 \le a[j] \le 4, \{j, 1, n\}\right]\Big]$;

$vars = $ Union[Table[eps[i], $\{i, 1, m\}$], Table[$a[j], \{j, 1, n\}$]]];

$f = $ Sqrt$\left[\dfrac{1}{m * \sum_{i=1}^{m}(\text{eps}[i]\text{^}2)}\right]$;

$\quad$ Return[NMinimize[$\{f, cons\}, vars$]];];

The matrix $\mathbf{M}$, where $\mathbf{M} = (\mathbf{M}_{ij})$, $1 \le i \le m$, $1 \le j \le n$, is an SPD matrix of the LEDs extracted from Table 1. The dimensions of the matrix are $m \times n$, where m is the selected spectral bandwidth and n is the number of selected diodes. For this matrix, the minimal standard deviation was equal to 0.019792.

Therefore, the maximal absolute declination was obtained for the coefficient $\varepsilon_{37} = 0.0565 = 5.65$ %. Notice that the constraints $-1 \le \varepsilon_i \le 1$, $i = \overline{1, m}$ in our mathematical model can be formulated as $-0.06 \le \varepsilon_i \le 0.06$, $i = \overline{1, m}$, since the maximal absolute declination never exceeded 6 % in our computations in MATHEMATICA.

The solutions obtained by means of this software showed considerable overlaps with the data received from our method (Table 5).

Comparison with an existing algorithmic solution

In order to evaluate potential merits of the newly-developed algorithm, a comparison was made with the algorithmic solution described in [15]. This previously-published algorithm solely depends on the minimization of parameter p (16). To make a meaningful comparison, the same set of assumptions were made as in [15]: individual LED spectra were taken as having a Gaussian

Table 5 LED coefficients obtained with MATHEMATICA software and our algorithmic solution for a "flat" curve in the spectral range 700 - 1070 nm

Spectral range 700 - 1070 nm			
LED	Coefficient	Mathematica	Our algorithm
L680	a_1	0.758	0.825
L690	a_2	0.020	0.002
L700	a_3	0.558	0.559
L710	a_4	0.260	0.261
L720	a_5	0.411	0.410
L735	a_6	0.560	0.560
L750	a_7	0.346	0.346
L760	a_8	0.328	0.327
L770	a_9	0.185	0.186
L780	a_{10}	0.480	0.479
L800	a_{11}	0.478	0.479
L810	a_{12}	0.401	0.401
L820	a_{13}	0	0
L830	a_{14}	0.354	0.354
L850	a_{15}	0.633	0.633
L870	a_{16}	0.240	0.240
L890	a_{17}	0.606	0.607
L910	a_{18}	0.280	0.279
L940	a_{19}	0.731	0.731
L970	a_{20}	0.012	0.012
L980	a_{21}	0.570	0.569
L1020	a_{22}	0.482	0.482
L1050	a_{23}	0.324	0.325
L1070	a_{24}	0.815	0.815
σ (standard deviation)		0.020860	0.020861
Maximum deviation (%)		5.68	5.69

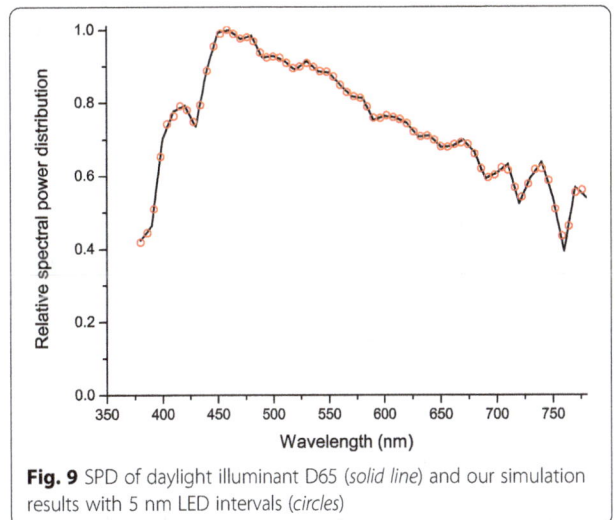

Fig. 9 SPD of daylight illuminant D65 (*solid line*) and our simulation results with 5 nm LED intervals (*circles*)

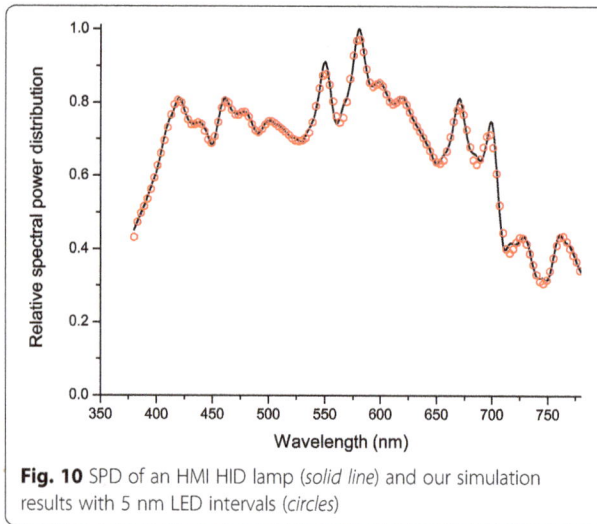

Fig. 10 SPD of an HMI HID lamp (*solid line*) and our simulation results with 5 nm LED intervals (*circles*)

shape with 20 nm FWHM. The simulations were performed for the visible spectrum (380 - 780 nm). The peek wavelengths for a given set of LEDs were chosen to be equidistant at 20, 10 and 5 nm. This effectively means that three sets of 20, 40 and 80 LEDs, respectively, were tested.

The validity of our optimization algorithm was tested for two well-known light spectra: CIE standard daylight illuminant D65 and HMI HID lamp. The simulation results with 5 nm LED intervals for the spectral range 380 – 780 nm are presented in Figs. 9 and 10. The figures show that the proposed algorithm simulated the spectra very accurately.

Table 6 shows a comparison between the results for parameter p of the two different algorithms. It is immediately apparent that the new algorithm yielded somewhat better results (lower value of p) for the 20 nm distance between peeks. For more densely populated sets of LEDs (10 nm and 5 nm inter-peak distance), our algorithm produced the same or slightly higher values of p for the CIE D65 and HMI HID lamps.

Based on the presented comparisons, we believe that the newly-developed algorithm offers some improvements, particularly in the case of sparsely populated sets of LEDs and/or target SPDs with more pronounced peeks.

Simulation of blackbody radiation

One of the SPDs of particular interest in our research was the spectrum of blackbody radiation in VNIR (700 -

1070 nm,) corresponding to blackbody temperatures above 800 °C [31, 32]. As previously mentioned, this kind of source could be useful for the calibration of optical pyrometers that operate in the range from 700 to 1070 nm.

Synthesis of Planck's curve in the given wavelength interval and for the temperatures of interest presented an additional challenge for the algorithm described in this paper. Namely, the spectra cover a very broad dynamic range with orders-of-magnitude different intensities at the endpoints. Our algorithm is limited in the sense that ub_a determines the dynamic range of the synthetized spectrum. To address this problem, the initial values of the lower (lb_a) and upper (ub_a) boundaries for the coefficients a_j were changed. Thus, for $t = 800$ °C boundaries they were $lb_a = 0$ and $ub_a = 4$, while for $t = 1300$ °C ub_a was increased to $ub_a = 30$. It was also determined that for this kind of target SPD, the optimal value of parameter β was $\beta = 0.992$.

Using the modifications mentioned in the previous paragraph, the newly-developed algorithm was applied to a set of real LEDs (using real SPDs for each LED from the set), in order to derive the intensity coefficients a_j and simulate Planck's law curve for temperatures between 800 °C and 1300 °C. Representative results of these simulations are presented in Fig. 11. Numerical results of the simulations for three different temperatures are given in Table 7, along with the values of a_j calculated using MATHEMATICA software. The similarity of the coefficients obtained by our algorithm and MATHEMATICA was yet another verification of the validity of our approach.

In order to verify that the synthesized blackbody spectra can indeed be used for calibrating optical pyrometers, the errors introduced by this non-ideal calibration source needed to be estimated. The basic assumption was that in the 800 - 1300 °C temperature range, the most common sensors used in pyrometry are PIN diodes with typical spectral sensitivity characteristics in the spectral range 600 - 1100 nm, as illustrated in Fig. 12 [33–35].

The output voltage of the PIN diode amplifier is proportional to the spectral radiance and for the spectral range (λ_1, λ_2) it depends on the blackbody temperature as:

$$\mathrm{U}(T) = C \times \int_{\lambda_1}^{\lambda_2} \mathrm{N}(\lambda, T) \mathrm{S}(\lambda) d\lambda \qquad (22)$$

Table 6 Parameter p for wavelength intervals over the spectral range from 380 nm to 780 nm

LED's peak wavelength intervals		Algorithm with function $\beta^{cln(c+1)}$			I. Fryc, S. W. Brown and Y. Ohno algorithm [15]		
		20 nm	10 nm	5 nm	20 nm	10 nm	5 nm
Parameter p for target light source	CIE D65	0.040	0.004	0.004	0.079	0.004	0.003
	HMI HID	0.042	0.011	0.010	0.074	0.008	0.007

The modeled LED distributions were based on normalized Gaussian functions with FWHM of 20 nm

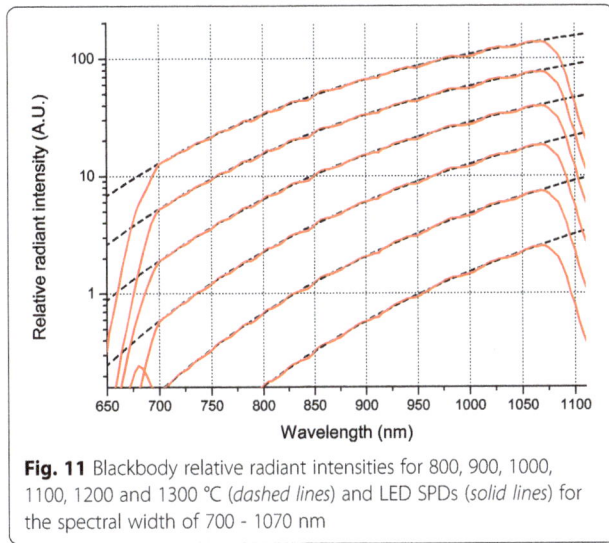

Fig. 11 Blackbody relative radiant intensities for 800, 900, 1000, 1100, 1200 and 1300 °C (*dashed lines*) and LED SPDs (*solid lines*) for the spectral width of 700 - 1070 nm

where: C is the proportionality constant which includes amplification and optical characteristics of a sensor and optical parts of the pyrometer, and $S(\lambda)$ is the spectral responsivity of the PIN diode. The quantity $N_\lambda(\lambda,T)$ is the spectral radiance in the given temperature range and can be derived from Planck's law:

$$N_\lambda(\lambda, T) = \frac{2hc^2}{\lambda^5} \frac{1}{e^{\frac{hc}{\lambda k_B T}} - 1}$$ (23)

By substituting the values of the ideal blackbody spectral radiance (23) with radiance values obtained from the simulation, it is possible to calculate the difference in temperature readings between the ideal blackbody and the synthesized calibration source.

Our calculations showed that the errors in temperature readout due to the non-ideality of the LED

Table 7 LED coefficients produced by MATHEMATICA and our algorithmic solutions for different temperatures in the spectral range 700 - 1070 nm

SPD of LEDs for Planck's law in the spectral range 700 - 1070 nm

LED	Coefficient	Temperature 800 °C		Temperature 1000 °C		Temperature 1200 °C	
		Mathematica	Our algorithm	Mathematica	Our algorithm	Mathematica	Our algorithm
L680	a_1	0.008	0.023	0.290	0.215	2.893	3.770
L690	a_2	0.004	0	0	0.012	0	0
L700	a_3	0.013	0.013	0.319	0.325	2.893	2.727
L710	a_4	0.010	0.010	0.171	0.168	1.497	1.570
L720	a_5	0.014	0.015	0.295	0.297	2.565	2.532
L735	a_6	0.031	0.030	0.542	0.542	4.407	4.417
L750	a_7	0.020	0.021	0.362	0.362	2.916	2.906
L760	a_8	0.030	0.029	0.478	0.479	3.593	3.592
L770	a_9	0.015	0.015	0.233	0.231	1.792	1.786
L780	a_{10}	0.054	0.054	0.834	0.835	6.048	6.051
L800	a_{11}	0.067	0.065	0.955	0.955	6.701	6.700
L810	a_{12}	0.074	0.074	1.008	1.008	6.753	6.754
L820	a_{13}	0	0	0	0	0	0
L830	a_{14}	0.076	0.076	1.003	1.002	6.548	6.547
L850	a_{15}	0.190	0.190	2.298	2.298	14.156	14.156
L870	a_{16}	0.072	0.072	0.930	0.930	5.803	5.804
L890	a_{17}	0.322	0.323	3.401	3.401	19.004	19.004
L910	a_{18}	0.156	0.156	1.681	1.682	9.371	9.371
L940	a_{19}	0.662	0.663	6.226	6.226	31.796	31.796
L970	a_{20}	0	0	0	0	0.029	0.029
L980	a_{21}	0.706	0.706	6.188	6.188	29.972	29.972
L1020	a_{22}	0.802	0.801	6.499	6.498	29.740	29.740
L1050	a_{23}	0.571	0.571	4.617	4.617	20.981	20.981
L1070	a_{24}	2.273	2.273	16.032	16.032	66.493	66.492
σ (standard deviation)		0.022123	0.022124	0.181485	0.181485	0.855000	0.855004
Maximum deviation (%)		4.63	4.65	4.79	4.79	5.00	5.00

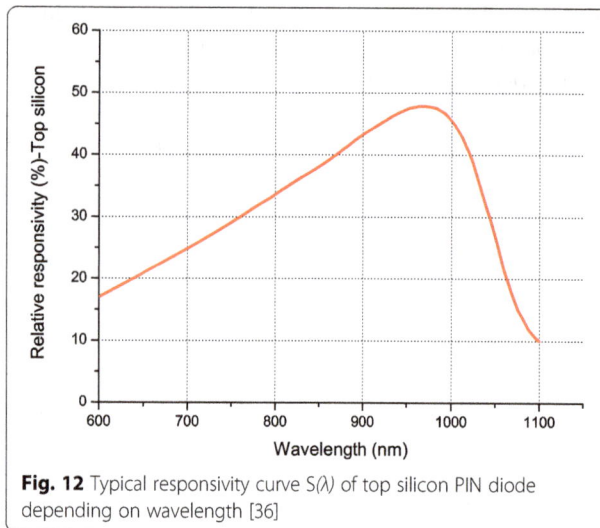

Fig. 12 Typical responsivity curve $S(\lambda)$ of top silicon PIN diode depending on wavelength [36]

source did not exceed 0.1 °C in the temperature range of interest.

Conclusion

This paper examined the feasibility of using a set of LED sources whose cumulative output simulated the "flat" spectrum as well as the blackbody radiation spectrum in the 700 - 1070 nm interval and in the 800 - 1300 °C temperature range. Our research team intends to use the actual composite LED source primarily to calibrate optical pyrometers. We developed a novel algorithm that calculates the intensities of each individual LED in the composite source, in order to achieve the desired output spectral profile. Apart from the "flat" and blackbody spectra, the algorithm was tested on various other target spectral profiles. Especially the "flat" spectrum which is very important as it might enable evaluation and direct measurements of spectral responses of various types of optical sensors and systems. We demonstrated that the proposed algorithm compares favorably with other methods for shaping the output spectral profile of tunable LED light sources. It provides an efficient theoretical base for practical realization of calibration sources.

Abbreviations
CIE: International Commission on Illumination; FWHM: Full width at half maximum; HID: High intensity discharge; HMI: Hydrargyrum medium-arc iodide; LED: Light emitting diode; SPD: Spectral power distribution; VNIR: Very near infrared region

Acknowledgment
The light spectra data of the CIE standard daylight illuminant D65 and HMI HID lamp were obtained from the National Institute of Standards and Technology (NIST), courtesy of Irena Fryc, private communication – Bialystok University of Technology, Poland.

Funding
This ongoing research is funded was funded by the Ministry of Education, Science and Technological Development under project No. 171035.

Authors' contributions
ML developed the proposed algorithm for tunable light sources and carried out all the simulations. VL programmed the algorithm in C and the graphical view of the algorithm in C#. IB proposed the specification and the initial design of the tunable light source. BK provided expert advice. IS programmed the algorithm using Mathematica software. MV coordinated the work and contributed to writing. All the authors have read and approved the final manuscript.

Competing interests
The authors declare that they have no competing interests.

Author details
[1]University of Kragujevac, Faculty of Technical Sciences, Cacak 32000, Serbia. [2]University of Belgrade, Faculty of Physics, Belgrade 11000, Serbia. [3]University of Nis, Faculty of Science and Mathematics, Nis 18000, Serbia.

References
1. Ries, H., Leike, I., Muschaweck, J.: Optimized additive mixing of colored light-emitting diode sources. Opt. Eng. **43**, 1531–1536 (2004)
2. Li, Y.-L., Shah, J.M., Leung, P.-H., Gessmann, T., Schubert, E.F.: Performance characteristics of white light sources consisting of multiple light emitting diodes. Proc. SPIE **5187**, 178–184 (2004)
3. Shang, P. Y., Tang, C. W., Huang, B. J.: Charaterizing LEDs for mixture of colored LED light sources, in 2006 Int. Conf. Electron. Mater. Packag. EMAP 2006, (IEEE, Kowloon, 2006)
4. Lu, S.S.L.W., Zhang, T., He, S.M., Zhang, B., Li, N.: Light-emitting diodes for space applications. Opt. Quant. Electron. **41**, 883–893 (2009)
5. Finlayson, G., Mackiewicz, M., Hurlbert, A., Pearce, B., Crichton, S.: On calculating metamer sets for spectrally tunable LED illuminators. J. Opt. Soc. Am. A Opt. Image Sci. Vis. **31**, 1577–1587 (2014)
6. Hirvonen, J.M., Poikonen, T., Vaskuri, A., Kärhä, P., Ikonen, E.: Spectrally adjustable quasi-monochromatic radiance source based on LEDs and its application for measuring spectral responsivity of a luminance meter. Meas. Sci. Technol. **24**, 115201–115208 (2013)
7. Mackiewicz, M., Crichton, S., Newsome, S., Gazerro, R., Finlayson, G. D., Hurlbert, A.: Spectrally tunable led illuminator for vision research, in Conf. Colour Graph. Imaging, Vision, CGIV 2012, (Society for Imaging Science and Technology, Amsterdam, 2012)
8. Kolberg, D., Schubert, F., Lontke, N., Zwigart, A., Spinner, D.M.: Development of tunable close match LED solar simulator with extended spectral range to UV and IR. Energy Procedia **8**, 100–105 (2011)
9. Kasalica, B.V., Belca, I.D., Stojadinovic, S.DJ., Zekovic, LJ.D., Nikolic, D.: Light-emitting-diode-based light source for calibration of an intensified charge-coupled device detection system intended for galvanoluminescence measurements. Appl. Spectrosc. **60**, 1090–1094 (2006)
10. O'Hagan, W.J., McKenna, M., Sherrington, D.C., Rolinski, O.J., Birch, D.J.S.: MHz LED source for nanosecond fluorescence sensing. Meas. Sci. Technol. **13**, 84–91 (2002)
11. Fryc, I., Brown, S.W., Ohno, Y.: A spectrally tunable LED sphere source enables accurate calibration of tristimulus colorimeters. Proc. SPIE **6158**, 61580E (2004)
12. Brown, S.W., Santana, C., Eppeldauer, G.P.: Development of a tunable LED-based colorimetric source. J. Res. Natl. Inst. Stand. Technol. **107**, 363–371 (2002)
13. Fryc, I., Brown, S.W., Eppeldauer, G.P., Ohno, Y.: LED-based spectrally tunable source for radiometric, photometric, and colorimetric applications. Opt. Eng. **44**, 111308–111309 (2005)
14. Burgos, F. J., Perales, E., Herrera-Ramírez, J. A., Vilaseca, M., Martínez-Verdú, F. M., Pujol, J.: Reconstruction of CIE standard illuminants with an LED-based spectrally tuneable light source, in 12th Int. AIC Congr. AIC **2013**, (2013)
15. Fryc, I., Brown, S.W., Ohno, Y.: Spectral matching with an LED-based spectrally tunable light source. Proc. SPIE **5941**(59411I), 300–308 (2005)
16. Brown, S.W., Rice, J.P., Neira, J.E., Johnson, B.C., Jackson, J.D.: Spectrally tunable sources for advanced radiometric applications. J. Res. Natl. Inst. Stand. Technol. **111**, 401–410 (2006)
17. Yuan, S.J., Huimin Yan, K.: LED-based spectrally tunable light source with optimized fitting. Chinese Opt. Lett. **12**, 32301 (2014)

18. Hsu, C.-W., Hsu, K.-F., Hwang, J.-M.: Stepless tunable four-chip LED lighting control on a black body radiation curve using the generalized reduced gradient method. Opt. Quant. Electron. **48**, 1–8 (2016)

19. Lasertechnik, R.: Roithner Lasertechnik - LEDs. 2014, <http://www.roithner-laser.com/led.html> Accessed 9 Apr 2014

20. Van Benthem, M.H., Keenan, M.R.: Fast algorithm for the solution of large-scale non-negativity-constrained least squares problems. J. Chemom. **18**, 441–450 (2004)

21. Chalmers, A., Soltic, S.: Light source optimization: spectral design and simulation of four-band white-light sources. Opt. Eng. **51**, 044003–1 (2012)

22. Kim, H., Park, H., Eldén, L.: Non-negative tensor factorization based on alternating large-scale non-negativity-constrained least squares, in Proc. 7th IEEE Int. Conf. Bioinforma. Bioeng. (BIBE, Boston, 2007)

23. Bro, R., Jong, S.: A fast non-negativity-constrained least squares algorithm. J. Chemom. **11**, 393–401 (1997)

24. Rokhlin, V., Tygert, M.: A fast randomized algorithm for overdetermined linear least-squares regression. Proc. Natl. Acad. Sci. **105**, 13212–13217 (2008)

25. Lampe, J., Voss, H.: Large-scale Tikhonov regularization of total least squares. J. Comput. Appl. Math. **238**, 95–108 (2013)

26. Golub, G.H., Hansen, P.C., O'Leary, D.P.: Tikhonov regularization and total least squares. Siam J. Matrix Anal. Appl. **21**, 185–194 (1999)

27. Wei, Y., Zhang, N., Ng, M.K., Xu, W.: Tikhonov regularization for weighted total least squares problems. Appl. Math. Lett. **20**, 82–87 (2007)

28. Beck, A., Ben-Tal, A.: On the solution of the Tikhonov regularization of the total least squares problem. SIAM J. Optim. **17**, 98–118 (2006)

29. Nash, J.C.: The (Dantzig) simplex method for linear programming. Comput. Sci. Eng. **2**, 29–31 (2000)

30. Forrest, J.J., Goldfarb, D.: Steepest-edge simplex algorithms for linear programming. Math. Program. **57**, 341–374 (1992)

31. Planck, M.: The theory of heat radiation. Search **30**, 85–94 (1914)

32. Chandrasekhar, S.: Radiative Transfer, in Energy (Dover Publication Inc, New York, 1960)

33. Photonics H.: Characteristics and use of infrared detectors, Small, 43 (2004)

34. Bellotti, E., D'Orsogna, D.: Numerical analysis of HgCdTe simultaneous two-color photovoltaic infrared detectors. IEEE J. Quantum Electron. **42**, 418–426 (2006)

35. Rogalski, A.: Infrared detectors: an overview. Infrared Phys. Technol. **43**, 187–210 (2002)

36. OSI Optoelectronics: Two Color Sandwich Detectors. Silicon Photodiodes, 2015, <http://www.osioptoelectronics.com/standard-products/silicon-photodiodes/two-color-sandwich-detectors.aspx> Accessed 2 Apr 2016.

7

Phase functions as solutions of integral equations

Margarita L. Shendeleva

Abstract

A phase function is an important characteristic of a scattering medium. A method to derive new analytic phase functions is proposed. The relation between a phase function and an angle-averaged single-scattering intensity, derived earlier [M. L. Shendeleva, J. Opt. Soc. Am. A **30**, 2169 (2013)], is considered as an integral equation for a phase function. This equation is classified as an Abel integral equation of the first kind, whose solution is known. Two phase functions newly derived with this method are presented.

Keywords: Radiative transfer, Phase function, Successive scattering orders, Integral equation, Abel, Single scattering, Anisotropy, Henyey-Greenstein

Background

The radiative transfer equation (RTE), which models light propagation in scattering media, contains two unknown functions: the radiance and the scattering phase function. Usually, the phase function is modeled separately and then inserted into the RTE. A common approach to such modelling relies on the use of Mie theory [1], which models, using the Maxwell equations, the light scattering from a single spherical particle. Mie theory was also extended to particles of other shapes [2]. In aerosols and soft tissue, various sorts of averaging over particle size distributions are applied. After averaging, the Mie phase function appears to become much smoother and can be well approximated by a co-called analytic or parametric phase function.

A widely used one-parametric phase function is that derived by Henyey and Greenstein (HG) [3]:

$$p^{HG}(\mu) = \frac{1-g^2}{4\pi(1+g^2-2g\mu)^{3/2}}, \tag{1}$$

where g is a parameter in the range $[-1, 1]$ and μ is the cosine of the scattering angle. The phase function is normalized such that

$$2\pi \int_{-1}^{1} p(\mu)d\mu = 1. \tag{2}$$

A convenient property of the HG phase function is that the parameter g is identically equal to the anisotropy (or asymmetry) factor g^{ani}, defined as

$$g^{ani} = \frac{\int_{-1}^{1} \mu p(\mu)d\mu}{\int_{-1}^{1} p(\mu)d\mu}. \tag{3}$$

Therefore, $g^{ani} = g$ for the HG phase function, where $|g^{ani}| \leq 1$.

A generalization of the HG phase function was introduced by Reynolds and McCormick [4] as

$$p^{RM}(\mu) = \frac{C}{\left(1+k^2-2k\mu\right)^{\alpha+1}}, \tag{4}$$

where C is the normalization factor,

$$C = \frac{\alpha k \left(1-k^2\right)^{2\alpha}}{\pi\left[(1+k)^{2\alpha}-(1-k)^{2\alpha}\right]}, \tag{5}$$

for $\alpha \neq 0$ and $C = k/\{2\pi ln[(1+k)/(1-k)]\}$ for $\alpha = 0$. Here α and k are real parameters, where $|k| \leq 1$.

Correspondence: shendeleva@gmail.com
Institute of Physics, 46 Prospekt Nauki, Kiev 03680, Ukraine

For $\alpha = 1/2$ the Reynolds–McCormick function reduces to the HG phase function, and, for $\alpha = 0$, it reduces to the ellipsoidal phase function

$$p^{el}(\mu) = \frac{k}{2\pi ln\left(\frac{1+k}{1-k}\right)\left(1 + k^2 - 2k\mu\right)}, \qquad (6)$$

where k is a real parameter related to the anisotropy factor as

$$g^{ani} = \frac{\left(1 + k^2\right)}{2k} - \frac{1}{ln[(1+k)/(1-k)]}, \qquad (7)$$

where $|g^{ani}| \le 1$.

More flexibility for modelling phase functions is obtained by combining two HG phase functions or a HG phase function with an isotropic phase function or with a delta function, yielding two-parameter phase functions [5] or three parameter phase functions [6].

As was pointed out by Selden [7], the main features of a phase function typically comprise a narrow forward lobe (corona), a broad diffuse background, and a narrow backscattering peak (glory). Various analytic phase functions that model these three major components were proposed by Cornette and Shanks [8], Liu [9], Draine [10] and many others (see the review of Sharma [11]). It should be noted that the influence of the choice of the analytic phase function in photon transport introduces errors in the determination of optical parameters that are difficult to evaluate [12].

Another approach to obtaining a phase function is based on understanding that the phase function can be found from the RTE itself, provided the radiance is known at some points. Such inverse problems for solving the RTE were considered by Zaneveld and Pak [13], Case [14], and McCormick [15]. The approach includes the decomposition of the phase function in a series of Legendre polynomials $P_n(\mu)$,

$$p(\mu) = \frac{1}{4\pi}\sum_{n=0}^{N}b_nP_n(\mu), \qquad (8)$$

where N is finite. The radiance is also extended to a series of Legendre polynomials and, from the RTE, one eventually obtains a system of equations for unknown coefficients b_n.

In this paper, we apply an inverse procedure of a different kind. First, we expand the radiance in successive scattering orders and then exploit the relation between the first-order angle-averaged scattering intensity and the phase function. The first part of this problem was solved, using the successive order expansion developed by Paasschens [16], in Shendeleva [17]. Here we focus on the second part. Consider first a few examples illustrating the relation between the first-order scattering intensity and the phase function.

For the isotropic phase function,

$$p^{iso} = 1/(4\pi), \qquad (9)$$

the first-order angle-averaged intensity generated by an instantaneous point source is [16]

$$I_1^{iso} = \frac{H(vt-r)e^{-vt/l_s}}{4\pi vtrl_s}ln\left(\frac{vt+r}{vt-r}\right), \qquad (10)$$

where r is the distance from the point source, t is the time from the moment of photon emission, v is the speed of light in the medium, l_s is the scattering length, and H (x) is the Heaviside step function that equals zero for $x < 0$ and one for $x \ge 0$.

For the linear phase function

$$p^{lin}(\mu) = (1 + 3g\mu)/(4\pi), \qquad (11)$$

where parameter $|g| \le 1/3$, the angle-averaged single-scattering intensity has been found as [17, 18]

$$I_1^{lin} = \frac{e^{-vt/l_s}H(1-u)}{4\pi rvtl_s}\left\{\left(1 + 3gu^2\right)ln\left(\frac{1+u}{1-u}\right) - 6gu\right\}, \qquad (12)$$

where $u = r/(vt)$.

Generally, there is a one-to-one correspondence between a phase function and a first-order scattering intensity. For a given phase function, the first-order angle-averaged intensity (also called a single-scattering intensity) can be found from Eq. (21) of Shendeleva [17]. In this paper, we consider an inverse problem: Given the fist-order angle-averaged intensity, find the corresponding phase function. The integral equation for this purpose is derived in the next section. Fortunately, this equation happens to be an integral equation of the Abel type, whose solutions are known [19]. Sections 3 and 4 provide examples of the application of this equation.

Derivation of the integral equation

The time-dependent RTE for radiance $L(\overrightarrow{r}, t, \hat{s})$ is considered in the form

$$\frac{1}{v}\frac{\partial L(\overrightarrow{r}, t, \hat{s})}{\partial t} + \hat{s}\nabla L(\overrightarrow{r}, t, \hat{s}) + (\mu_s + \mu_a)L(\overrightarrow{r}, t, \hat{s})$$

$$= \mu_s\int_{4\pi}p(\hat{s}\cdot\hat{s}')L(\overrightarrow{r}, t, \hat{s}')d\hat{s}' + \frac{1}{v}S(\overrightarrow{r}, t), \qquad (13)$$

where $p(s \cdot \hat{s}')$ is a phase function, normalized such that $\int_{4\pi}p(\hat{s}\cdot\hat{s}')d\hat{s}' = 1$, where $\hat{s}'$ and $\hat{s}$ are directions before and after a scattering event, respectively, and μ_s and μ_a are the scattering and absorption coefficients, respectively, related to the scattering length $l_s = 1/\mu_s$ and to the absorption length $l_a = 1/\mu_a$. An instantaneous point source is represented by delta functions,

$$S(\vec{r}, t) = \frac{N_0 \hbar \omega v}{4\pi} \delta(r)\delta(t), \quad (14)$$

where $\hbar\omega$ is photon's energy, and N_0 is the number of photons emitted at $t = 0$. Note that, in the following, we consider a non-absorbing case, since absorption enters into the solution through the exponential factor $Exp(-\mu_a vt)$ (16).

The multiple collision approach uses the decomposition of the radiance in successive scattering orders

$$L(\vec{r}, t, \hat{s}) = \sum_{N=0}^{\infty} L_N(\vec{r}, t, \hat{s}), \quad (15)$$

where the term with $N = 0$ describes the unscattered radiance, the term with $N = 1$ corresponds to the radiance scattered once, and so forth. Correspondingly, for the angle-averaged intensity, which is defined as

$$I(\vec{r}, t) = \frac{1}{4\pi} \int_{4\pi} L(\vec{r}, t, \hat{s}) d\hat{s}, \quad (16)$$

we have the expansion

$$I(\vec{r}, t) = \sum_{N}^{\infty} I_N(\vec{r}, t), \quad (17)$$

where $I_N(\vec{r}, t)$ is an angle-averaged intensity of the N-s order defined as $I_N = \int_{4\pi} L_N(\vec{r}, t, \hat{s}) d\hat{s}/(4\pi)$.

For the first-order scattering, the relation between a phase function and first-order intensity is obtained as [17]

$$I_1(r, t) = \frac{e^{-vt/l_s}}{l_s} \int_{-1}^{1} \frac{p(\mu) d\xi}{(vt)^2 + r^2 - 2vtr\xi}, \quad (18)$$

where μ is the cosine of the scattering angle θ and ξ is the cosine of the angle α between the direction from the source to the observation point and the direction from the scattering point to the observation point (as shown in Fig. 1 of Shendeleva [17]). Using the notation $r = 2c$, $vt = 2a$, and $c/a = r/(vt) = u$, we can express cosines μ and ξ as

$$\xi = \frac{c - x}{a - ux}, \quad (19)$$

$$\mu = \frac{2c^2 - a^2 - u^2 x^2}{a^2 - u^2 x^2}, \quad (20)$$

where x is a coordinate along the x-axis in the range $[-a, a]$.

Changing the integration variable in Eq. (18) from ξ to μ, we transform Eq. (18) to

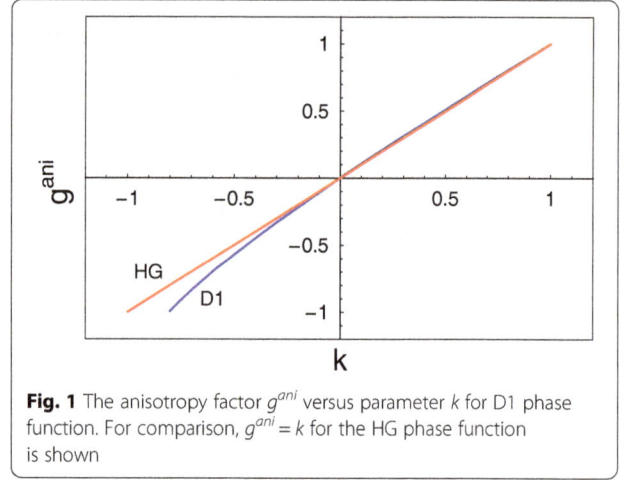

Fig. 1 The anisotropy factor g^{ani} versus parameter k for D1 phase function. For comparison, $g^{ani} = k$ for the HG phase function is shown

$$I_1(r, t) = \frac{e^{-vt/l_s}}{vtrl_s} \int_{-1}^{X} \frac{p(\mu) d\mu}{\sqrt{1-\mu}\sqrt{X-\mu}}, \quad (21)$$

with $\chi = 2u^2 - 1$, where χ varies in the range $[-1, 1]$.

Eq. (21) takes the form

$$\int_{-1}^{X} \frac{p(\mu) d\mu}{\sqrt{1-\mu}\sqrt{X-\mu}} = F(\chi) \quad (22)$$

and thus can be classified as an Abel integral equation of the first kind. The solution of this equation is known to be [19]

$$p(\mu) = \frac{\sqrt{1-\mu}}{\pi} \int_{-1}^{\mu} \frac{F'(\tau) d\tau}{\sqrt{\mu-\tau}} + \frac{F(-1)\sqrt{1-\mu}}{\sqrt{1+\mu}}, \quad (23)$$

where $F'(\tau)$ means the derivative with respect to the argument. Note also the normalization conditions in Eq. (23). From the normalization of the phase function given by Eq. (2), one obtains the normalization for $F(\chi)$,

$$\int_{-1}^{1} F(\tau) d\tau = \frac{1}{\pi}, \quad (24)$$

and therefore a single-scattering intensity is normalized as

$$4\pi \int_{0}^{vt} I_1(r, t) r^2 dr = \frac{vt}{l_s} e^{-vt/l_s}, \quad (25)$$

which is consistent with the normalization obtained by Shendeleva [17].

Examples of solutions
Ellipsoidal phase function
To test the method, consider an ellipsoidal phase function given by Eq. (6). By direct substitution of this phase function in Eq. (18), one can find that the ellipsoidal phase function corresponds to the single-scattering intensity [17]

$$I_1^{el}(r,t) = \frac{ke^{-\frac{vt}{l_s}}}{2\pi r v t l_s (1-k) ln\left(\frac{1+k}{1-k}\right)\sqrt{1+k^2-2k\chi}}$$

$$ln\left(\frac{\sqrt{1+k^2-2k\chi}+(1-k)\sqrt{(\chi+1)/2}}{\sqrt{1+k^2-2k\chi}-(1-k)\sqrt{(\chi+1)/2}}\right),$$

(26)

where $\chi = 2r^2/(vt)^2 - 1$.

Vice versa, considering the integral equation (22) with function

$$F^{el}(\chi) = \frac{k}{2\pi(1-k)ln\left(\frac{1+k}{1-k}\right)\sqrt{1+k^2-2k\chi}}$$

$$ln\left(\frac{\sqrt{1+k^2-2k\chi}+(1-k)\sqrt{(\chi+1)/2}}{\sqrt{1+k^2-2k\chi}-(1-k)\sqrt{(\chi+1)/2}}\right),$$

(27)

one obtains, after some manipulations, the ellipsoidal phase function (6).

The following considers the integral equations (22) with the right-hand side of the form

$$F^{Da}(\chi) = \frac{K}{(1+k^2-2k\chi)^a}ln\left(\frac{\sqrt{1+k^2-2k\chi}+(1-k)\sqrt{(\chi+1)/2}}{\sqrt{1+k^2-2k\chi}-(1-k)\sqrt{(\chi+1)/2}}\right),$$

(28)

where K is the normalization factor that depends on parameter k. The superscript Da indicates that expression $(1+k^2-2k\chi)$ in the denominator enters in degree a. Consider the following two particular cases.

Phase function D1

In this case, $a = 1$ and the right-hand side of Eq. (22) takes the form

$$F^{D1}(\chi) = \frac{K}{1+k^2-2k\chi}ln\left(\frac{\sqrt{1+k^2-2k\chi}+(1-k)\sqrt{(\chi+1)/2}}{\sqrt{1+k^2-2k\chi}-(1-k)\sqrt{(\chi+1)/2}}\right),$$

(29)

where K is found from the normalization condition (24) as

$$K = 1/\int_{-1}^{1}\frac{\pi}{1+k^2-2k\chi}ln\left(\frac{\sqrt{1+k^2-2k\chi}+(1-k)\sqrt{\chi+1}/\sqrt{2}}{\sqrt{1+k^2-2k\chi}-(1-k)\sqrt{\chi+1}/\sqrt{2}}\right)d\chi.$$

(30)

Taking the derivative, one obtains

$$\left(F^{D1}(\tau)\right)' = \frac{2kK}{(1+k^2-2k\tau)^2}ln\left(\frac{\sqrt{1+k^2-2k\tau}+(1-k)\sqrt{\tau+1}/\sqrt{2}}{\sqrt{1+k^2-2k\tau}-(1-k)\sqrt{\tau+1}/\sqrt{2}}\right)$$

$$+\frac{\sqrt{2}(1-k)K}{\sqrt{1+\tau}(1-\tau)(1+k^2-2k\tau)^{3/2}}.$$

(31)

Thus, the solution for normalized phase function p^{D1} is found to be

$$p^{D1}(\mu) = \frac{\sqrt{1-\mu}}{\pi^2}\int_{-1}^{\mu}\left[\frac{2k}{(1+k^2-2k\tau)^2}ln\left(\frac{\sqrt{1+k^2-2k\tau}+(1-k)\sqrt{\tau+1}/\sqrt{2}}{\sqrt{1+k^2-2k\tau}-(1-k)\sqrt{\tau+1}/\sqrt{2}}\right)\right.$$

$$\left.+\frac{\sqrt{2}(1-k)}{\sqrt{1+\tau}(1-\tau)(1+k^2-2k\tau)^{3/2}}\right]\frac{d\tau}{\sqrt{\mu-\tau}}$$

$$/\int_{-1}^{1}ln\left(\frac{\sqrt{1+k^2-2k\tau}+(1-k)\sqrt{\tau+1}/\sqrt{2}}{\sqrt{1+k^2-2k\tau}-(1-k)\sqrt{\tau+1}/\sqrt{2}}\right)\frac{d\tau}{(1+k^2-2k\tau)}.$$

(32)

It should be noted that $p^{D1}(\mu)$ takes on negative values for $k < -0.8$; therefore, the range of parameter k should be restricted to $-0.8 < k < 1$.

The anisotropy factor for this function is calculated as

$$g^{ani} = \frac{\int_{-1}^{1}\mu p(\mu)d\mu}{\int_{-1}^{1}p(\mu)d\mu} = \frac{\int_{-1}^{1}(1-\tau)(1+3\tau)F'(\tau)d\tau}{4\int_{-1}^{1}F(\tau)d\tau}.$$

(33)

The plot of the anisotropy factor versus parameter k is shown in Fig. 1. Note that, for $-0.2 < k < 1$, the anisotropy factor closely follows the linear dependence $g^{ani} = k$, which is characteristic of the HG phase function. In contrast to the HG phase function, the phase function D1 has a simple analytic form for the single-scattering angle-averaged intensity. It is shown in the Appendix that this fact can be used to approximate the single-scattering intensity for the HG phase function by the single-scattering intensity for the D1 phase function. A plot for the phase function $p^{D1}(\mu)$ versus the scattering angle θ is shown in Fig. 2.

Phase function D4

Consider the integral equation (22) with the function

$$F^{D4}(\chi) = \frac{Q}{(1+k^2-2k\chi)^4}ln\left(\frac{\sqrt{1+k^2-2k\chi}+(1-k)\sqrt{(\chi+1)/2}}{\sqrt{1+k^2-2k\chi}-(1-k)\sqrt{(\chi+1)/2}}\right).$$

(34)

The notation D4 means that expression $(1+k^2-2k\chi)$ enters the denominator in degree 4.

Here, Q is the normalizing factor, which can be analytically calculated as follows. Changing back to the

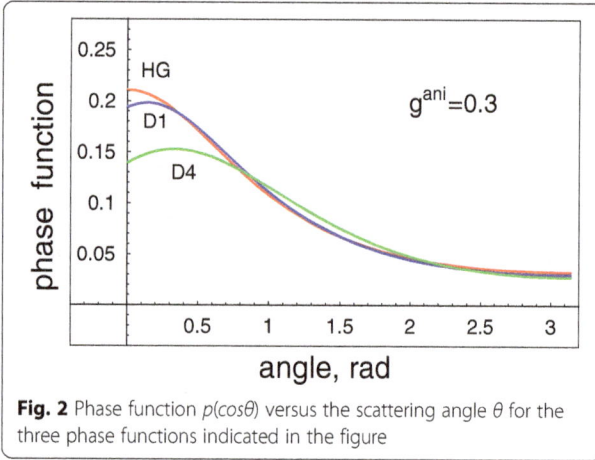

Fig. 2 Phase function $p(cos\theta)$ versus the scattering angle θ for the three phase functions indicated in the figure

variable $u^2 = (\chi + 1)/2$ in Eq. (34), one can write the normalization condition as

$$\int_0^1 \frac{Q ln\left(\frac{\sqrt{(1+k)^2-4ku^2}+(1-k)u}{\sqrt{(1+k)^2-4ku^2}-(1-k)u}\right)u\,du}{\left((1+k)^2-4ku^2\right)^4} = \frac{1}{4\pi}. \quad (35)$$

Evaluating the integral with the use of Mathematica, we obtain

$$Q = \frac{45(1-k)^6(1+k)^6}{4\pi\left(45 + 60k + 158k^2 + 60k^3 + 45k^4\right)}. \quad (36)$$

Therefore, the single-scattering intensity corresponding to $F^{D4}(\chi)$ can be written in the form

$$I_1^{D4} = \frac{45(1-k)^6(1+k)^6 e^{-vt/l_s}H(1-u)}{4\pi vtrl_s \left(45 + 60k + 158k^2 + 60k^3 + 45k^4\right)\left[(1+k)^2-4ku^2\right]^4}$$
$$ln\left(\frac{\sqrt{(1+k)^2-4ku^2}+(1-k)u}{\sqrt{(1+k)^2-4ku^2}-(1-k)u}\right). \quad (37)$$

Calculating the derivative

$$\left(F^{D4}(\tau)\right)' =$$
$$= \frac{8kQ}{(1+k^2-2k\tau)^5} ln\left(\frac{\sqrt{1+k^2-2k\tau}+(1-k)\sqrt{\tau+1}/\sqrt{2}}{\sqrt{1+k^2-2k\tau}-(1-k)\sqrt{\tau+1}/\sqrt{2}}\right)$$
$$+ \frac{\sqrt{2}(1-k)Q}{(1-\tau)\sqrt{1+\tau}(1+k^2-2k\tau)^{9/2}}, \quad (38)$$

one obtains the phase function in the form

$$p^{D4}(\mu) = \frac{Q\sqrt{1-\mu}}{\pi} \int_{-1}^{\mu} \left[\frac{8k}{(1+k^2-2k\tau)^5}\right.$$

$$ln\left(\frac{\sqrt{1+k^2-2k\tau}+(1-k)\sqrt{(\tau+1)/2}}{\sqrt{1+k^2-2k\tau}-(1-k)\sqrt{(\tau+1)/2}}\right)$$

$$\left. + \frac{\sqrt{2}(1-k)}{(1-\tau)\sqrt{1+\tau}(1+k^2-2k\tau)^{9/2}}\right]\frac{d\tau}{\sqrt{\mu-\tau}}, \quad (39)$$

where Q is given by Eq. (36).

Figure 3 shows the anisotropy factor g^{ani} versus parameter k. It can be seen that the curve $g^{ani}(k)$ is closely approximated by the curve

$$g^{ani} = 1 - (1-k)^2/(1+k)^2 \quad (40)$$

for $0 \le k \le 1$. A polar plot for the phase function $p^{D4}(\mu)$, and also for $p^{D1}(\mu)$, is shown in Fig. 4. It should be noted that, for polar plots, it is convenient to make the change of variable $\tau + 1 = (1+\mu)(1-y^2)$, which transforms the integral in Eq. (39) to the integral with constant limits

$$p^{D4}(\mu) = \frac{2Q\sqrt{1-\mu^2}}{\pi}\int_0^1 \left[\frac{8k}{[(1+k)^2-2k(1+\mu)(1-y^2)]^5}\right.$$

$$ln\left(\frac{\sqrt{(1+k)^2-2k(1+\mu)(1-y^2)}+(1-k)\sqrt{\frac{(1+\mu)(1-y^2)}{2}}}{\sqrt{(1+k)^2-2k(1+\mu)(1-y^2)}-(1-k)\sqrt{\frac{(1+\mu)(1-y^2)}{2}}}\right)$$

$$\left. + \frac{\sqrt{2}(1-k)}{\sqrt{1+\mu}\sqrt{1-y^2}[1-\mu+y^2(1+\mu)]((1+k)^2-2k(1+\mu)(1-y^2))^{9/2}}\right]dy. \quad (41)$$

The use of the phase function D4 will be shown elsewhere.

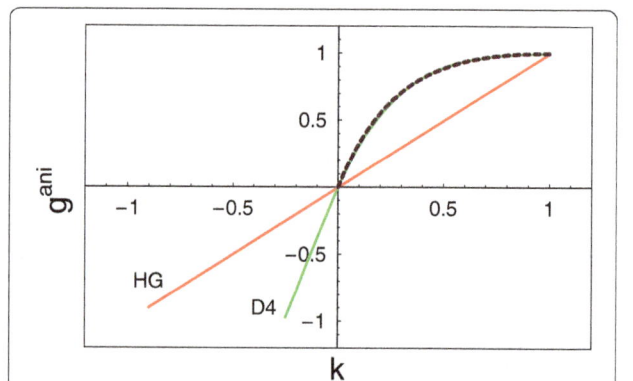

Fig. 3 The anisotropy factor g^{ani} versus parameter k for the D4 phase function and the HG phase function. The curve $g^{ani} = 1 - (1-k)^2/(1+k)^2$ for $0 < k < 1$ is indicated by the dashed line

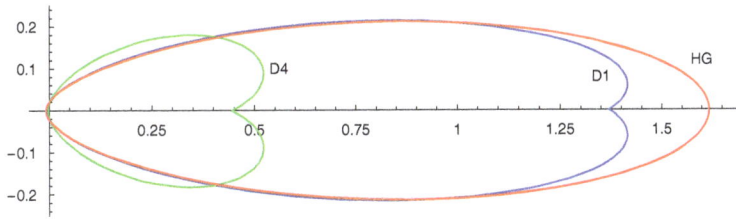

Fig. 4 Polar plot for the D1, D4 and HG phase functions for anisotropy factor $g^{ani} = 0.71$

Conclusion

We have derived an integral equation that relates a phase function and an angle-averaged intensity for the first-order scattering, Eq. (21). The solution of this equation, classified as an Abel integral equation of the first kind, can be readily written down as given by Eq. (23). Then, we consider several examples of application of this equation. The first example is merely illustrative, since it concerns the ellipsoidal phase function for which the first-order intensity is known. This example is useful in the sense that it gives an idea of what the first-order intensity can look like. In the two subsequent examples, small modifications of the fist-order intensity allow us to derive two new one-parameter phase functions. The first one, denoted by D1, has the useful property that its anisotropy factor is practically identical to the parameter of the phase function (in the range $-0.2 < k < 1$). The same property is characteristic of the HG phase function. Although the function D1 is more complicated than the HG phase function, it has a very simple first-order scattering intensity. The second phase function, D4, has a simple algebraic relation (in the range $0 < k < 1$) between the anisotropy factor and the parameter of the phase function, as given by Eq. (40). These examples show that, through small modifications of the single-scattering intensity, one can derive new phase functions with useful properties.

Appendix

Approximation of the single-scattering intensity for the HG phase function

For the HG phase function, the single-scattering intensity is derived in the form [17, 20]

$$I_1^{HG}(r,t) = \frac{(1-g^2)e^{-vt/l_s}}{2\pi l_s} \int_0^1 \frac{(v^2t^2-r^2\lambda^2)^{1/2}d\lambda}{\left[(v^2t^2-r^2\lambda^2)(1-g)^2 + 4g(v^2t^2-r^2)\right]^{3/2}}.$$

(42)

Reilly and Warde [20] derived this expression for use in non–line-of-sight communications. For practical purposes, it would be useful to find an approximate analytical expression for this integral. Here, we propose approximating this expression by the first-order intensity found for the D1 phase function, since it has $k \approx g^{ani}$ (for $-0.2 < k < 1$), similar to $g = g^{ani}$ for HG. Thus, we obtain

$$I_1^{HG}(r,t) \approx \frac{K(g)e^{-vt/l_s}}{vtrl_s[(1+g)^2-4gu^2]} ln\left(\frac{\sqrt{(1+g)^2-4gu^2}+(1-g)u}{\sqrt{(1+g)^2-4gu^2}-(1-g)u}\right),$$

(43)

for $-0.2 < g < 1$. Here, $K(g)$, defined by Eq. (30), depends only on g and, therefore, the dependence of $I_1^{HG}(r,t)$ on r and t is obtained in analytical form. Moreover, for $K(g)$, we find an approximation $K(g) \approx 0.05g + 0.08$ for $-0.2 < g < 1$. As a spinoff, we also obtain an approximation for the integral

$$\int_0^1 \frac{(1-u^2\lambda^2)^{1/2}d\lambda}{\left[(1-u^2\lambda^2)(1-g)^2 + 4g(1-u^2)\right]^{3/2}} \approx \frac{2\pi(0.05g+0.08)}{u(1-g^2)\left[(1+g)^2-4gu^2\right]}$$

$$ln\left(\frac{\sqrt{(1+g)^2-4gu^2}+(1-g)u}{\sqrt{(1+g)^2-4gu^2}-(1-g)u}\right),$$

(44)

where $-0.2 < g < 1$ and $0 < u < 1$. Denoting the left-hand side of the above equation by Υ_L and the right-hand side by Υ_R, we calculate the relative error of this approximation as $\Delta = (\Upsilon_R - \Upsilon_L)/\Upsilon_L$. The relative error Δ for various parameters g in shown in Fig. 5. It can be seen that the error is biggest for $u = 0$. The error, which is bigger for large g, is decreasing in absolute value with increasing u and, in the region $0.8 < u < 1$, it is less than 5%.

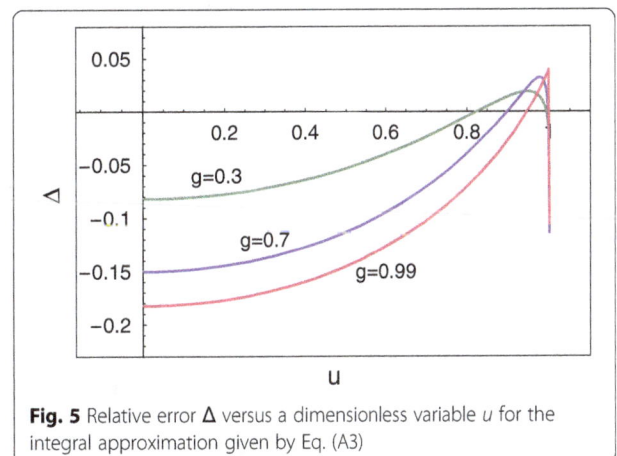

Fig. 5 Relative error Δ versus a dimensionless variable u for the integral approximation given by Eq. (A3)

Competing interests
The author declares that she has no competing interests.

References
1. Mie, G.: Contributions to the optics of turbid media, particularly of colloidal metal solutions. [in German]. Annal. Phys. **25**(N3), 377–445 (1908)
2. Bohren, C.F., Huffman, D.R.: Absorption and Scattering of Light by Small Particles. Wiley, New-York (1983)
3. Henyey, L.G., Greenstein, J.L.: Diffuse radiation in the galaxy. Astrophys. J. **93**, 70–83 (1941)
4. Reynolds, L.O., McCormick, N.J.: Approximate two-parameter phase function for light scattering. J. Opt. Soc. Am. **70**, 1206–1212 (1980)
5. Irvin, W.M.: Multiple scattering by large particles. Astrophys. J. **142**, 1563–1575 (1965)
6. Kattawar, G.W.: A three-parameter analytic phase function for multiple scattering calculations. J. Quant. Spectrosc. Radiat. Transf. **15**, 839–849 (1975)
7. Selden, A.C.: Attenuation and impulse response for multiple scattering of light in atmospheric clouds and aerosols. Appl. Opt. **45**, 3144–3151 (2006)
8. Cornette, W.M., Shanks, J.G.: Physically reasonable analytic expression for the single-scattering phase function. Appl. Opt. **31**, 3152–3160 (1992)
9. Liu, P.: A new phase function approximating to Mie scattering for radiative transport equations. Phys. Med. Biol. **39**, 1025–1036 (1994)
10. Draine, B.T.: Scattering by interstellar dust grains. I. Opt. Ultraviolet Astrophys. J **598**, 1017–1025 (1992)
11. Sharma, S.K.: A review of approximate analytic light-scattering phase functions. In: Kokhanovsky, A.A. (ed.) Light Scattering Reviews, vol. 9, pp. 53–100. Springer, Berlin (2015)
12. Selden, A.C.: Photon transport parameters of diffuse media with highly anisotropic scattering. Phys. Med. Biol. **49**, 3017–3027 (2004)
13. Zaneveld, J.R.V., Pak, H.: Some aspects of the axially symmetric submarine daylight field. J. Geophys. Res. **77**, 2677–2680 (1972)
14. Case, K.M.: Inverse problem in transport theory. Phys. Fluids **16**, 1607–1611 (1973)
15. McCormick, N.J.: Ocean optics phase-function inverse equations. Appl. Opt. **31**, 4958–4961 (2002)
16. Paasschens, J.C.J.: Solution of time-dependent Boltzmann equation. Phys. Rev. E **56**, 1135–1141 (1997)
17. Shendeleva, M.L.: Single-scattering solutions to radiative transfer in infinite turbid media. J. Opt. Soc. Am. A **30**, 2169–2174 (2013)
18. Fomenko, V.N., Shvarts, F.M., Shvarts, M.A.: Exact description of photon migration in anisotropically scattering media. Phys. Rev. E **61**, 1990–1995 (2000)
19. Bitsadze, A.V.: Integral Equations of First Kind. World Scientific, Singapore (1995)
20. Reilly, D.M., Warde, C.: Temporal characteristics of single-scatter radiation. J. Opt. Soc. Am. **69**, 464–470 (1979)

8

Scattering from metamaterial coated nihility sphere

A. Ghaffar[1*], M. M. Hussan[1], Majeed A. S. Alkanhal[2] and Sajjad ur Rehman[2]

Abstract

Background: To show the Back-Scattering efficiencies of nihility sphere become non-zero, scattering from metamaterial coated nihility sphere has been carried out in the presented work.

Methods: The field phasors are expanded in terms of spherical wave vector function along with assumed scattering coefficients. Boundaries conditions are applied at each interface, i.e., free space- metamaterial coating and metamaterial coating-nihility sphere core.

Results: Scattering coefficients are obtained by using boundary conditions at each interface. The scattering efficiencies are obtained in graphical form by varying both permittivity and permeability of metamaterial using MATHEMATICA software. Under some special conditions obtained results are compared with already published literature to show the correctness of present formulations.

Conclusion: It is observed that Back-Scattering efficiencies of nihility sphere becomes non-zero when we introduce a metamaterial coating layer on it. It is also noticed that the scattering efficiencies are dependent on coating thickness as well as permittivity and permeability of metamaterials.

Keywords: Nihility, Sphere, Metamaterial, Scattering, Forward-scattering efficiency, Back-scattering and Extinction efficiency

Background

A lot of engineers, researchers and professional's belonging to optical society pay a vital consideration towards those materials which were artificially designed, which we often call as metamaterial. Because metamaterials are tremendously good to tune and control the electromagnetic properties at vide range of frequencies which is not possible for naturally occurring materials. Many researchers used metamaterials as filters, phase shifters, perfect reflectors, electromagnetic invisibility cloak, and wave guiders [1–5]. Some characteristic metamaterials which were being considered widely in already published work are plasma, Perfect Electromagnetic Conductor(PEMC), chiral [6], chiral nihility and split ring resonator (SRR) [7–9]. The nihility medium was considered as "the electromagnetically nilpotent" and has the most surprising impression in the field of electromagnetics and optics, in this material the relative permeability and relative permittivity both have zero magnitude.

* Correspondence: aghaffar16@uaf.edu.pk
[1]Department of Physics, University of Agriculture, Faisalabad, Pakistan
Full list of author information is available at the end of the article

Nihility material introduced by Lakhtakia [10, 11] attracted many researchers because wave cannot propagate in nihility medium. Lakhtakia postulated that nihility medium is a medium whose permittivity and permeability both are null valued [10–13]. Due to null valued permittivity and permeability this medium does not allow the wave to propagate. Maxwell equation under nihility conditions reduced into below form

$$\nabla \times \boldsymbol{E} = 0 \tag{1}$$

$$\nabla \times \boldsymbol{H} = 0 \tag{2}$$

Many researchers used nihility material and performed many experiments regarding nihility waveguides, electromagnetic/plane wave scattering from different objects made up of nihility material [11, 12, 14–18]. Lakhtakia derived the mathematical solutions for scattering of electromagnetic wave from sphere made up of nihility and concluded that its Back-Scattering efficiency is exactly zero, he also reported that both Forward-scattering and Extinction efficiencies of nihility sphere showed larger magnitude than perfect electric and perfect magnetic conductors [12]. In

addition Lakhtakia and Geddes proposed the analytical solution of plane wave scattering from infinite long nihility
cylinder [11]. Ahmad *et al.*, analytically solved the plane
wave scattering by considering infinite long nihility cylinder
coated with metamaterials and deduced that metamaterial
layer can tune the scattering efficiencies [19]. The scattering
characteristics of plane wave scattering from infinite cylinder made up of nihility covered with chiral layer also has
been studied by Ahmed and Naqvi [20]. Sobia *et al.*, analytically formulated the scattering of the electromagnetic wave
from chiral coated nihility cylinder embedded in chiral
media [21]. Yaqoob *et al.*, used anisotropic plasma as a
coating material and solved the scattering problem from
nihility cylinder placed in chiral medium, and deduced that
chiral and anisotropic plasma parameters can be used to
tune the scattering efficiencies [22].

In the present paper, the problem of electromagnetic
wave scattering from nihility sphere coated with metamaterial, i.e., Double Positive (DPS), Double Negative (DNG),
Epsilon Negative (EMG) and Mu negative (MNG) are
formulated mathematically and computationally. For the
simplicity of mathematical formulations coating layer has
been considered of uniform thickness. In order to expand
the incident, the scattered and the transmitted electromagnetic field, spherical wave vector function are used. For
mathematical solutions, the proposed geometry of subjected problem was sliced into three regions free space as
region 0, metamaterial coating as region 1 and Nihility
sphere as region 2, shown in the Fig. 1. By applying the
boundaries conditions at each interface a set of eight equations are obtained which are used to find the scattering coefficients. These obtained scattering coefficients are used to
calculate the scattering efficiencies, which were then
compared with already published literature to show the
accuracy of present formulation. In this paper the time
dependence $e^{-j\omega t}$ was considered throughout the mathematical formulation.

Analytical Formulations and Method

The propose geometry of the presented problem of
plane wave scattering from nihility sphere coated with
DPS, DNG, MNG or ENG material is depicted in Fig. 1.
Where b and a represents the radius of sphere with coating and without coating respectively. The outer medium i.e.
$\rho \geq b$ is free space having wave number $k_0 = \omega\sqrt{\epsilon_0\mu_0}$ is
represented by region 0. The coating medium $a < \rho < b$ with
wave number $k_1 = \omega\sqrt{\epsilon_1\mu_1}$ is termed as region 1. Region
$\rho < a$ with wavenumber $k_2 = \omega\sqrt{\epsilon_2\mu_2}$ is represented by region 2.

In spherical coordinate system (r, θ, ϕ) the spherical
wave vector functions are given as.

$$M_{\sigma mn\gamma}^{(l)} = \nabla \times \left[r Y_{\sigma mn}(\theta, \phi) R_n^l(k_\gamma r) \right] \quad (3)$$

$$N_{\sigma mn\gamma}^{(l)} = \frac{1}{k_\gamma} \left[\nabla \times M_{\sigma mn\gamma}^{(l)} \right] \quad (4)$$

Where $Y_{\sigma mn}(\theta, \phi)$ represents spherical harmonic, the
most significant property of spherical harmonics is its parity. Here σ represents the parity of spherical harmonics,
which is even when $\sigma = e$ and odd when $\sigma = o$. The radial
function $R_n^l(k_\gamma r)$ transform to the spherical Bessel $J_n(k_\gamma r)$,
Spherical Neumann function $n_n(k_\gamma r)$ and spherical Hankel
function $h_n(k_\gamma r)$ corresponding to $l = 1, 2, 3$ respectively.
Where subscript γ represents the appropriate wave number i.e., $\gamma = 0, 1, 2$ Represents the region 0 (free space)
wave number and region 1 wave number k_1 and so on.

A Plane wave traveling in $+z$ direction with its electric
field polarized in the positive x direction is incident on
metamaterial coated nihility sphere. The incident electromagnetic field in terms of spherical vector wave functions is given as.

$$E_{\text{inc}}(r) = E_0 \sum_{n=1}^{\infty} i^n \frac{(2n+1)}{n(n+1)} \left(M_{\text{oln}}^{(1)} - i N_{\text{eln}}^{(1)} \right) \quad (5)$$

$$H_{\text{inc}}(r) = -\frac{k_0}{\omega\mu_0} E_0 \sum_{n=1}^{\infty} i^n \frac{(2n+1)}{n(n+1)} \left(M_{\text{eln}}^{(1)} + i N_{\text{oln}}^{(1)} \right)$$
$$\quad (6)$$

Where k_0 and μ_0 is the wavenumber and permeability
of free space.

The scattered field can be written as

$$E_{\text{sc}}(r) = E_0 \sum_{n=1}^{\infty} i^n \frac{(2n+1)}{n(n+1)} \left(i a_n N_{\text{eln}}^{(3)} - b_n M_{\text{oln}}^{(3)} \right) \quad (7)$$

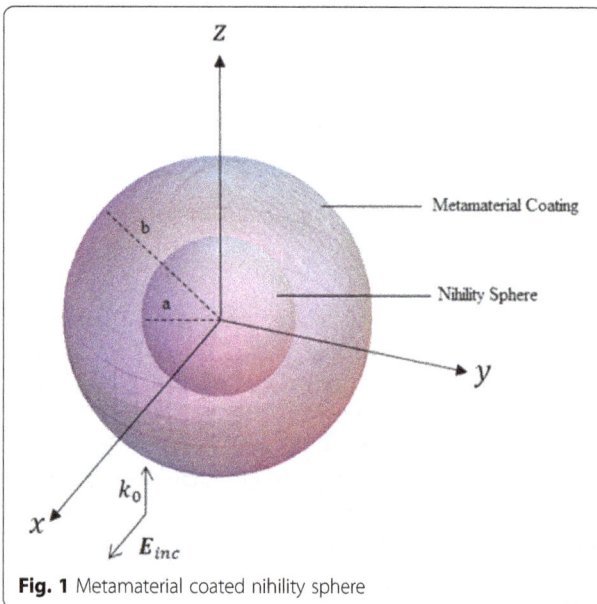

Fig. 1 Metamaterial coated nihility sphere

$$H_{sc}(r) = \frac{k_0}{\omega\mu_0} E_0 \sum_{n=1}^{\infty} i^n \frac{(2n+1)}{n(n+1)} \left(ib_n N_{oln}^{(3)} + a_n M_{eln}^{(3)} \right)$$

(8)

Electromagnetic field that transmitted in region 1 can be represented as

$$E_I(r) = E_0 \sum_{n=1}^{\infty} i^n \frac{(2n+1)}{n(n+1)} \left(c_n M_{oln}^{(1)} + d_n M_{oln}^{(2)} - ie_n N_{eln}^{(1)} \right.$$
$$\left. - if_n N_{eln}^{(2)} \right)$$

(9)

$$H_I(r) = \frac{-k_1}{\omega\mu_1} E_0 \sum_{n=1}^{\infty} i^n \frac{(2n+1)}{n(n+1)} \left(e_n M_{eln}^{(1)} + f_n M_{eln}^{(2)} - ic_n N_{oln}^{(1)} \right.$$
$$\left. - id_n N_{oln}^{(2)} \right)$$

(10)

Here k_1 represents the wavenumber in region 1 having μ_1 its permeability. Electromagnetic Field present in region 2 is

$$E_{II}(r) = E_0 \sum_{n=1}^{\infty} i^n \frac{(2n+1)}{n(n+1)} \left[g_n M_{oln}^{(1)} - ih_n N_{eln}^{(1)} \right] \quad (11)$$

$$H_{II}(r) = \frac{-k_2}{\omega\mu_2} E_0 \sum_{n=1}^{\infty} i^n \frac{(2n+1)}{n(n+1)} \left[l_n M_{eln}^{(1)} + il_n N_{oln}^{(1)} \right]$$

(12)

The analytically solved boundary conditions at both interface are listed below

$$\left.\begin{array}{ll} E_\theta^{inc} + E_\theta^{scat} = E_\theta^I & \rho = b \\ H_\theta^{inc} + H_\theta^{scat} = H_\theta^I & \rho = b \\ E_\theta^I = E_\theta^{II} & \rho = a \\ H_\theta^I = H_\theta^{II} & \rho = a \end{array}\right\}$$

(13)

By using field Eqs. 5, 6, 7, 8, 9, 10, 11 and 12 in the Eq. 13 set of eight equations are obtained in terms of scattering coefficients.

$$h_n(r_0)b_n + j_n(r_1)c_n + n_n(r_1)d_n = j_n(r_0)$$ (14)

$$r_1[r_0 h_n(r_0)]' a_n + r_0[r_1 j_n(r_1)]' e_n + r_0[r_1 n_n(r_1)]' f_n$$
$$= r_1[r_0 j_n(r_0)]'$$

(15)

$$\eta_0^{-1} h_n(r_0)a_n + \eta_{1}^{-1} j_n(r_1)e_n + \eta_{1}^{-1} n_n(r_1)f_n$$
$$= \eta_{0}^{-1} j_n(r_0)$$

(16)

$$\eta_0^{-1} r_1[r_0 h_n(r_0)]' b_n - \eta_{1}^{-1} r_0[r_1 j_n(r_1)]' c_n - \eta_{1}^{-1} r_0[r_1 n_n(r_1)]' d_n$$
$$= \eta_{0}^{-1} r_1[r_0 j_n(r_0)]'$$

(17)

$$j_n(r_2)c_n + n_n(r_2)d_n - j_n(r_3)g_n = 0$$ (18)

$$r_3[r_2 j_n(r_2)]' e_n + r_3[r_2 n_n(r_2)]' f_n - r_2[r_3 j_n(r_3)]' h_n = 0$$

(19)

$$\eta_1^{-1} j_n(r_2)e_n + \eta_{1}^{-1} n_n(r_2)f_n - \eta_{2}^{-1} j_n(r_3)h_n = 0$$ (20)

$$\eta_1^{-1} r_3[r_2 j_n(r_2)]' c_n + \eta_{1}^{-1} r_3[r_2 n_n(r_2)]' d_n - \eta_{2}^{-1} r_2[r_3 j_n(r_3)]' g_n = 0$$

(21)

Where the impedance is defined as $\eta_n = \sqrt{\frac{\mu_n}{\epsilon_n}}$. These equations are solved for scattering coefficients a_n and b_n which are then used to solve the scattering efficiency, Forward-scattering efficiency, Back-scattering and Extinction efficiency as given below where $k_0 b = r_0$, $k_1 b = r_1$, $k_1 a = r_2$, $k_2 a = r_3$

The energy flow (Poynting vector) was obtained by implementing the scattered field solution and can be written as [23]

$$S_i = \frac{1}{2} Re(E_i \times H_i^*)$$

$$S_s = \frac{1}{2} Re(E_s \times H_s^*)$$

$$S_{ext} = \frac{1}{2} Re(E_i H_s^* - E_s H_i^*)$$

Where S_i and S_s are the corresponding Poynting vector associated with incident and scattered field respectively, while S_{ext} represents the Poynting vector induced due to interaction between incident and scattered

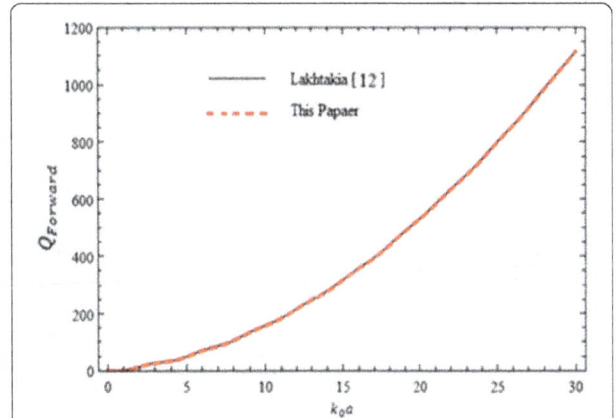

Fig. 2 Forward-scattering efficiency as a function of $k_0 a$ when coating layer has been removed by setting coating layer parameter equals to core i.e. nihility b = a, $\epsilon_2 = \epsilon_1$, $\mu_2 = \mu_1$

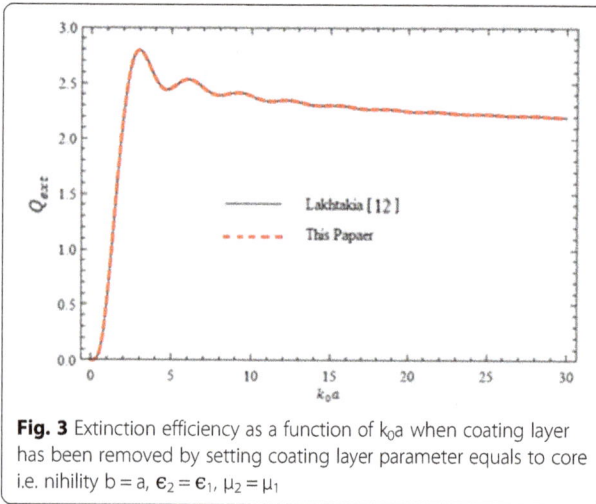

Fig. 3 Extinction efficiency as a function of k_0a when coating layer has been removed by setting coating layer parameter equals to core i.e. nihility b = a, $\epsilon_2 = \epsilon_1$, $\mu_2 = \mu_1$

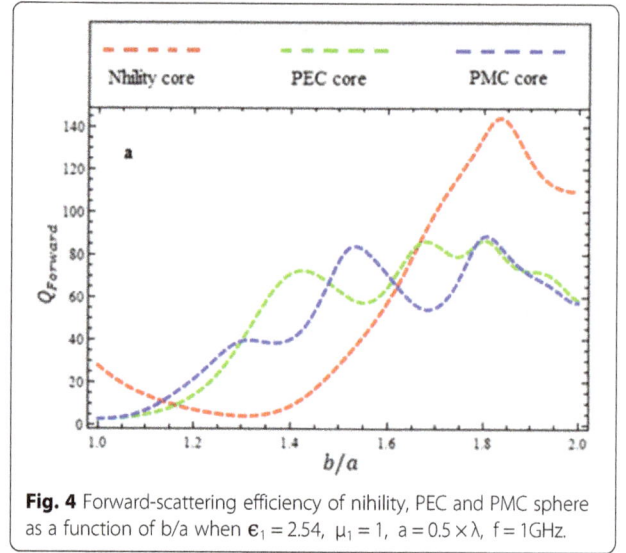

Fig. 4 Forward-scattering efficiency of nihility, PEC and PMC sphere as a function of b/a when $\epsilon_1 = 2.54$, $\mu_1 = 1$, $a = 0.5 \times \lambda$, $f = 1$ GHz.

electromagnetic waves. By following standard Mie theory, if we integrate the above equation over a large sphere, various scattering efficiencies can be obtained [24]. This yields the extinction efficiency and scattering efficiencies (forward scattered and backscattered) [23, 24] as presented in Eqs. 22, 23 and 24.

$$Q_{ext} = \frac{2}{r_0^2} \sum_{n=1}^{\infty} (2n+1)\Re(a_n + b_n) \tag{22}$$

$$Q_{forward} = \frac{1}{r_0^2} \left| \sum_{n=1}^{\infty} (2n+1)(a_n + b_n)^2 \right|^2 \tag{23}$$

$$Q_{backward} = \frac{1}{r_0^2} \left| \sum_{n=1}^{\infty} (-1)^n (2n+1)(b_n - a_n)^2 \right|^2 \tag{24}$$

By using values of scattering coefficients a_n and b_n in above equation nihility condition is applied i.e., $\epsilon_2 = 0$ and $\mu_2 = 0$, and results are obtained.

Results and discussions

In the previous section, mathematical formulations of plane wave scattering from nihility sphere coated with metamaterial have been calculated. In order to gain the more detail of the presented work and also to check the correctness of mathematical formulations Forward-scattering and Extinction efficiencies are plotted under some special conditions. The obtained results are then compared with already published literature [12] which depicted the accuracy of the presented formulation. When we removed the coating layer

by considering $b = a$, $\epsilon_2 = \epsilon_1$, $\mu_2 = \mu_1$ and by applying nihility boundary conditions i.e., $\epsilon_1 \rightarrow 0$, $\mu_1 \rightarrow 0$ the presented problem reduced to electromagnetic wave scattering from nihility sphere. Then by using these specified conditions Forward-scattering and Extinction efficiencies are plotted by using MATHEMATICA software and compared with already published work which are in great agreement as shown in Figs. 2 and 3.

To compare the obtained results of nihility core coated with metamaterial with PEC and PMC sphere coated with metamaterial conversion conditions presented in Table 1 were enforced.

Table 1 Material conversion values

Material	ϵ_1	μ_1
Nihility	≈ 0	≈ 0
PEC	$\approx \infty$	≈ 0
PMC	≈ 0	$\approx \infty$

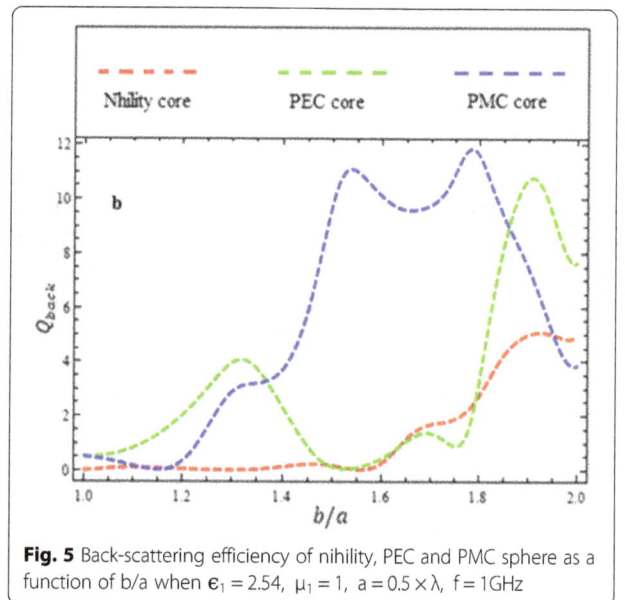

Fig. 5 Back-scattering efficiency of nihility, PEC and PMC sphere as a function of b/a when $\epsilon_1 = 2.54$, $\mu_1 = 1$, $a = 0.5 \times \lambda$, $f = 1$ GHz

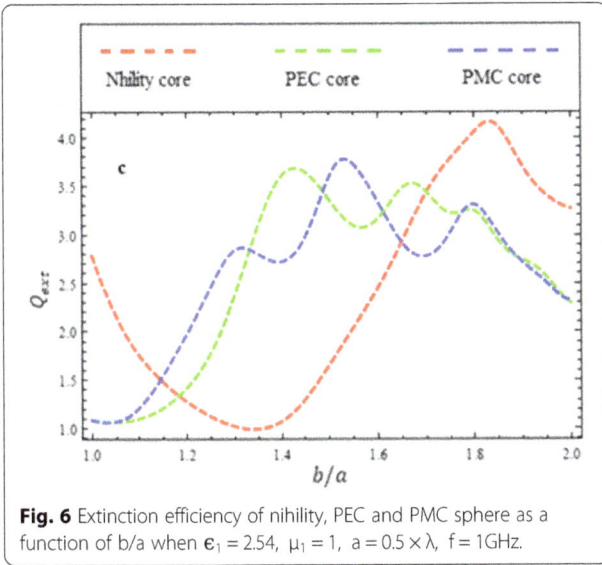

Fig. 6 Extinction efficiency of nihility, PEC and PMC sphere as a function of b/a when $\epsilon_1 = 2.54$, $\mu_1 = 1$, $a = 0.5 \times \lambda$, $f = 1GHz$.

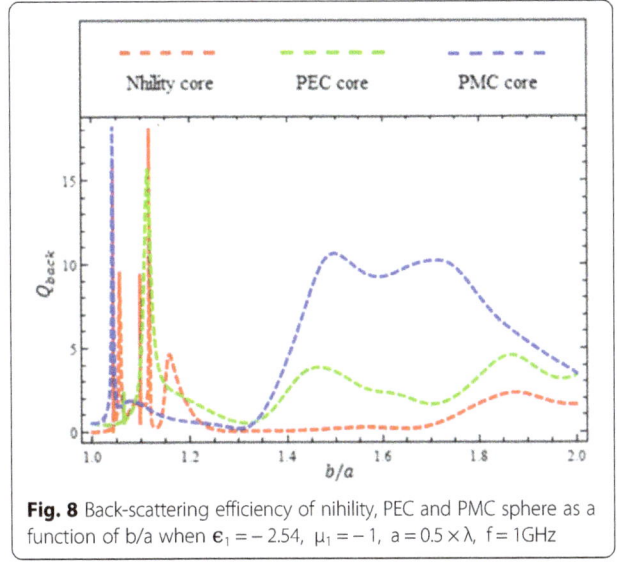

Fig. 8 Back-scattering efficiency of nihility, PEC and PMC sphere as a function of b/a when $\epsilon_1 = -2.54$, $\mu_1 = -1$, $a = 0.5 \times \lambda$, $f = 1GHz$

DPS Coating

Figure 4 represents the comparison between Forward-Scattering efficiencies of nihility, PEC and PMC sphere when coated with DPS metamaterial as a function of coating layer thickness with following parameters $\epsilon_1 = 2.54$, $\mu_1 = 1$, $a = 0.5 \times \lambda$, $f = 1GHz$. By analyzing Fig. 4 we deduced that the Forward-scattering efficiency of nihility sphere is higher than that of PEC sphere and PMC sphere at higher values of coating thickness and lower at lower values of coating thickness. Figure 5 shows the compared results of Back-Scattering efficiencies versus coating layer thickness when nihility, PEC and PMC sphere covered with layer of DPS metamaterial by inserting following parameters $\epsilon_1 = 2.54$, $\mu_1 = 1$, $a = 0.5 \times \lambda$, $f = 1GHz$. Figure 5 reflects that the Back-Scattering efficiency of nihility sphere is lower (but not zero as in uncoated

case [12]) than PEC sphere and PMC sphere when coated with DPS metamaterial layer. In Fig. 6 the Extinction efficiencies of different sphere core i.e., nihility, PEC and PMC are coated with DPS metamaterial and are comparatively plotted against the coating thickness by considering the same parameters as that used in Figs. 4 and 5. By analyzing Fig. 6, we reported that the extinction efficiency of nihility sphere is higher as compared to PEC and PMC cases at higher values of coating thickness and vice versa.

DNG Coating

Forward-scattering, Back-scattering and Extinction efficiencies of nihility sphere are plotted along with PEC and PMC sphere when coated with DNG metamaterial when $\epsilon_1 = -2.54$, $\mu_1 = -1$, $a = 0.5 \times \lambda$, $f = 1GHz$ as shown

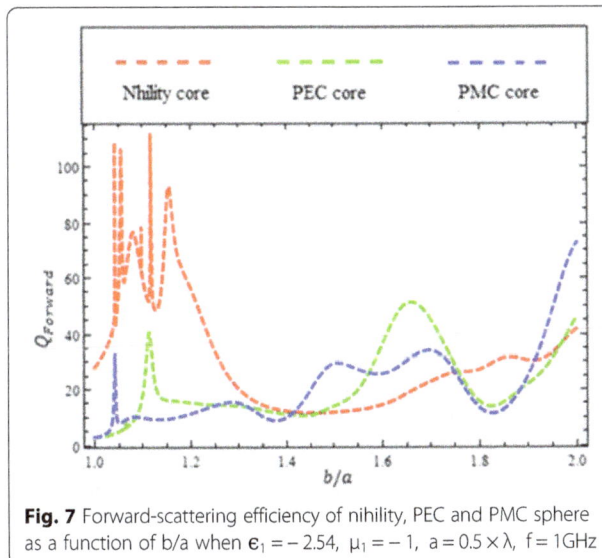

Fig. 7 Forward-scattering efficiency of nihility, PEC and PMC sphere as a function of b/a when $\epsilon_1 = -2.54$, $\mu_1 = -1$, $a = 0.5 \times \lambda$, $f = 1GHz$

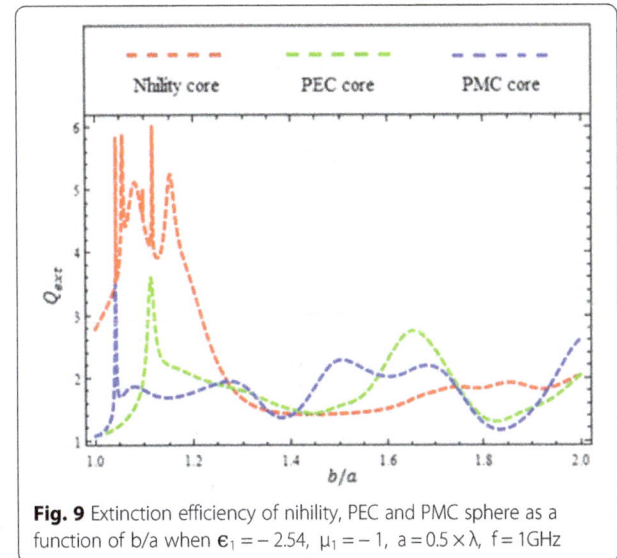

Fig. 9 Extinction efficiency of nihility, PEC and PMC sphere as a function of b/a when $\epsilon_1 = -2.54$, $\mu_1 = -1$, $a = 0.5 \times \lambda$, $f = 1GHz$

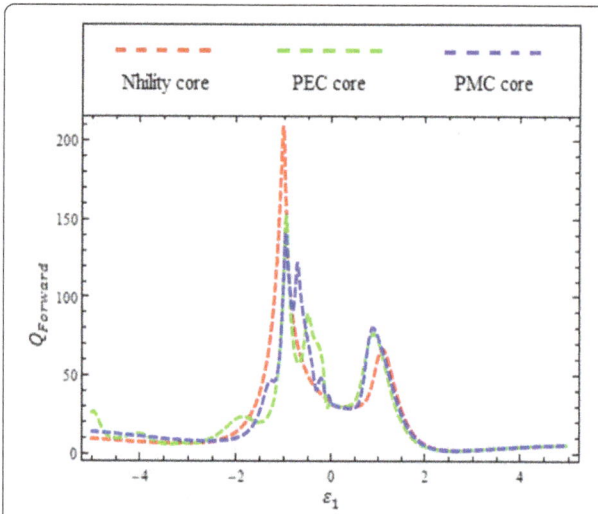

Fig. 10 Comparison between Forward-Scattering efficiencies of nihility, PEC and PMC sphere as a function of ϵ_1 when $\mu_1 = -1$, $a = 0.4 \times \lambda$, $b = 1.5 \times a$, $f = 1GHz$

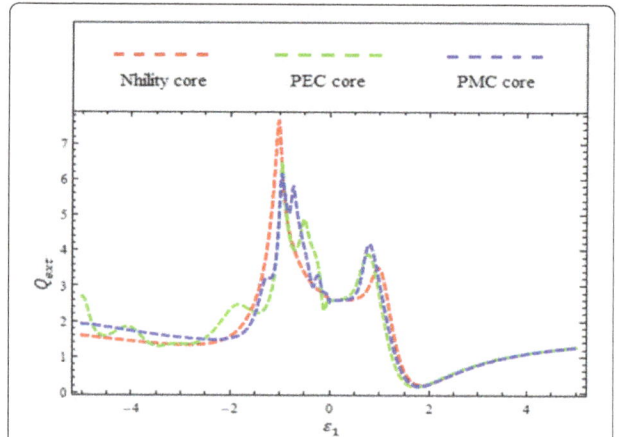

Fig. 12 Comparison between Extinction efficiencies of nihility, PEC and PMC sphere as a function of ϵ_1 when $\mu_1 = -1$, $a = 0.4 \times \lambda$, $b = 1.5 \times a$, $f = 1GHz$

in Figs. 7, 8 and 9 respectively. From Figs. 7, 8 and 9 we concluded that when the nihility sphere is coated with DNG metamaterial the behavior of Forward-scattering, Back-scattering and Extinction efficiencies are more sensitive than DPS coated nihility sphere and Forward-Scattering, Back-scattering and Extinction efficiencies depict higher magnitude then PEC and PMC at lower values of coating thickness and vice versa.

ENG and MNG Coating

Forward-scattering, Back-Scattering and Extinction efficiencies of nihility sphere when coated with metamaterial by substituting following parameters: $\mu_1 = -1$, $a = 0.4 \times \lambda$, $b = 1.5 \times a$, $f = 1GHz$ were compared with PEC and PMC sphere as shown in Figs. 10, 11 and 12. From Fig. 10 we

concluded that the both Forward-scattering and Extinction efficiencies of nihility core depict lower magnitude than PEC and PMC sphere when we considered only MNG ($\epsilon_1 = 0$ to 5) coating while, show higher magnitude than PEC and PMC when coating has been considered of DNG ($\epsilon_1 - 2$ to -1). It was further deduced that the Back-scattering efficiency of nihility is lower than PEC and PMC, which can be confirmed from Fig. 11.

Forward-scattering, Back-Scattering and Extinction efficiencies of nihility sphere as a function of μ_1 when coated with metamaterial by considering following parameters $\epsilon_1 = -2.54$, $a = 0.4 \times \lambda$, $b = 1.5 \times a$, $f = 1GHz$ are compared with PEC and PMC sphere as shown in Figs. 13, 14 and 15 respectively. By analyzing Figs. 13 and 15 it is reported that the magnitude of both Forward-Scattering and Extinction efficiencies corresponding to nihility sphere depicts higher magnitude than PEC

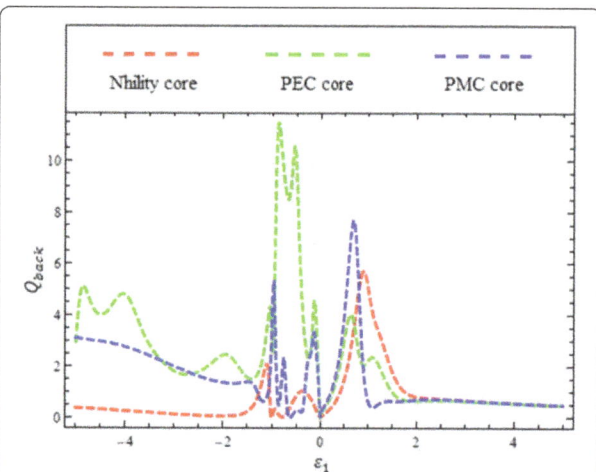

Fig. 11 Comparison between Back-Scattering efficiencies of nihility, PEC and PMC sphere as a function of ϵ_1 when $\mu_1 = -1$, $a = 0.4 \times \lambda$, $b = 1.5 \times a$, $f = 1GHz$

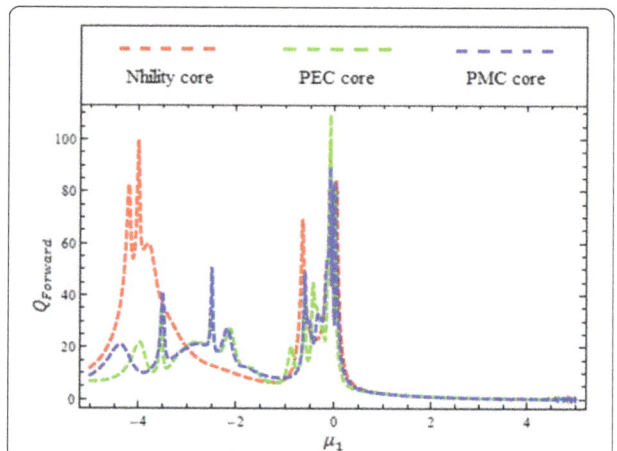

Fig. 13 Comparison between Forward-Scattering efficiencies of nihility, PEC and PMC sphere as a function of μ_1 when $\epsilon_1 = -2.54$, $a = 0.4 \times \lambda$, $b = 1.5 \times a$, $f = 1GHz$

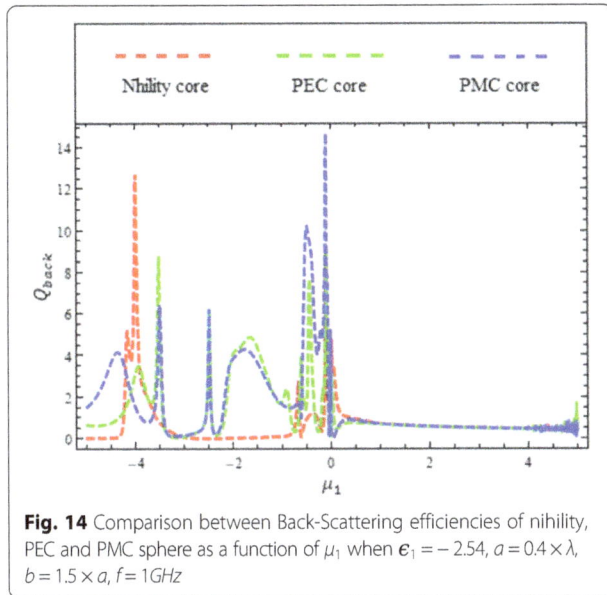

Fig. 14 Comparison between Back-Scattering efficiencies of nihility, PEC and PMC sphere as a function of μ_1 when $\epsilon_1 = -2.54$, $a = 0.4 \times \lambda$, $b = 1.5 \times a$, $f = 1GHz$

and PMC sphere only when the value of μ_1 ranging between -5 to -4, while show lower magnitude value as compared to PEC and PMC when $\mu_1 = -4$ to 0. The Back-Scattering efficiency of nihility sphere recorded maximum when $\mu_1 = -4$, and minimum as compared to PEC and PMC at other values of μ_1, which can be verified from Fig. 14. From these Figs, it was deduced that Forward-scattering and Extinction efficiencies of nihility depict higher magnitude than PEC and PMC in MNG coating case (at some specific values of permeability) and lower in ENG coating case as shown in Figs. 13, 14 and 15.

Conclusions

The problem of electromagnetic wave scattering from metamaterial coated nihility sphere by using extended classical wave theory have been mathematically formulated and analyzed. It is observed that the Forward-Scattering, Back-scattering and Extinction efficiencies strongly depend on coating thickness. When we consider the DPS coating, these efficiencies depict higher magnitude than PEC and PMC sphere at higher values of coating thickness, while in DNG case scattering efficiencies following increasing pattern at lower values of coating thickness. It is furthermore reported that in DNG coating case, we observed that lower scattering efficiencies magnitude of Nihility sphere as compared to PEC and PMC at higher values of coating thickness. Based on the obtained results, we predicted that, scattering efficiencies can be controlled and tuned more effectively by producing metamaterial coating layer. These results may be very helpful in the defense technologies and optical illusion devices.

Abbreviations
DNG: Double Negative; DPS: Double Positive; EMG: Epsilon Negative; MNG: Mu negative; PEC: Perfect Electric Conductor; PMC: Perfect Magnetic Conductor; SRR: Split Ring Resonator

Funding
Deanship of Scientific Research (DSR) at King Saud University for its funding of this research through the Research Group no RG-1436-001.

Authors' contributions
All authors contributed equally in all the sections of this work. All authors read and approved the final manuscript.

Competing interests
The authors declare that they have no competing interests.

Author details
[1]Department of Physics, University of Agriculture, Faisalabad, Pakistan. [2]Department of Electrical Engineering, King Saud University, Riyadh, Saudi Arabia.

References
1. Capolino, F: Applications of metamaterials. CRC press, USA (2009)
2. Engheta, N, and Ziolkowski, RW: Metamaterials: physics and engineering explorations. John Wiley & Sons, USA (2006)
3. Gil, M., Bonache, J., Martin, F.: Metamaterial filters: A review. Metamaterials **2**(4), 186–197 (2008)
4. Ding, F, and Yang, H: EM scattering by objects coated with DNM. In: Microwave, Antenna, Propagation and EMC Technologies for Wireless Communications (MAPE), 2013 IEEE 5th International Symposium on. IEEE, USA (2013)
5. Geng, Y-L: Mie Scattering by a Conducting Sphere Coated Uniaxial Single-Negative Medium. Int J Antennas Propagation. **2012**(856476), 1–6 (2012)
6. Zhang, S., et al.: Negative refractive index in chiral metamaterials. Phys Rev Lett **102**(2), 023901 (2009)
7. Afzaal, M., et al.: Scattering of electromagnetic plane wave by an impedance strip embedded in homogeneous isotropic chiral medium. Opt Commun **342**, 115–124 (2015)

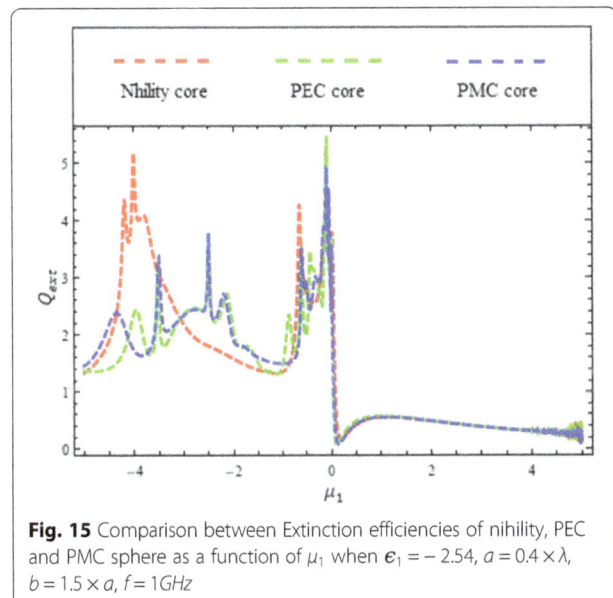

Fig. 15 Comparison between Extinction efficiencies of nihility, PEC and PMC sphere as a function of μ_1 when $\epsilon_1 = -2.54$, $a = 0.4 \times \lambda$, $b = 1.5 \times a$, $f = 1GHz$

8. Ghaffar, A., Alkanhal, M.A.: Electromagnetic waves in parallel plate uniaxial anisotropic chiral waveguides. Opt Mater Express **4**(9), 1756–1761 (2014)
9. Sihvola, A.: Metamaterials in electromagnetics. Metamaterials **1**(1), 2–11 (2007)
10. Lakhtakia, A: An electromagnetic trinity from "negative permittivity" and "negative permeability". arXiv preprint physics/0112003. Int J Infrared Millimeter Waves **22**(12), 1731–1734 (2001)
11. Lakhtakia, A., Geddes Iii, J.B.: Scattering by a nihility cylinder. AEU-Int J Electron Commun **61**(1), 62–65 (2007)
12. Lakhtakia, A.: Scattering by a nihility sphere. Microwave Opt Technol Lett **48**(5), 895–896 (2006)
13. Lakhtakia, A.: Radiation pressure efficiencies of spheres made of isotropic, achiral, passive, homogeneous, negative-phase-velocity materials. Electromagnetics **28**(5), 346–353 (2008)
14. Capretti, A., et al.: Comparative study of second-harmonic generation from epsilon-near-zero indium tin oxide and titanium nitride nanolayers excited in the near-infrared spectral range. ACS Photonics **2**(11), 1584–1591 (2015)
15. Liberal, I., Engheta, N.: Nonradiating and radiating modes excited by quantum emitters in open epsilon-near-zero cavities. Sci Adv **2**(10), e1600987 (2016)
16. Liberal, I, and Engheta, N: Selected features of metamaterials with near-zero parameters. In: Electromagnetic Theory (EMTS), 2016 URSI International Symposium on. IEEE, USA (2016)
17. Liberal, I., Engheta, N.: Zero-Index Platforms: where light defies geometry. Opt Photonic News **27**(7), 26–33 (2016)
18. Dong, J.-F., Li, J.: Characteristics of guided modes in uniaxial chiral circular waveguides. Prog Electromagnetics Res **124**, 331–345 (2012)
19. Ahmed, S., Naqvi, Q.A.: Scattering of electromagnetic waves by a coated nihility cylinder. J Infrared Millimeter Terahertz Waves **30**(10), 1044–1052 (2009)
20. Ahmed, S., Naqvi, Q.A.: Electromagnetic scattering from a chiral-coated nihility cylinder. Prog Electromagnetics Res Lett **18**, 41–50 (2010)
21. Shoukat, S., et al.: Scattering from a coated nihility circular cylinder placed in chiral metamaterial. Optik-Int J Light Electron Opt **125**(15), 3886–3890 (2014)
22. Yaqoob, M., et al.: Scattering of electromagnetic waves from a chiral coated nihility cylinder hosted by isotropic plasma medium. Opt Mater Express **5**(5), 1224–1229 (2015)
23. Bohren, CF, and Huffman, DR: Absorption and Scattering of Light by Small Particle, John Wiley & Sons, USA (1983)
24. Stratton, J: Electromagnetic Theory, McGrow-Hill, New-York. London (1941)

9

Robust and precise algorithm for aspheric surfaces characterization by the conic section

Petr Křen

Abstract

Background: A new algorithm for precise characterisation of rotationally symmetric aspheric surfaces by the conic section and polynomial according to the ISO 10110 standard is described.

Methods: The algorithm uses only the iterative linear least squares. It uses fitting the surface form in a combination with terms containing its spatial derivatives that represent infinitesimal transformations of form.

Results: The algorithm reaches sub-nanometre residuals even though the aspheric surface is translated and rotated in the space.

Conclusion: he algorithm is computationally robust and an influence of local surface imperfections can be easily reduced by use of a criterion for residuals.

Keywords: Aspheric lens, Robust algorithm, Least squares fitting, Metrology, ISO 10110

Background

Aspheric surfaces are recently widely used in industry. One of their applications is aspheric lens that often needs its precise characterisation of form. The description of the aspheric surface by the conic section with a polynomial correction is common in ray tracing software and in producer specifications of aspheric lens. The conic section surface fitting with a polynomial correction was addressed by several authors [1–5]. Also alternative descriptions of aspheric surfaces were introduced e.g. in [6, 7]. Nevertheless, a simple and robust algorithm is still needed to evaluate the conic section from measurement data in the ISO 10110-12 form. The coordinate system is shown in the Fig. 1.

The design shape of aspheric surfaces is often described by the z-coordinate as a function of the distance r from z-axis in the form

$$z = c(R, k, r) + \sum_{i=2}^{n} A_{2i} r^{2i} \qquad (1)$$

where function c describes the conic section given by function

Correspondence: pkren@cmi.cz
Czech Metrology Institute, Okružní 31, CZ63800 Brno, Czech Republic

$$c(R, k, r) = \frac{r^2}{R\left(1 + \sqrt{1-(1+k)\dfrac{r^2}{R^2}}\right)} \approx \frac{1}{2R}r^2 + \frac{1+k}{8R^3}r^4 + \frac{(1+k)^2}{16R^5}r^6 + \dots$$

(2)

where R is the radius of curvature at the vertex and k is the conic constant ($k < -1$ hyperbolic, $k = -1$ parabolic, $k > -1$ elliptical, $k = 0$ spherical surfaces). The correction of surface is given by the even-power polynomial with coefficients A.

Method
Radius of curvature

The expansion of (2) shows that the radius of curvature R at the vertex (r is small) can be obtained from linear least squares (i.e. L2-norm) simply using the first coefficient of an even-power polynomial

$$z = c(R, k, r) + \sum_{i=2}^{n} A_{2i} r^{2i} \approx \sum_{i=1}^{n} q_{2i} r^{2i}. \qquad (3)$$

The higher-order terms of Taylor series at the vertex are negligible and the even-power polynomial with the 18th power

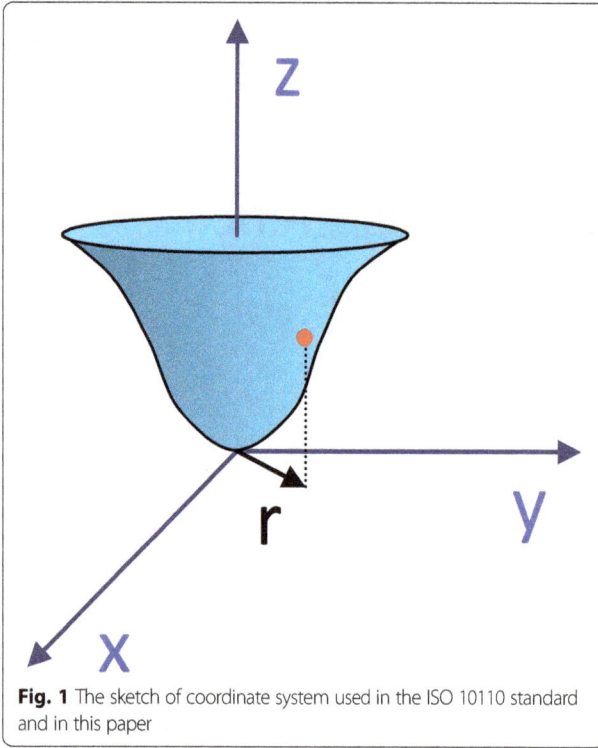

Fig. 1 The sketch of coordinate system used in the ISO 10110 standard and in this paper

$$z = \tilde{q}_2 r^2 + \tilde{q}_4 r^4 + \tilde{q}_6 r^6 + \tilde{q}_8 r^8 + \tilde{q}_{10} r^{10} + \tilde{q}_{12} r^{12}$$
$$+ \tilde{q}_{14} r^{14} + \tilde{q}_{16} r^{16} + \tilde{q}_{18} r^{18} \tag{4}$$

is sufficient for reduction of estimation error for R from data with a given range of r. However, high degree of polynomial could introduce numerical errors. They can be reduced by the following way. The linearity of problem allows making the second fit for residuals $\bar{z} = z - \tilde{z}$ for initial estimate $\tilde{z}$ obtained from $\tilde{q}_{2i}$ as

$$\bar{z} = \bar{q}_2 r^2 + \bar{q}_4 r^4 + \bar{q}_6 r^6 + \bar{q}_8 r^8 + \bar{q}_{10} r^{10} + \bar{q}_{12} r^{12}$$
$$+ \bar{q}_{14} r^{14} + \bar{q}_{16} r^{16} + \bar{q}_{18} r^{18}. \tag{5}$$

The final estimate of coefficients with reduced numerical error is then

$$q_{2i} = \tilde{q}_{2i} + \bar{q}_{2i} \tag{6}$$

thanks to the linearity (additivity) of the problem. The higher power terms are negligible for r close to the vertex. Thus the radius of curvature for rotational paraboloid is obtained as

$$R = \frac{1}{2q_2} \tag{7}$$

with relatively small error because the first term of expansion of (2) also does not depend on k. The error is below 10^{-6} in relative for examples from [3] and the 18th power polynomial, except the case 1 with relative error 0.002 for R. Thus it is a robust way to evaluate radius of curvature at the vertex. The polynomial with coefficients q_{2i} also describes the aspheric surface very well for medium precision applications (i.e. 1λ flatness of wavefront). Nevertheless, the conic section describes the aspheric form better with less number of coefficients. For example, the Taylor series of (2) for hyperbolic surface with high k converges slowly and thus the even-power polynomial must have more terms for the corresponding precision.

Conic constant

The conic constant k will be obtained by the following way. The initial values are obtained as

$$R_0 = \frac{1}{2q_2}, \quad k_0 = -1. \tag{8}$$

It corresponds to the parabolic solution from the previous section. The convergence of iterations is worse close to $k = -1$ because there is a small contribution from the conic section (the terms of expansion for function c). In the next step of algorithm, user selects between the hyperbolic region ($k < -1$) and the elliptic region ($k > -1$). If the algorithm output has large errors the second option could be selected automatically. The next values are then $R_1 = R_0$ and $k_1 = -201$ or $k_1 = -0.5$ respectively (The algorithm also works for oblate elliptical surfaces if k_1 is set as a larger positive number.). The value of k_1 for hyperbolic region can be selected closer to the value -1 (e.g. -3 because the most of commercial aspheric lenses have the conic constant above -3). Nevertheless, the value -201 was selected for demonstration purposes. Note that in some cases (e.g. the case 3 from [3]) the maximum radial distance of points r_{max} from axis z is large and it is not possible to calculate the conic section for selected k_1 (e.g. -0.5). In that case the value k_1-1 is divided by 2 until the k_1-1 is small enough to calculate the square root in the conic section function c as a real number. Now we calculate two values k_{2+} and k_{2-} (iteration index $i = 2$) by halving the interval as

$$k_{i\pm} = k_{i-1} \pm (k_{i-1} - k_{i-2})/2. \tag{9}$$

In the next step, we calculate a pair of differences for all data points using equation

$$\Delta z_{i\pm} = z - c(R_{i-1}, k_{i\pm}, r) \tag{10}$$

and fit them independently by the polynomials $p_{2j\pm} r^{2j}$ using the least squares with selected power larger than r^2

(e.g. up to r^{12} or corresponding to the aspheric lens specifications). The decision between these two fits is based on the lower residual sum of squares. Then the next k and R values are

$$k_i = k_{i+}, \quad R_i^{-1} = R_{i-1}^{-1} + 2p_{2+} \tag{11}$$

or

$$k_i = k_{i-}, \quad R_i^{-1} = R_{i-1}^{-1} + 2p_{2-} \tag{12}$$

and the iterations are repeated until the residuals are small enough. Then the output coefficients A_i are equal to the coefficients p_i and the final conic section parameters are k_i and R_i (Note that the described algorithm also works with negative values of R). The results for five cases from [3] (see also Table 1) are shown in Fig. 2. The convergence is good for all cases and thus the robustness of such algorithm is shown. Nevertheless, the final k values can differ by a few percent from the designed values.

The decision between k_{i+} and k_{i-} values in the algorithm must be 100% correct. However, an error can occur in some cases. This problem comes from the fact that the combination of conic section with even-power polynomial is underdetermined in parameters within the expected form error. I.e. even thought the relative error of k seems to be large, it cannot be evaluated more precisely from experimental form errors. Nevertheless, the final residuals are in sub-nanometre range even if the initial value k is changed (see Fig. 3). I.e. repeating the procedure with different initial k (in range within the multiple of 2) can improve the results. Also additional conic constants such as e.g.

$$\tilde{k}_{i\pm} = k_{i-1} \pm (k_{i-1}-k_{i-2})/2 \cdot 1.2 \tag{13}$$

can be used in each iteration step for the decision of minimal residuals to improve the result (Figs. 2 and 3 show this option). However, it is not necessary because the corresponding residuals are below the uncertainty of measurement that can be carried out. In the case of aspheric lens testing, the known designed value k_{tbt} (that should be calibrated) can be used. Then the initial values can be set e.g. $(k_0 + 1) = 0.99(k_{tbt} + 1)$ and $k_1 = k_{tbt}$ for 1% initial range and the obtained k is then much closer to the lens design value (see Figs. 2 and 3 for case 2). I.e. if

the initial parameters are closer to their final values then a less number of iteration steps is needed.

Rotations and translations

The algorithm from previous section and also from e.g. [3] solves the problem where the vertex of aspheric surface is in the origin of coordinates and the surface is not rotated. However, it is not the case for measurement results (3D data of [x, y, z] coordinates) that are generally in an arbitrary coordinate system. This problem can be solved by the following way for relatively small rotations (up to few tens of degrees) and translations (up to few tenths of optical element size).

Infinitesimal transformations could be used together $(\cos \alpha \approx 1)$ and thus we can apply linearized substitutions

$$x \rightarrow x + t_x - z \sin \alpha_x \cong x + t_x - z\alpha_x \tag{14}$$

$$y \rightarrow y + t_y - z \sin \alpha_y \cong y + t_y - z\alpha_y \tag{15}$$

$$z \rightarrow z + t_z + z \sin \alpha_x + z \sin \alpha_y \cong z + t_z + z\alpha_x + z\alpha_y \tag{16}$$

and equation $r^2 = x^2 + y^2$ into the even-power polynomial equation of r for z. This transformation introduces odd powers of x and y to this polynomial. The initial values of even-power polynomial q_{2i} are obtained by the linear least squares using equation

$$z = \sum_{i=1}^{9} q_{2i} r^{2i} + g_0 + g_1 x + g_2 y + g_3 x^3 + g_4 y^3 \tag{17}$$

The additional terms with coefficients g_j are used to partially compensate unknown arbitrary transformation in the initial stage. Nevertheless, these coefficients are not used in further calculations. Then the each iteration step consists of the least squares fitting for the following equation

$$z = \sum_{i=1}^{9} q_{2i} r^{2i} - t_z + t_x x s_1 + t_y y s_1 - \alpha_x x(1 + s_1 s_2) - \alpha_y y(1 + s_1 s_2) \tag{18}$$

where

$$s_1 = \sum_{i}^{9} 2i\hat{q}_{2i} r^{2i-2} \quad \text{and} \quad s_2 = \sum_{i}^{9} \hat{q}_{2i} r^{2i} \tag{19}$$

Table 1 Parameters of aspherical surfaces from [3]

Case	R	k	A_4	A_6	A_8	A_{10}	A_{12}
1	44.577884	−171.0312	2.316294E-4	3.495852E-8			
2	4.25	−0.863601	1.77613E-4	−1.55395E-5			
3	2.708638	−0.8968698	2.788402E-3	1.553377E-4	−7.281244E-6		
4	56.031	−3	−4.33E-6	−9.76E-9	−1.09E-12	−1.23E-14	
5	1.898836	−0.5603343	−6.8505495E-4	−4.1501354E-4	−4.4705513E-5	−1.8065968E-5	−2.1569936E-7

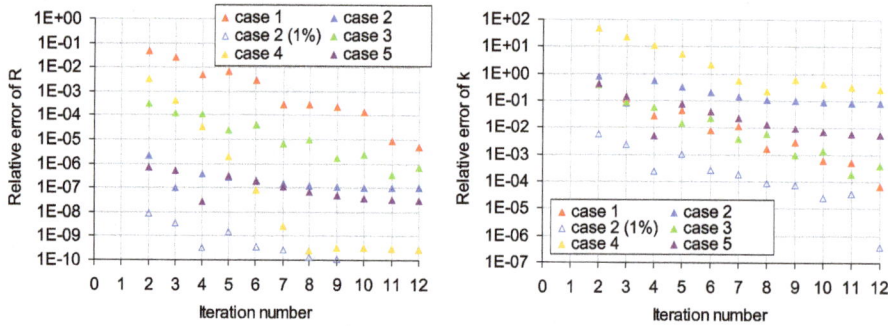

Fig. 2 The convergence of relative error of R and k for 5 cases from [3] as a function of the number of iteration index (the polynomial degree was selected the same as the degree of designed A_{2i} in each individual case). An example for case 2 with the initial value of k within 1% (hollow triangles)

are terms used to represent infinitesimal (derivative) linear and angular transformation contributions to the form derived from significant terms solving polynomial expression with substitutions (14)-(16). The even-power polynomial coefficients $\hat{q}_{2i}$ are taken from the previous iteration step (or they are equal to the initial coefficients q_{2i} in the beginning. The obtained translation coefficient t_z is corrected as $t_z - \overline{a}_x t_x - \overline{a}_y t_y$ (for better convergence of t_z), where coefficients $\overline{a}$ are total rotation angles from all previous iterations (and are equal to zero at the beginning of algorithm). The resulting translation coefficients t_x, t_y, t_z and rotation angles a_x, a_y are summed with corresponding translation coefficients $\overline{t}$ and angles $\overline{a}$ from the previous iteration. Then the total translation and rotation parameters (i.e. with overline) are used to translate and then to rotate all [x,y,z] values from their original position (to avoid cumulation of numerical errors) and these transformed data points are used in the next iteration step. These iterations are repeated until all transformations do not change with a sufficient precision. Thereafter, the algorithm from previous section is used to calculate aspheric surface parameters such as R and k in correct coordinate system.

The rate of convergence of these iterations is superlinear (better for smaller transformations) and faster for higher degree of even-power polynomial. The convergence is slower for larger data sets. Nevertheless, the polynomial degree of 18 ($i = 9$) is sufficient for iteration convergence of all cases that have been tested (It also includes parameters of various real aspheric lens from different producers.). In the case 1, which is an extreme hyperbolic case with k equal about -171 (not a real lens case), the position error is larger. However, it can be reduced by reduction of the radial range of data or by increasing of polynomial degree.

The good convergence that algorithm reached is shown in Fig. 4. The calculation of position and angles of rotations takes less than one tenth of second on standard computer for data set with a few thousands of points. Thus this algorithm can be used for a real time tracking and displaying of aspheric surface in desired coordinate system.

It should be also noted that in case of large dataset (millions of coordinates), the iterative process could be carried out with some randomly selected part of data (e.g. every thousandth) due to the low roughness of optical surfaces. It speeds up the calculations because the algorithm complexity is given by the linear least squares and thus its calculation time is linearly proportional to the number of data points. The results of such pre-

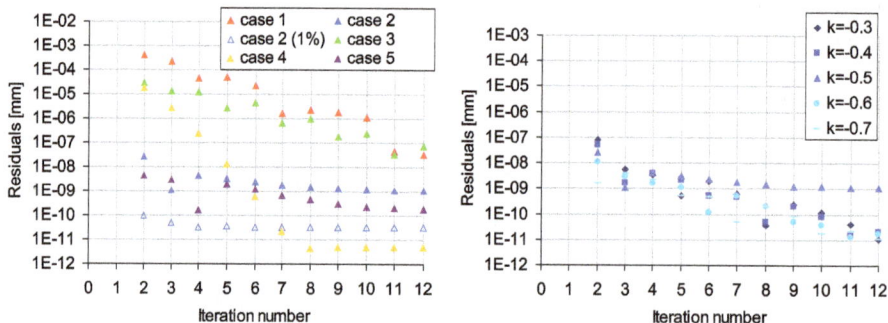

Fig. 3 The standard deviation of residuals for 5 cases from [3] and the effect of different initial k on convergence in the case 2 ($k = -0.5$ is the standard initial value for prolate elliptical surfaces)

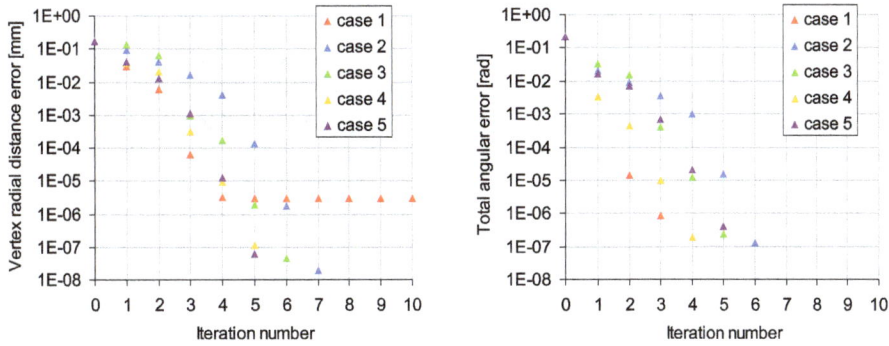

Fig. 4 Errors of the radial distances of vertex and the total angles of rotation as a function of iteration number for 5 cases of surfaces from [3] with initial translations 0.1 mm in all directions X, Y and Z and rotations 0.1 rad and 0.2 rad for axes X and Y respectively

calculation of transformation and form parameters are used for modification of the initial state of algorithm for subsequent calculation with the full dataset that will describe the form more precisely.

Surface imperfections

The local imperfections in data arise from measurement outliers or effect of dust, marks and/or scratches. These imperfections influence the fitting algorithm results although the area of such imperfection is often relatively small. The lens parameters should be evaluated more precisely without an influence of these data (e.g. the focal point of lens). The optical aperture is not so affected if such data from optically not usable areas are removed. The following algorithm with the residual criterion can be used to solve this problem effectively and to keep the robustness of the presented algorithm.

All iteration steps for evaluation of transformation parameters that was described in the previous section will contain the criterion for the correct data. If the residuals of given points will be greater than e.g. 10σ (where σ is the standard deviation in z-direction for each iteration step) of least squares residuals then these points will be removed from the data set.

The following example is applied on the case 2 from [3] that is aligned (transformed by the same translations and rotations) as in the previous section. The artificial imperfection is applied to the aspheric form as a difference with the Gaussian form

$$\delta z = h \exp\left(-\left((x-x_0)^2 + (y-y_0)^2\right)/s^2\right) \qquad (20)$$

where h is the height of imperfection, s is its width and $[x_0,y_0]$ are its coordinates. The results for different heights and for $s = 0.1$ mm, $x_0 = 0.5$ mm and $y_0 = 0.3$ mm are shown in Fig. 5.

We can clearly see that the error of estimated radius is proportional to the height of such imperfection if it is not filtered out (the filter factor is too large). However, the form is fitted correctly for lower filter factors (such as 10σ in this case). Nevertheless, the filter factor cannot be too small because too many points will be excluded from calculations. Thus some compromise must be made and the corresponding filter factor value is different for different aspheric surfaces and for different

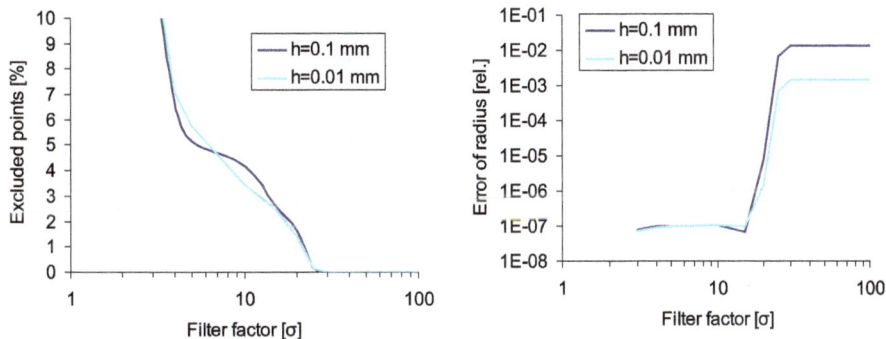

Fig. 5 Excluded fraction of points and relative error of estimated radius R as functions of filter factor (sigma multiples) for the case 2 from [3]. (Other parameters such as estimated angle of rotation are not shown because the transition between the correct and incorrect value is similar as for the radius R.)

imperfections. In the case of min-max algorithm (e.g. in [8]) the form error is large if 0% of points is excluded. However, it can be significantly smaller if some points are excluded (e.g. about 5%). The parameters obtained from such fitting with a reasonable filter factor can be used to display surface deviations for all points and the standardized form error can be evaluated from these residuals.

Another example of imperfection is measurement noise or roughness of surface. The noise in [x,y,z] data coordinates stops the algorithm convergence. Nevertheless, the resulting error of parameters corresponds to the uncertainty arising from this noise and the algorithm robustness is not affected.

The weighted least squares can be introduced for the conic constant iterative search to deal with the heteroscedasticity of residuals. We can use e.g. weights w corresponding to the projection of orthogonal least squares to the z-direction

$$w = \left(1 + \left(\frac{\partial z}{\partial r}\right)^2\right)^{-1/2} \qquad (21)$$

or to the local curvature of form (i.e. inversely proportional to local radius of curvature)

$$w = \frac{\partial^2 z}{\partial r^2}\left(1 + \left(\frac{\partial z}{\partial r}\right)^2\right)^{-3/2} \qquad (22)$$

containing the first and the second derivatives of surface analytically calculated from parameters in the previous iteration step. However, the change of weight from $w = 1$ has a negligible impact on the results.

Results and discussion

The comprehensive test of described algorithm is shown in Fig. 6. An aspheric surface was transformed to a coordinate system unknown for the computer program. Ten artificial imperfections (positive and negative in z-direction) were used and than automatically filtered-out by the fitting algorithm (8.5% of data were removed). The obtained residuals have standard deviation 3.3 pm and maximal error is 11 pm in z-direction (indicated as red colour on the edge of aspheric surface).

The results clearly showed that algorithm allows finding positions of aspheric lenses of unknown form with sub-nanometre accuracy and it was achieved only by least squares fitting. The fitted form for various aspheric surfaces reached residuals also at sub-nanometre level. The non-linear least squares methods (such as in [3]) allows less number of iterations. Nevertheless, linear least squares method, presented here, calculates each step faster and its convergence is robust as it was shown on various examples. All imperfections can be effectively filtered out and precise parameters of lens could be easily obtained. The wide range of algorithm properties such as speed, robustness and effective filtering enables its use in automatic processes.

Conclusions

The algorithm for the evaluation of parameters of rotationally symmetric aspheric surfaces was described and tested. The sub-nanometre precision of fitting was reached with this new robust and fast algorithm. The algorithm also does not need precise estimation of parameters for the initial iteration. It allows ISO 10110 characterization with the conic section parameters of surfaces that are rotated and translated as it is common in the data output from measurement devices and alignment markers on lens are not needed. The additional criterion for residuals can effectively

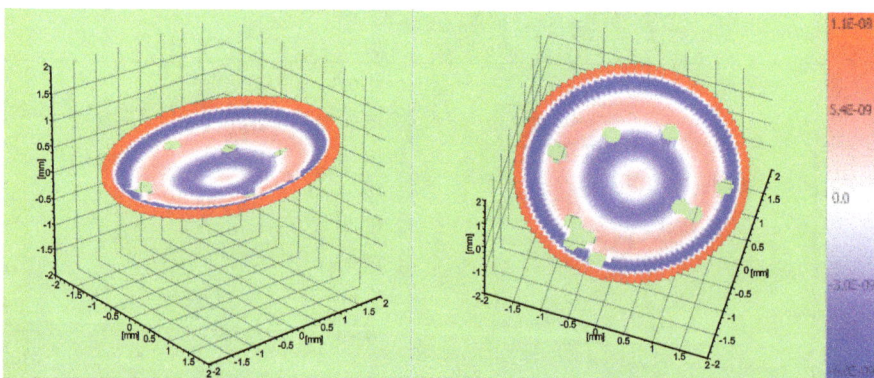

Fig. 6 The fitting errors of algorithm for the case 2 from [3] indicated as colours of points (two views in [mm]). The asphere is translated by 0.1 mm in all directions X, Y and Z and rotated by 0.1 rad and 0.2 rad for axes X and Y respectively. Ten randomly placed Gaussian imperfections were artificially added (with $h = 0.05$ mm or $h = -0.05$ mm and with $s = 0.05$ mm). The filter with factor 10σ filtered out these imperfections (holes in the surface)

remove unuseful data to keep results for the main part of aperture unaffected by local imperfections. Moreover, the part of algorithm that finds the correct coordinate system and removes outliers can be used independently on the conic section fitting part. Thus the algorithm will be useful for the optical community for precise characterisation and testing of aspheric lens.

Acknowledgments
I would like to thank Pavel Mašika for comments and help with testing.

Funding
This work was done in the EMPIR project 15SIB01 FreeFORM. The EMPIR initiative is co-founded by the European Union's Horizon 2020 research and innovation programme and the EMPIR Participating States. The CMI participation in the project is co-funded by the Ministry of Education, Youth and Sports of the Czech Republic (8B16008).

Competing interests
The author declares that he has no competing interests.

References
1. Zhang, Z.: Parameter estimation techniques: a tutorial with application to conic fitting. Image. Vis. Comp. **15**, 59–76 (1997)
2. Gugsa, S., Davies, A.: Monte Carlo analysis for the determination of the conic constant of an aspheric micro lens based on a scanning white light interferometric measurement. Proc. SPIE **5878**, 92–102 (2005)
3. Sun, W., McBride, J.W., Hill, M.: A new approach to characterising aspheric surfaces. Precis. Eng. **34**, 171–179 (2010)
4. El-Hayek, N., Nouira, H., Anwer, N., Gibaru, O., Damak, M.: A new method for aspherical surface fitting with large-volume datasets. Precis. Eng. **38**, 935–947 (2014)
5. Piratelli-Filho A., Anwer N., Souzani C.M., Devedzic G., Arencibia R.V.: Error evaluation in reverse engineering of aspherical lenses. 17th International Congress of Metrology. 13007 (2015)
6. Forbes, G.W.: Shape specification for axially symmetric optical surfaces. Opt. Exp. **15**, 5218–5226 (2007)
7. Park, H.: A solution for NURBS modelling in aspheric lens manufacture. Int. J. Adv. Manuf. Technol. **23**, 1–10 (2014)
8. Zhang, X., Jiang, X., Scott, P.J.: A minimax fitting algorithm for ultra-precision aspheric surfaces. J. Phys.: Conf. Ser. **311**, 012031 (2011)

10

MMI filters configuration for dual-wavelength generation in a ring cavity erbium-doped fibre laser

Ricardo I. Álvarez-Tamayo[1], José G. Aguilar-Soto[1], Manuel Durán-Sánchez[1,2], José E. Antonio-López[3], Baldemar Ibarra-Escamilla[1*] and Evgeny A. Kuzin[1]

Abstract

Background: Dual wavelength laser generation has been of constant interest due to their applications in optical communications and teraheartz generation. A novel configuration for Dual-wavelength laser generation based on the use of a couple of multimode interference (MMI) filters is demonstrated in a ring cavity Erbium-doped fibre laser.

Methods: The MMI filters consist of a segment of no-core fibre spliced between two SMF-28 single mode fibre segments. The MMI filters configured as a Mach-Zehnder interferometer, are used as transmission spectral filters for simultaneous generation of two laser wavelengths. An optical attenuator is used to adjust the intra-cavity losses for dual-wavelength laser generation.

Results: Laser emission at 1540.4 and 1554 nm for a wavelength separation of ~13.6 nm is obtained. The laser wavelengths output power stability variations with the applied pump power is also experimentally discussed.

Conclusions: The use of MMI filters in the proposed dual-wavelength laser filter is experimentally demonstrated as a reliable device for dual-wavelength generation in fibre lasers.

Keywords: Fibre lasers, Multimode interference filter, Dual-wavelength laser, Erbium-doped fibre

Background

Spectral filters in optical fibre based on single-mode-multimode-single-mode (SMS) fibre structure have been of significant interest as reliable optical devices because of their many advantages such as compatibility for all-fibre integration, low cost, low insertion loss and ease to fabricate. The operation principle of the SMS fibre structure is based on a self-imaging phenomenon described for slab waveguides by Soldano and Penings [1]. The filtering aspects of the SMS fibre structure was discussed by Mohammed et al. [2]. The SMS filter exhibits a narrow transmission spectral width with a leading wavelength peak. Moreover, its transmission wavelength peak can be tuned by external parameters such as temperature, strain and liquid refractive index [3, 4]. This features make the SMS fibre structure

attractive for fibre-optical sensing [5–7] and tuneable fibre laser applications [8–10]. The performance of the SMS structure as a band-pass filter [8, 9] make it a reliable device for wavelength selection and tuning of ring cavity fibre lasers. Furthermore, when the MMI filter is properly designed, the transmission peak wavelength can be easily selected in terms of the multimodal fibre (MMF) refractive index, core diameter and length [2, 3].

As the dual-wavelength laser emission has generated a considerable interest in recent years, the obtaining of two simultaneous laser wavelengths by using a single cavity has been attractive in different areas such as optical fibre sensing, optical communications, microwave and terahertz generation. However, EDF is a homogeneous gain medium at room temperature, which leads to a strong mode competition for the generated laser lines. Therefore, several techniques to achieve dual-wavelength laser emission where the use of optical filters as a reliable method for cavity losses adjustment have been reported [11–15], among which can

---

* Correspondence: baldemar@inaoep.mx
[1]Optics Department, Instituto Nacional de Astrofísica, Óptica y Electrónica, Luis Enrique Erro 1, Puebla 72824, Mexico
Full list of author information is available at the end of the article

be mentioned fibre Bragg gratings (FBG), Mach-Zehnder interferometer (MZ) and Sagnac interferometer. Recently, tuneable Erbium- and Ytterbium-doped fibre lasers using MMI filters have been reported [8–10]. However, the potential of using MMI filters for dual-wavelength laser generation, has been underexploited. To our knowledge, only few researches of dual-wavelength fibre lasers in which a SMS structure is used for dual-wavelength generation have been reported [16–18]. However, the use of a single SMS structure in these investigations, limits the laser lines wavelength separation and make difficult the wavelength selection and the cavity losses adjustment.

In this paper, we demonstrate stable dual-wavelength laser emission of an EDF laser based on the use of two SMS filters. The filters with SMS structure are disposed in a MZ interferometer configuration for separately laser wavelength selection. The dual-wavelength laser lines separation is ~13.6 nm with laser lines wavelength at 1540.4 and 1554 nm. The wavelength is marginally modified with the increase of the pump power. The reliability of using two MMI filters for separately wavelength selection of stable dual-wavelength laser operation is experimentally demonstrated.

Methods

MMI dual-wavelength filter principle and characterization

Figure 1a shows the schematic of the SMS structure used in our experiments. A segment of multimode no-core fibre (NCF) is spliced between two segments of standard single mode fibre (SMF-28).

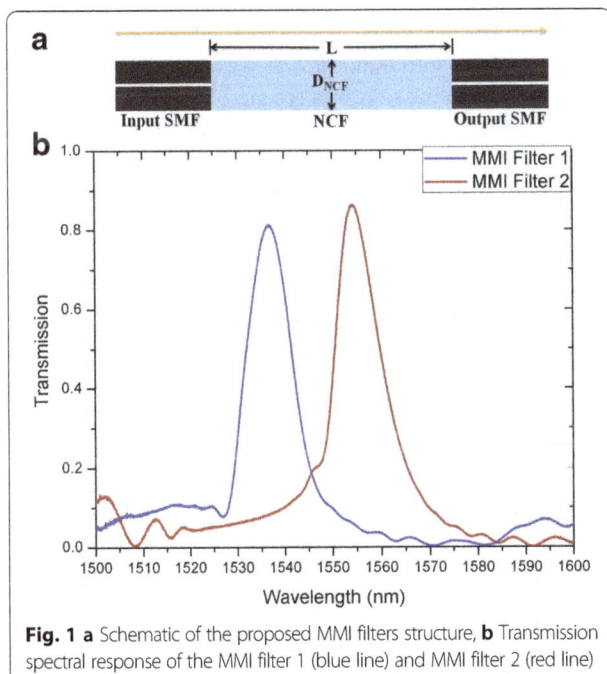

Fig. 1 a Schematic of the proposed MMI filters structure, **b** Transmission spectral response of the MMI filter 1 (blue line) and MMI filter 2 (red line)

The SMS structure has a narrow band-pass response with a wavelength peak based on the multimode-interference self-imaging effect. The modes are excited in the NCF when a field is launched from the input SMF. The interference of these excited modes produces self-images at periodic intervals along the NCF. The central wavelength of the SMS filter can be expressed as [4]:

$$\lambda_0 = p \frac{n_{NCF} D_{NCF}^2}{L}, \qquad (1)$$

where p is the self-image number, n_{NCF} is the NCF effective refractive index, D_{NCF} is the diameter of the NCF corresponding to the effective width of the fundamental mode, and L is the NCF length. The homemade MMI filters were constructed to obtain the fourth self-image by using a NCF (in which air acts as cladding) with 125 μm diameter. The NCF lengths of 58.7 and 58 mm, to obtain transmission wavelength peaks of 1537.5 and 1554.3 nm (for MMI filter 1 and MMI filter 2 respectively) were calculated with Eq. (1). Fig. 1b shows the transmission of the constructed MMI filters. To measure the transmission, we used a LED source with emission in a wavelength range from 1465 to 1650 nm, as input signal. The transmission of the MMI filters was estimated by the following method: Initially, three spliced segments of SMF-28 fibre were mounted in a metal plate to avoid instability in the estimation, and the fibre ends were connected to bare fibre adaptors. In one of the adaptors the LED source was connected and the input signal spectrum was recorded with an OSA at the other connector. Then, the central segment of SMF-28 fibre was replaced with one of the MMI filters. Likewise, the output signal of the LED source due to the MMI was measured. The transmission of both separately MMI shown in Fig. 1b was estimated as the measured MMI output signal divided by the measured LED source output signal. As it can be observed, the transmission spectrum of the MMI filter 1 (blue line) exhibits a wavelength peak at ~1538.6 nm and for the MMI filter 2 (red line) a wavelength peak is observed at ~1554.3 nm. The FWHM of both MMI filters is around 11 nm. The transmission losses for the MMI filter 1 and MMI filter 2 are 19 % and ~14 %, respectively.

Figure 2 shows the configuration and the spectral response of the proposed dual-wavelength filter with two MMI filters (MMI-DWF) used to obtain dual wavelength laser emission in a ring cavity fibre laser. To achieve both transmissions with independent performance, the MMI filters were disposed between two 50/50 fibre couplers (coupler 1 and coupler 2) in a MZ interferometer configuration as it is shown in Fig. 2a. The filter is mounted on a metal plate to avoid instability of the laser performance due to mechanical deformation.

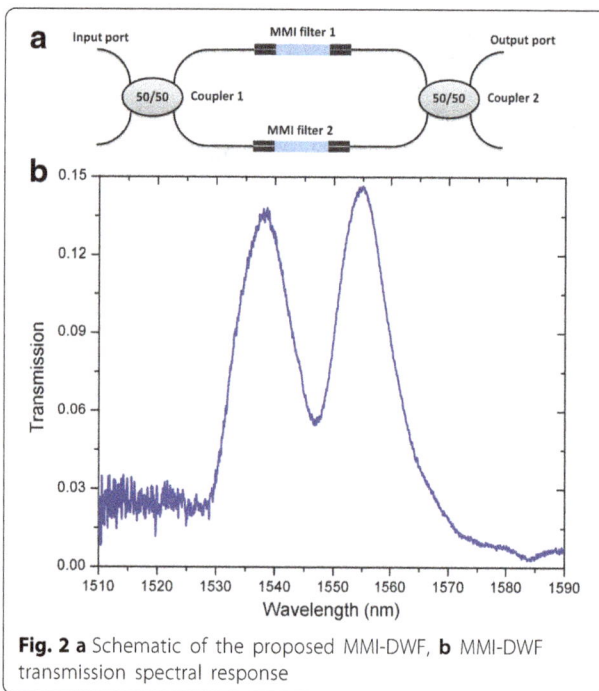

Fig. 2 a Schematic of the proposed MMI-DWF, **b** MMI-DWF transmission spectral response

The interference modulation is avoided due to the long path difference between the two arms of the MZ interferometer. Figure 2b shows the transmission of the MMI-DWF. The transmission spectrum exhibits two wavelength peaks at 1538.5 and 1554.6 nm corresponding to the MMI filter 1 and the MMI filter 2, respectively. As it can be observed, the longer wavelength peak presents slightly higher transmission than the shorter wavelength peak, in accordance with the individual MMI filters transmission spectra shown in Fig. 1b.

Results and discussions

The experimental setup for the dual-wavelength EDF laser using the proposed MMI-DWF is shown in Fig. 3. A 3-m length EDF (MetroGain M-12 980/125) with

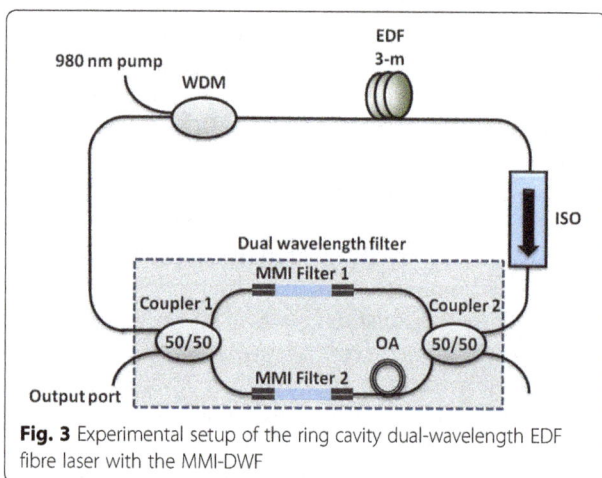

Fig. 3 Experimental setup of the ring cavity dual-wavelength EDF fibre laser with the MMI-DWF

absorption of 20 dB/m at 1531 nm and numerical aperture (NA) of 0.24 was used as a gain medium. The EDF is pumped by a 980 nm single mode laser diode with maximal output power of 120 mW through a 980/1550 nm wavelength-division multiplexer (WDM). A polarization independent optical isolator (ISO) is placed to force unidirectional ring laser operation. As it was shown in Fig. 2b, the longer wavelength peak of the MMI-DWF exhibits a higher transmission than the shorter wavelength. For intra-cavity losses adjustment to achieve stable dual-wavelength laser operation, an optical attenuator (OA) was inserted in the MMI-DWF arm containing the MMI filter 2 to add losses by curvature on the MMI filter with higher transmission. The coupler 1 output port taken as the laser output is used to measure and analyze the laser spectrum by an optical spectrum analyzer (OSA, Yokogawa AQ6375) with scanning range from 1200 to 2400 nm and spectral resolution of 0.05 nm.

Figure 4 shows the EDF laser output spectra at pump power of 100 mW. Figure 4a shows the simultaneous dual-wavelength laser emission with equal output powers at laser wavelengths $\lambda_1 = 1540.4$ nm and $\lambda_2 = 1554$ nm, where λ_1 and λ_2 are the generated laser wavelengths due to the MMI filter 1 and the MMI filter 2, respectively. The cavity losses adjustment performed by the OA requires similar amplification gain at the generated wavelengths. The MMI filters wavelengths were chosen where the EDF amplification spectrum profile exhibits a flat zone proper for cavity losses balancing to reach dual wavelength laser generation.

Figure 4b shows repeated measurements of the dual-wavelength laser output with a launched pump power of 100 mW. The OA was adjusted to obtain dual wavelength laser operation with equal power wavelengths. Once simultaneous laser emission was reached, the OA was fixed to prevent instability by deformation. A set of ten measurements with a 2 min interval were obtained at room temperature in which thermal dependence of the MMI-DWF is not noticeable. The peak power variation for each generated wavelength was less than 0.3 %. Therefore, dual-wavelength laser emission stability is observed.

Figure 5 shows the output power of the generated laser lines as a function of the pump power. The measurements were obtained for pump powers from 40 to 120 mW with an interval of 20 mW. A set of 10 measurements with interval of 2 min was performed at each pump power. With a pump power of 80 mW, dual-wavelength laser emission with equal powers was initially set by adjusting the OA. For the subsequent measurements with different pump powers, the same OA adjustment was used. As it can be observed at pump power of 40 and 60 mW, the output power of λ_1 is higher than for λ_2 and the output power of both laser

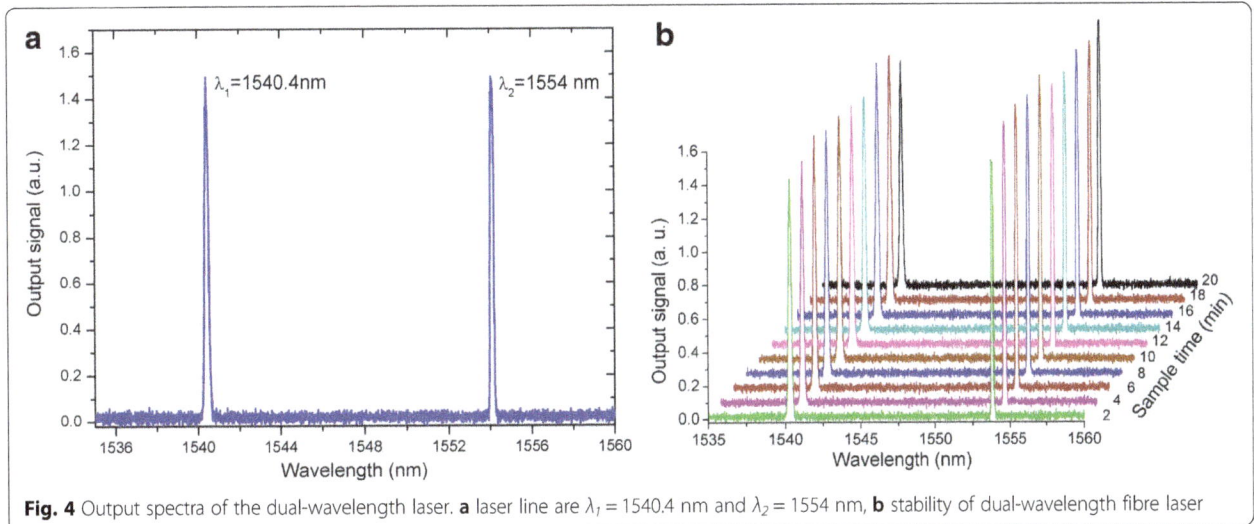

Fig. 4 Output spectra of the dual-wavelength laser. **a** laser line are $\lambda_1 = 1540.4$ nm and $\lambda_2 = 1554$ nm, **b** stability of dual-wavelength fibre laser

lines remains stable. At pump power of 80 and 100 mW stable dual-wavelength laser operation with equal power is observed. At pump power of 120 mW, the power became unstable.

Figure 6 shows the laser wavelengths behaviour as a function of the pump power. The measurements were performed with the same settings used to obtain the measurements shown in Fig. 5. With a pump power of 80 mW an initial adjustment in the OA was performed to achieve dual-wavelength laser operation with equal power. As is shown in Fig. 6a, laser lines displacement toward shorter wavelengths is observed with increment of the pump power. The wavelength displacement for the longer wavelength is larger than for the shorter wavelength, resulting in a variation of the wavelength separation between generated laser lines. With a pump power of 40 mW the separation between the laser wavelengths is ~13.4 nm whereas a

separation of ~14.2 nm is observed with a pump power of 120 mW. Figure 6b shows the wavelength displacement $\Delta\lambda$ as a function of the pump power where λ_1 is the shorter wavelength and λ_2 the longer wavelength. $\Delta\lambda$ is the wavelength shift from the initial generated peak wavelengths with a pump power of 40 mW. The wavelength displacement for λ_2 is larger than the observed for λ_1, where the maximal wavelength displacements for λ_1 and λ_2 are -0.7 and -1.42 nm, respectively.

The laser generation depends on the EDF amplification spectrum gain and on the filter transmission at the specific wavelength. The laser wavelength which will be generated is defined by the maxima of the product of the filter transmission and the EDF amplification. However, each MMI filter exhibits a narrow wavelength range of ~ 2 nm were the transmission maxima is approximately the same. Therefore, when the pump power is changed, the EDF amplification spectrum undergoes a profile modification that leads to a laser line generation in a slightly shifted wavelength. We attributed the larger shift of the longer wavelength to the fact of λ_2 is generated in a wavelength range in which the EDF amplification spectrum exhibits a downward slope of gain which leads to a more pronounced wavelength shift when pump power is varied.

Conclusions

In this paper, stable dual-wavelength laser emission using two MMI filters (configured as a MZ interferometer) was experimentally demonstrated. The novel filter structure was used for individual wavelengths selection and cavity losses adjustment to achieve dual-wavelength laser operation of an EDF ring cavity laser. With a pump power of 100 mW, dual-wavelength laser emission at 1540.4 and 1554 nm with a wavelengths separation of 13.6 nm was achieved. The laser wavelengths output power stability variations with the

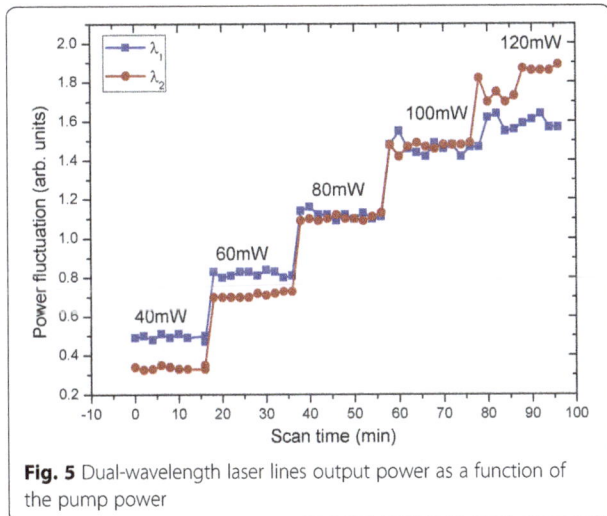

Fig. 5 Dual-wavelength laser lines output power as a function of the pump power

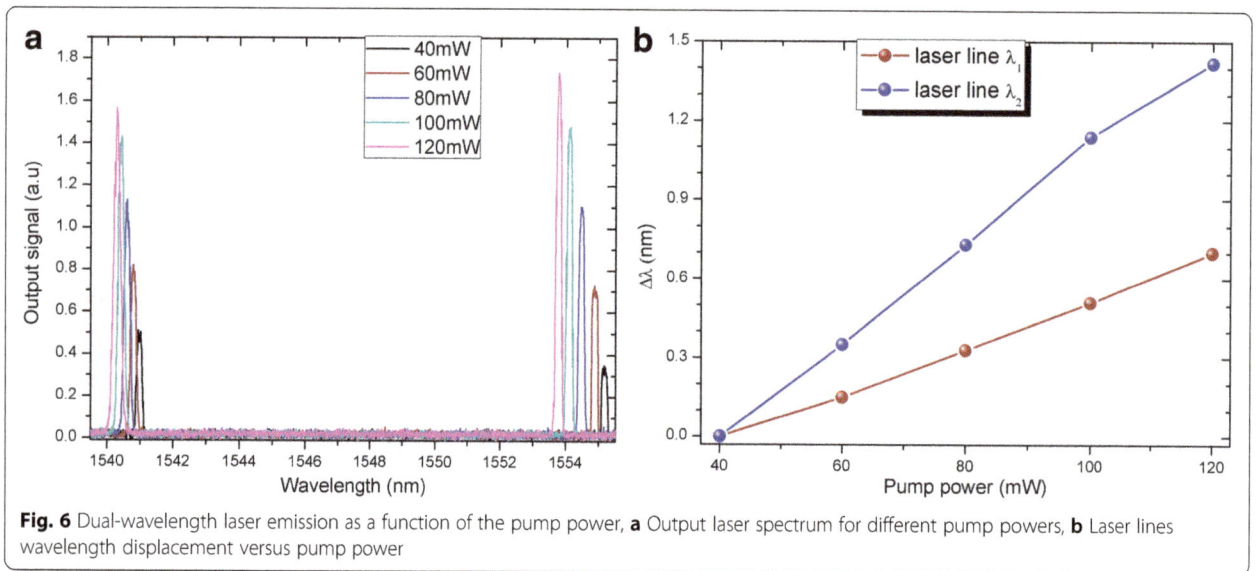

Fig. 6 Dual-wavelength laser emission as a function of the pump power, **a** Output laser spectrum for different pump powers, **b** Laser lines wavelength displacement versus pump power

applied pump power was also experimentally discussed. The wavelength of the generated laser lines slightly shifts toward shorter wavelengths with an increase of the applied pump power. The obtained results demonstrate the reliability of the proposed MMI-DWF for dual-wavelength laser generation in an all-fibre ring cavity fibre laser with potential applications in optical communications and sensing.

Acknowledgements
Manuel Durán-Sánchez was supported by Cátedras-CONACyT project 2728. R. I. Álvarez-Tamayo wants to thanks CONACyT postdoctoral fellow 160248. B. Ibarra-Escamilla was supported by CONACyT Grants No. 237855 and No. 255284.

Authors' contributions
RIAT and MDS conceived the work. MDS did the implementation, obtained the experimental results and contributed to results analysis. RIAT did results analysis and contributed to writing. JGAS did the MMI filters design and analysis. JEAL developed and fabricated the no-core fibre and contributed to writing. BIA and EAK coordinated the work, did result analysis and contributed to writing. All authors read and approved the final manuscript.

Competing interests
The authors declare that they have no competing interests.

Author details
[1]Optics Department, Instituto Nacional de Astrofísica, Óptica y Electrónica, Luis Enrique Erro 1, Puebla 72824, Mexico. [2]CONACyT Research Fellow - Instituto Nacional de Astrofísica, Óptica y Electrónica, Luis Enrique Erro 1, Puebla 72824, Mexico. [3]CREOL, The College of Optics and Photonics, University of Central Florida, Orlando, FL 32816-2700, USA.

References
1. Soldano, LB, Pennings, ECM: Optical multi-mode interference devices based on self imaging: Principles and applications. J Lightw Technol. **13**, 615–627 (1995)
2. Mohammed, WS, Smith, PWE, Gu, X: All-fibre multimode interference bandpass filter. Opt. Lett. **31**, 2547–2549 (2006)
3. Wang, P, Brambilla, G, Ding, M, Semenova, Y, Wu, Q, Farrell, G: Investigation of single-mode-multimode-single-mode and single-mode-tapered-multimode-single-mode fiber structures and their application for refractive index sensing. J. Opt. Soc. Am. B **28**, 1180–1186 (2011)
4. Antonio-Lopez, JE, Castillo-Guzman, A, May-Arrioja, DA, Selvas-Aguilar, R, LiKamWa, P: Tunable multimode-interference bandpass fiber filter. Opt. Lett. **35**, 324–326 (2010)
5. Taue, S, Matsumoto, Y, Fukano, H, Tsuruta, K: Experimental analysis of optical fiber multimode interference structure and its application to refractive index measurement. Japanese J Appl Phys **51**, 04DG14 (2012)
6. André, RM, Biazoli, CR, Silva, SO, Marques, MB, Cordeiro, CMB, Frazão, O: Multimode interference in tapered single mode-multimode-single mode fiber structures for strain sensing applications. Proceedings of SPIE, OFS2012 22nd International Conference on Optical Fibre Sensors, 84213B, (2012)
7. Antonio-Lopez, JE, Castillo-Guzman, A, May-Arrioja, DA, Selvas-Aguilar, R, LiKamWa, P: Fiber-optic sensor for liquid level measurement. Opt. Lett. **36**, 3425–3427 (2011)
8. Castillo-Guzman, A, Antonio-Lopez, JE, Selvas-Aguilar, R, May-Arrioja, DA, Estudillo-Ayala, J, LiKamWa, P: Widely tunable erbium-doped fiber laser based on multimode interference effect. Opt. Express **18**, 591–597 (2010)
9. Mukhopadhyay, PK, Gupta, PK, Singh, A, Sharma, SK, Bindra, KS, Oak, SM: Note: Broadly tunable all-fiber ytterbium laser with 0.05 nm spectral width based on multimode interference filter. Rev Sci Instr **85**, 056101 (2014)
10. Selvas, R, Torres-Gomez, I, Martinez-Rios, A, Alvarez-Chavez, JA: Wavelength tuning of fiber lasers using multimode interference effects. Opt. Express **13**, 9439–9445 (2005)
11. Liu, Z, Liu, Y, Du, J, Yuan, S, Dong, X: Switchable triple-wavelength erbium-doped fiber laser using a single fiber Bragg grating in polarization-maintaining fiber. Opt. Comm. **279**, 168–172 (2007)
12. Han, Y, Lee, JH: Switchable dual wavelength erbium-doped fiber laser at room temperature. Microwave Opt Technol Lett. **49**, 1433–1435 (2007)
13. Meng, Y, Zhang, S, Wang, X, Du, J, Li, H, Hao, Y, Li, X: Tunable double-clad ytterbium-doped fiber laser based on a double-pass Mach–Zehnder interferometer. Opt Laser Eng. **50**, 303–307 (2012)
14. Álvarez-Tamayo, RJ, Durán-Sánchez, M, Pottiez, O, Kuzin, EA, Ibarra-Escamilla, B, Flores-Rosas, A: Theoretical and experimental analysis of tunable Sagnac high-birefringence loop filter for dual-wavelength laser application. Appl Optics. **50**, 253–260 (2011)
15. Durán-Sánchez, M, Álvarez-Tamayo, RI, Pottiez, O, Ibarra-Escamilla, B, Hernández-García, JC, Beltrán-Pérez, G, Kuzin, EA: Actively Q-switched dual-wavelength laser with double-cladding Er/Yb-doped fiber using a Hi-Bi Sagnac interferometer. Laser Phys. Lett. **12**, 025102 (2015)
16. Ma, L, Kang, Z, Qi, Y, Jian, S: Tunable dual-wavelength fiber laser based on an MMI filter in a cascaded Sagnac loop interferometer. Laser Phys. **24**, 045102 (2014)
17. Antonio-Lopez, JE, Sanchez-Mondragon, JJ, LiKamWa, P, May-Arrioja, DA: Tunable Dual-Wavelength Erbium-Doped Fibre Ring Laser. Frontiers in Optics 2012/Laser Science XXVIII, OSA Technical Digest (online) (Optical Society of America, 2012), FW3A.32 (2012)
18. Zhang, P, Wang, T, Ma, W, Dong, K, Jiang, H: Tunable multiwavelength Tm-doped fibre laser based on the multimode interference effect. Appl Optics. **54**, 4667–4671 (2015)

11

Analysis of dispersion effect on a NRZ-OOK terrestrial free-space optical transmission system

Mohamed Bouhadda[1*], Fouad Mohamed Abbou[2], Mustapha Serhani[1], Fouad Chaatit[2] and Ali Boutoulout[1]

Abstract

Background: In this paper, the impact of the dispersion effect, due to atmospheric pressure and temperature, on NRZ-OOK terrestrial free-space optical transmission system is investigated. An expression for the dispersion parameter in FSO atmospheric channel is derived.

Results: The results show that the variation of the refractive index along the transmission path induces fluctuations of group velocity dispersion of the optical pulse resulting in broadening of the pulse duration. Simulation results show that at a propagation distance of 7.5 km, the broadening ratio for input pulse duration of 300 fs is approximately 2.39. Further, at a propagation distance of 7.5 km, the remaining fraction of energy is approximately 40 % for a 300 fs input pulse duration. However, by increasing the transmitter input power, the effect of dispersion could be reduced. Namely, for a reference BER of 10^{-9}, the maximum distance that it could be achieved is about 1. 461 km for an input power of 1 mW, while it is about 2.694 km for an input power of 4 mW.

Conclusions: The results indicate that the effect of dispersion resulting from pressure and temperature increases with the propagation distance, which induces a high BER. However, the results show that it is possible to reach longer propagation distances with a lower BER by increasing the input power.

Keywords: Dispersion, Pulse broadening ratio, NRZ-OOK, BER, FSO

Background

Recently, free space optical communication technology has attracted much research because it has been successfully used in various applications such as satellite communication, deep-space probes and terrestrial communication. The free space optical communication offers remarkable advantages over the radio waves transmission, namely; high data transmission, unlicensed transmission, reduced interference and high security. Further, the capacity of FSO communication system has been successfully increased in recent years. In particular, an optical time division multiplexing system operating at 1.28 Tbit/s data transmission over a single-mode channel has been established [1]. According to [2], through free-space optical wireless systems, up to 2.5 Gbit/s of data, voice and video communications can be transmitted. FSO communication provides line of sight (LOS) communication thanks to its narrow transmit beamwidth and works in visible and IR spectrum. Furthermore, FSO communication systems are classified into terrestrial and space optical links which include building-to-building, ground-to-satellite, satellite-to-ground, satellite-to-satellite and satellite-to-airborne platforms (see [3, 4, 5]). Typical terrestrial communication wavelengths such as 808, 1064 or 1550 nm are applicable because they fall within the atmospheric transmission window in the absorption spectrum. As a result, the atmospheric loss due to absorption for these wavelengths turns out to be negligible as noted in [6, 7]. However, and due the variation of the atmospheric pressure and temperature, the refractive index undergoes random fluctuations along the transmission path. This induces fluctuations of group velocity dispersion of the optical pulse, and results in either, broadening or compressing the pulse duration. The Pulse broadening limits the bit rate of optical link, and induces inter-symbol interference between adjacent

---

* Correspondence: mohamed.bouhadda@usmba.ac.ma
[1]Moulay Ismail University, MACS Laboratory, Meknès, Morocco
Full list of author information is available at the end of the article

pulses, which increases, bit error rate of the free space optical communication system.

In this paper, we propose an analytical expression for temporal pulse broadening, and we investigate the effects of atmospheric pressure and temperature on temporal broadening and study the effect of atmospheric dispersion on NRZ-OOK terrestrial free-space optical transmission system. The paper is organized as follows. In Theoretical analysis section, we present the Theoretical analysis needed for the study. In Results and discussions section, we discuss and analyze the obtained results. Conclusion section concludes the paper.

Methods

Dispersion phenomena can drastically affect the propagation of an optical beam by random fluctuations of the refractive index due to temperature and pressure variations along the optical propagation path. Based on the work presented in [8], the refractive index in the visible light and infrared domain can be described by the following expression

$$n = 1 + 77.6\left(1 + 7.52 \times 10^{-3}\lambda^{-2}\right)\frac{P_h}{T_h} \times 10^{-6} \quad (1)$$

Here λ is the optical wavelength in μm, P_h is the atmospheric pressure in Millibar, and T_h is the temperature of the atmosphere in Kelvin. The gradient of standard atmospheric temperature as function of height can be expressed as [9];

$$T_h = \begin{cases} 288\text{--}6.5h10^{-3} & for\ 0km{\leq}h < 11Km \\ 216,5 & for\ 11km{\leq}h < 20Km \\ 216,5 + (h\text{--}20000)10^{-3} & for\ 20km{\leq}h < 32Km \end{cases} \quad (2)$$

For the gradient of standard atmospheric pressure as function of height is given by [9];

$$P_h = 1013\left(\frac{(288\text{--}0.006h)}{288}\right)^{5.255} \quad (3)$$

Where h is the altitude in meters. Further, when a group of optical waves with narrow range of wavelengths co-propagate along the optical propagation path, their resultant lightwave packet travels at the group velocity group v_g defined by:

$$v_g = \frac{c}{n-\lambda\frac{dn}{d\lambda}} \quad (4)$$

Where c is the speed of light in vacuum. Using $L = v_g\tau_{FSO}$, where L is link length of the FSO medium, we can obtain an expression of the pulse delay τ_{FSO} as

$$\tau_{FSO} = \frac{L}{C}\left(1 + 77.6\left(1 + 22.56 \times 10^{-3}\lambda^{-2}\right)\frac{P_h}{T_h} \times 10^{-4}\right) \quad (5)$$

Further, considering an optical laser source with rms spectral width $\Delta\lambda$, the rms pulse broadening due to FSO medium can be derived as:

$$\sigma_{FSO} = \Delta\lambda\left|\frac{d\tau_{FSO}}{d\lambda}\right| = 3501.31\Delta\lambda\frac{L}{c}\lambda^{-3}\frac{P_h}{T_h}10^{-7} \quad (6)$$

Hence, the FSO medium dispersion coefficient can be expressed as:

$$D_{FSO} = -3501,31\frac{\lambda^{-3}}{c}\frac{P_h}{T_h}10^{-7} \quad (7)$$

Hence, the third order β_3 and the second order β_2 derivatives of propagation constant β can be expressed as:

$$\beta_2 = \frac{\lambda^{-1}}{2\pi c^2}3501,31\frac{P_h}{T_h}10^{-7} \quad (8)$$

$$\beta_3 = \frac{10503,93}{4\pi^2 c^3}\frac{P_h}{T_h}10^{-7} \quad (9)$$

β_2 is the group velocity dispersion (GVD), is known to be the primary source of pulse broadening [10]. The frequency dependence of the group velocity results in pulse broadening because different spectral components of the pulse disperse during propagation due to frequency chirps generated by the GVD induced phase shift.

Further, by ignoring the channel losses induced by scattering and absorption in a terrestrial FSO link, the received signal power P_r at a distance L with a transmitter signal power P_t can be written as

Table 1 System parameters

Parameters	Value
Transmission Wavelength (λ)	1550 nm
Distance (L)	1–10 km
Transmitter power (P_t)	1–50 mW
Optical Efficiency of Transmitter τ_t	0.75
Optical Efficiency of Receiver τ_r	0.75
Full transmitting divergence angle θ	$2*10^{-3}$rad
Receiver Diameter	1 cm
Electron Charge (q)	1.6×10^{-19} C
PIN Load Resistance (R)	1kΩ
Boltzmann Constant (k)	1.38×10^{-23} J.k
Temperature (T)	298 K
Dark Current (I_d)	10 nA
Responsivity (R_d)	0.6A/W˙
Bandwidth (B)	0.5GHz

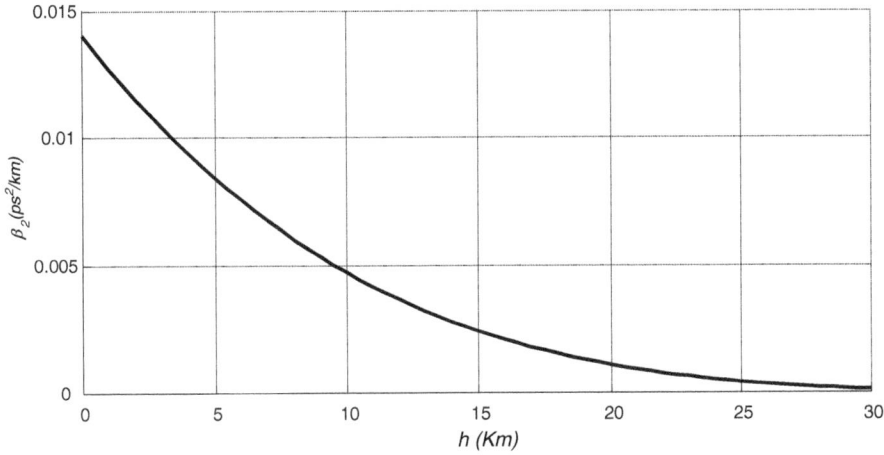

Fig. 1 The second order dispersion coefficient β_2 as a function of the altitude

$$P_r = P_t \left(\frac{D}{L\theta}\right)^2 \tau_t \tau_r \tag{10}$$

Where D is the receiver diameter, θ is the full transmitting divergence angle, and τ_r and τ_t are the optical efficiencies of the transmitter and the receiver respectively. In order to evaluate the FSO performance in the presence of dispersion, the SNR and BER are considered. For a PIN photodiode receiver, the signal-to-noise ratio (SNR) can be written as

$$\frac{S}{N} = \frac{(P_r R_d)^2}{i_d^2 + i_{th}^2 + i_{sh}^2} \tag{11}$$

Where i_d^2 is the detector dark noise, i_{th}^2 is the thermal noise and i_{sh}^2 is the shot noise. The noise sources are expressed mathematically by:

$$i_d^2 = 2qBI_d \tag{12}$$

$$i_{sh}^2 = 2qBI_p \tag{13}$$

$$i_{th}^2 = \frac{4kTB}{R} \tag{14}$$

Where $I_p = P_r R_d$ is the average photocurrent, R_d is the receiver responsivity, q is the charge of an electron, B represents the bandwidth, T is the absolute photodiode temperature (K), and R is the PIN load resistor, and k is the Boltzmann's constant. The NRZ-OOK Bit Error Rate (BER) of a FSO link can be expressed as

$$\text{BER} = \frac{1}{2}\text{erfc}\left(\frac{1}{2\sqrt{2}}\sqrt{\frac{S}{N}}\right) \tag{15}$$

In the next section, simulation results will be discussed to analyze the effect of dispersion on FSO optical

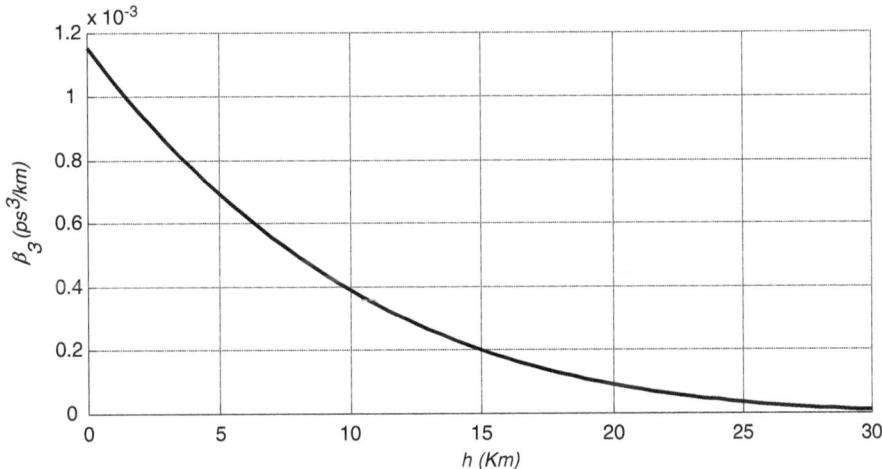

Fig. 2 The third order dispersion coefficient β_3 as a function of the altitude

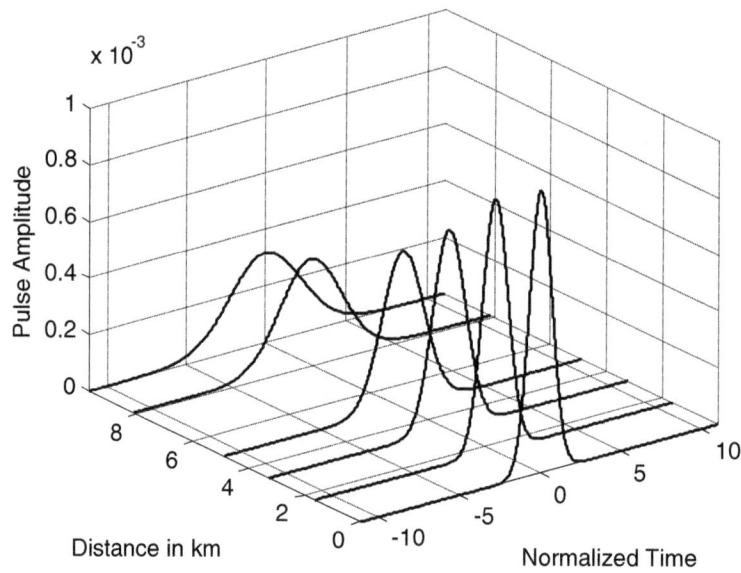

Fig. 3 Pulse propagation at $z = 10$ km and with $\beta_2 = 0.002$ ps²/km

wireless communication system employing NRZ-OOK modulation technique. The simulation parameters are defined in Table 1.

Results and discussions

Following the theoretical analysis presented early on, the effect of dispersion due to atmospheric pressure and temperature on a terrestrial free-space optical communication system is investigated. Figures 1 and 2 show the curves of second and third order dispersion coefficients as a function of the altitude. Clearly, it can be seen that both β_2 and β_3 are decreasing with the altitude. From Fig. 3, it is obvious that the GVD induced pulse broadening increases linearly with the propagation distance and therefore imposes limitation on the FSO link. Further, the pulse broadening ratio as a function of propagation distance, for different input pulses is depicted in Fig. 4. It is clear from Fig. 4 that the broadening ratio increases with the propagation distance. At a propagation distance of 7 km, the values of the broadening ratios for the three different input pulses with $T_0 = 300$ fs, $T_0 = 400$ fs, and $T_0 = 500$ fs, are

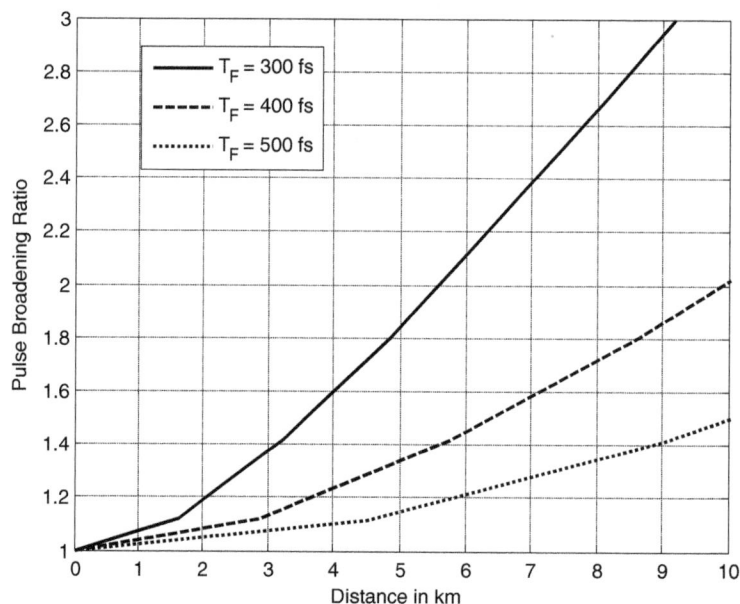

Fig. 4 Broadening ratio as a function of propagation distance for different input pulses

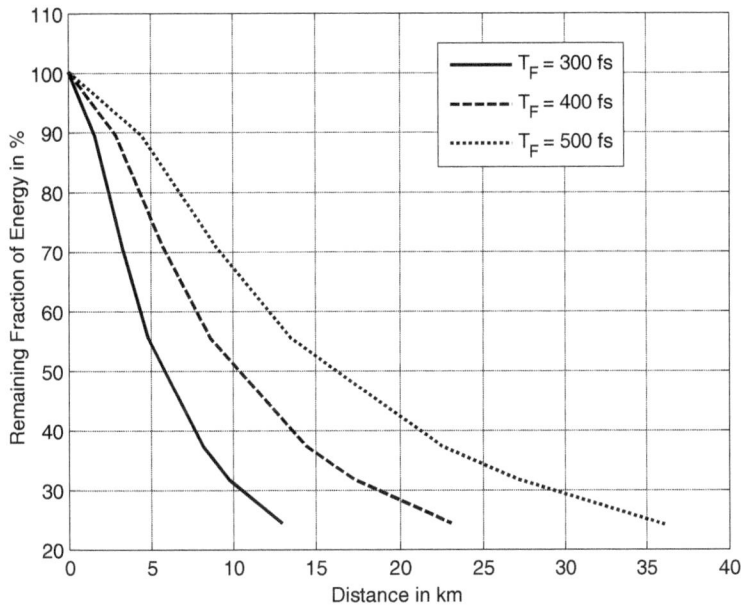

Fig. 5 Remaining fraction of energy as a function of the propagation distance

found to be approximately 2.39, 1.59 and 1.29 respectively. This is obvious as short pulsed are most sensitive to dispersion effect.

Further as shown in Fig. 5, the pulse remaining fraction of energy decreases with the distance due to the attenuation induced by dispersion. For example, at a propagation distance of 7.5 km, the remaining fraction of energy is approximately 40 % for a 300 fs input

pulse. Thus, it is quite obvious that at a large link distance, it is difficult to maintain sufficient pulse energy. However, from Fig. 6, the BER curves for NRZ-OOK modulation format for different values of input power show that by increasing the transmitter input power, the effect of dispersion could be reduced and therefore it would be possible to achieve longer propagation distance with significant lower BER.

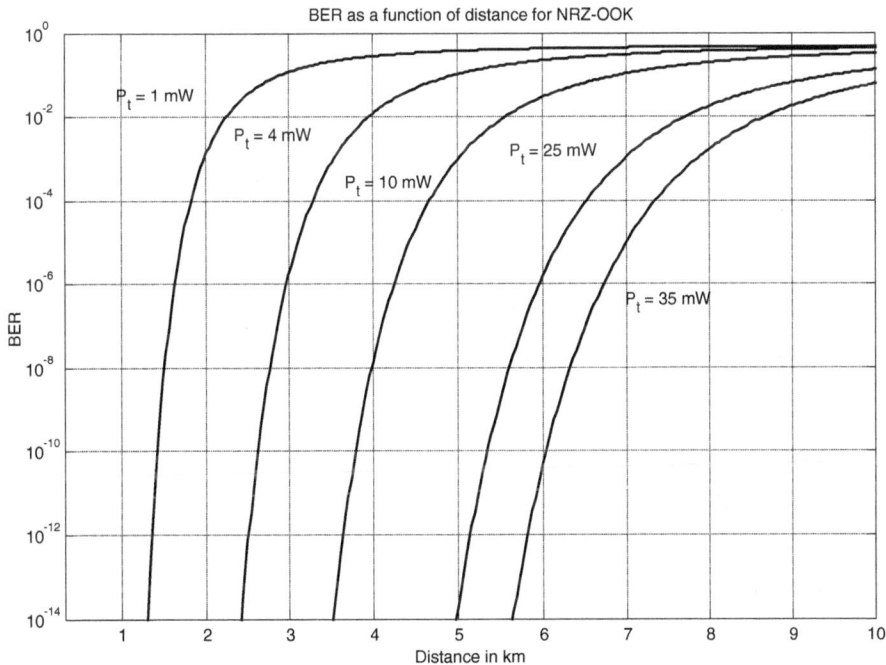

Fig. 6 NRZ-OOK BER versus link distance for different values of transmitter power

Conclusion

The effect of dispersion due to atmospheric pressure and temperature on a terrestrial free-space optical communication system is semi-analytically analyzed. A general expression for the medium dispersion coefficient due to pressure and temperature is derived. It is clear that the dispersion effect due to pressure and temperature increases with the propagation distance. At a propagation distance of 7.5 km, the remaining fraction of energy is approximately 40 % for a 300 fs input pulse. Further, performance results show that the dispersion induced pulse broadening limits the link distance and induces high BER. However, by increasing the transmitter input power, the effect of dispersion could be reduced and therefore it would be possible to achieve longer propagation distance with significant lower BER.

Acknowledgement
The authors would like to extend their special thanks and appreciations to the Al Akhawayen University and Moulay Ismail University, Morroco for supporting this work.

Authors' contributions
MB and FMA participated in the development of the mathematical model andcarried out the simulation. MB and FMA, MS, FC and AB contributed in the analysis of the results. All authors helped to draft the manuscript. All authors have read and approved the final manuscript.

Competing interests
The authors declare that they have no competing interests.

Author details
[1]Moulay Ismail University, MACS Laboratory, Meknès, Morocco. [2]Al Akhawayen University, School of Sciences and Enginnering, Ifrane, Morocco.

References
1. Howlader, MK, Jung, J: Inter-symbol interference due to the atmospheric turbulence for free-space optical communication system. In: IEEE International Conference on Communications, p. 5046. (2007)
2. FSONA Systems Corp: Unveils 2-5-Gbps free-space optical systems. (2012)
3. Ghassemlooy, Z, Popoola, WO: Terrestial Free-Space Optical Communications, ch. 17. In: Fares, SA, Adachi, F (eds.). pp. 356–392. European Union, Rijeka (2010). ISBN 978-953-307-042-1
4. Sharma, V, Kumar, N: Improved analysis of 2.5 Gbps-inter-satellite link (ISL) in inter-satellite optical wireless communication (ISOWC) system. Opt. Commun. **286**, 99–102 (2014)
5. Majumdar, AK, Ricklin, JC: Free-Space Laser Communications: Principles and Advanced, springer science+business media. LLC, New York (2008)
6. Henninger, H, Wilfert, O: An Introduction to Free-space Optical Communications. Radio Eng. **19**(2), 203–212 (2010)
7. Alkholidi, A, Altowij, K: Effect of Clear Atmospheric Turbulence on Quality of Free Space Optical Communications in Western Asia, Das, N. (ed.) Optical Communications Systems, ISBN: 978-953-51-0170-3, InTech (2012). doi:10.5772/35186
8. Andrews, LC, Phillips, RL: Laser beam propagation through random media, 2nd edn. SPIE Optical Engineering Press, Bellingham (2005)
9. Manual of the ICAO Standard Atmosphere Doc 7488/3, International Civil Aviation Organization, 3rd edn. (1993)
10. Govind, P: Agrawal Fiber-Optic Communication Systems, 2nd edn. (1997)

12

Nonlinear optical response of Mg/MgO structures prepared by laser ablation method

Fahimeh Abrinaei

Abstract

Background: Investigation of new materials plays an important role in advancing the field of optoelectronics.

Methods: In this work, Mg/MgO microstructures were prepared by Nd-YAG laser (λ= 1064 nm) ablation of magnesium target in acetone. For the first time, the nonlinear optical properties of square Mg/MgO microstructures were investigated by using the Z-scan technique with nanosecond Nd-YAG laser at 532 nm.

Results: The XRD analysis approved the formation of Mg/MgO microstructures. The energy band gap of Mg/MgO microstructures was calculated to equal 2.3 eV from UV-Vis spectrum. The ablated materials were ejected into acetone as structures with an average size of 1-1.5 µm. The nonlinear absorption coefficient, β, and nonlinear refractive index, n_2, for Mg/MgO microstructures at the laser intensity of 1.1×10^8 W/cm2 were measured to be 1. 15×10^{-8} cm/W and 8.2×10^{-13} cm^2/W, respectively. In order to investigate size particles and liquid medium effects, the nonlinear optical parameters, β and n_2 of Mg/MgO nanostructures synthesized by laser ablation of magnesium target in isopropanol also were calculated and it was found these parameters are an order of magnitude larger than the values for the β and n_2 of Mg/MgO microstructures synthesized in acetone. The third-order nonlinear optical susceptibility, $\chi^{(3)}$, of Mg/MgO microstructures and nanostructures were measured in order of 10^{-6} and 10^{-5} esu, respectively.

Conclusions: The results show that Mg/MgO structures synthesized in acetone and isopropanol have negative nonlinearity as well as good nonlinear absorption at 532 nm and these magnesium-based structures have the potential applications in the nonlinear optical devices.

Keywords: Laser materials processing, Mg/MgO microstructures, Mg/MgO nanostructures, Nonlinear optics, Z-scan technique

Background

Materials with the large nonlinear optical (NLO) response, good capabilities of processing, environmental constancy and ultrafast signal switching are necessary for potential applications in optical signal processing, optical limiting (OL), and optical devices. Designing of new materials with large nonlinearities is a promising line of the current optoelectronics. So, composite particles that are attractive to build various photonic devices provide many unique opportunities to achieve a large optical nonlinearity [1].

In recent years, many researchers have noted to magnesium microstructures and nanostructures due to their novel properties to distinguish them from bulk materials and diverse applications in fields of propellant, battery, composite fillers, etc. Magnesium has several very promising properties for applications in plasmonics and specially in switchable plasmonic metamaterials [2]. As reported by Sterl et al., magnesium nanoparticles exhibit a pronounced plasmonic response at throughout the whole visible wavelength range. Therefore, it can be an ideal alternative to established materials for UV plasmonics such as aluminum [3]. Magnesium is an excellent candidate for weight critical structural applications for its impressive low mass density. A good high-temperature crawl, a high damping capacity,

Correspondence: f_abrinaey@yahoo.com
Department of physics, East Tehran Branch, Islamic Azad University, Tehran, Iran

and good dimensional stability are the properties that make magnesium perfect for industrial applications [4–9]. Magnesium is a candidate for the fuel cell technologies. It is reversible, abundant and low-cost element possesses the large capacity for hydrogen storage (7.6 wt.%) [10].

Magnesium oxide (MgO) nanostructures and microstructures have attracted more and more regard in recent years. MgO is an important inorganic material possesses a wide band gap energy ($E_g \sim 7.8$ eV) [11]. MgO is a useful oxide material for industrial applications. MgO has been employed such as an additive in heavy fuel oils, toxic waste remediation, catalyst supports, refractory materials and adsorbents, catalysis, reflecting and anti-reflecting coatings, superconducting and ferroelectric thin films as the substrate, etc [12, 13]. For medical uses, MgO is used for the sedation of heartburn, sore stomach, and for bone instauration [14, 15]. Recently, researchers have discovered that MgO nanoparticles are appropriate for application in tumor treatment [16]. MgO nanoparticles are suitable for application as antibacterial agent [17].

Kurth et al. reported the initial oxidation of magnesium at oxygen partial pressures and showed that the initially formed oxide has a higher Mg/O ratio (>1.3) than bulk MgO. Also, they indicated that the band gap values of the oxide layers are considerably smaller than the value expected for bulk MgO (2.5 eV vs. 7.8 eV) [18]. Canney et al. studied electronic band structure of thin magnesium and magnesium oxide films, experimentally and theoretically. They showed that the investigated band structures of Mg and MgO have the characteristic of a metallic and ionic solid, respectively [19]. Phuoc et al. synthesized Mg/MgO nanocrystallites by laser ablation of magnesium in acetone and isopropanol with a Q-switched Nd–Yag laser operating at 1064 nm and investigated structural properties of the formed Mg/MgO nanostructures [20]. Abrinaei et al. reported on the study of structural properties of Mg/MgO nanoparticles synthesized by Ng-YAG ($\lambda = 1064$ nm) laser as well as CVL (Cooper Vapor Laser) in isopropanol [21]. In the recent months, Gutierrez et al. reported an investigation of the effect of an oxide shell on the ultraviolet plasmonic behavior of Ga, Mg, and Al nanostructures. They studied the plasmonic response of Mg/MgO nanostructures and effect of oxidation on the UV-plasmonic response of Mg/MgO spherical or hemisphere-on-substrate nanostructures [22].

This work analyzes the Mg/MgO microstructures obtained by Nd-YAG ($\lambda = 1064$ nm) laser ablation of solid magnesium target in acetone media, through the study of their X-ray diffraction powder pattern, SEM image, Fourier transform infrared spectroscopy (FTIR) spectrum and UV-VIS spectrum. The nonlinear absorption and refraction of these microstructures are investigated by using the Z-scan technique with Nd-YAG laser at 532 nm. Furthermore, the current research discusses the OL

properties of Mg/MgO colloids. To compare the case, linear and NLO properties of spherical and platelet-like Mg/MgO nanostructures have been calculated in this work, as well. The synthesis of these nanostructures by Nd-YAG ($\lambda = 1064$ nm) laser ablation of magnesium target in isopropanol media and investigation of their structural properties have been already reported in the researcher's previous work [21]. To the best of the author's knowledge, this is the first investigation of NLO properties of Mg/MgO structures induced by the Z-scan method.

Methods
Material synthesis
Mg/MgO microstructures were prepared by laser ablation of Mg metal plate (99.99%) in acetone. As shown in Fig. 1, the target was put at the bottom of a glass vessel filled with 5 ml acetone. The laser beam was focused on the surface of the magnesium target. The liquid depth above the target surface was several millimeters. An acoustic-optically Q-switched Nd-YAG laser ($\lambda = 1064$ nm) with 240 ns pulse duration, was used and adjusted to operate at 210 Hz repetition rate. The laser pulse energy applied on the target was 3.5 mJ, leading to a fluence of about 91 J cm^{-2}. The thickness of the target was 1 mm. The irradiation time in the experiments was 45 min. During the ablation in acetone, the spark plumes become larger with a cracking noise. Small bubbles were observed and suggest that acetone was pyrolized during the laser ablation process. Upon irradiation of a laser beam, the color of the solution turned light metallic gray. Structures were not stable and the rapid agglomeration even pending the laser ablation process. These agglomerations precipitated at the bottom of the container, rapidly.

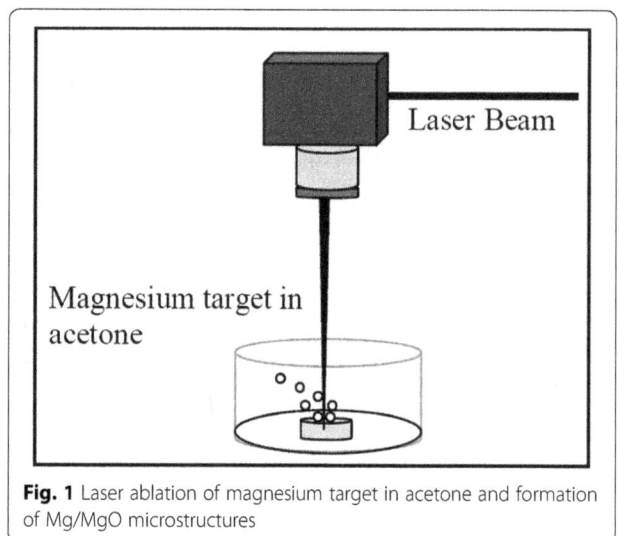

Fig. 1 Laser ablation of magnesium target in acetone and formation of Mg/MgO microstructures

Structural characterization

After laser ablation, a small amount of colloidal solution was transferred into a quartz cell for UV–VIS spectroscopy analysis. The optical absorbance of the sample was measured by using a high-resolution spectrophotometer, Camspec, ModelM350 in the wavelength range of 200–1000 nm.

X-ray diffraction patterns of structures were recorded by evaporation of the colloidal solutions onto a glass substrate at room temperature. The crystal structure of sample was analyzed by a Philips diffractometer (STADI MP) with Cu K_α radiation, angle step size of 0.02°, and count time of 1.0 s per step.

An amount of wet precipitate was transferred to a quartz vessel placed in the open air to evaporate the liquid in order to acquire some powder for FTIR investigations. The FTIR spectrum is prepared by a Therno Nicollet NEXUS 870 FT_IR model spectrophotometer.

Some drops of Mg/MgO colloidal solution have also been deposited onto an Aluminum plate in order to perform SEM analysis with a TS5136MM microscope operating at 30 kV.

Z-scan setup

By using the Z-scan setup can be measured the sign and magnitude of the nonlinear refractive index (n_2) and nonlinear absorption coefficient (β), simultaneously. The Z-scan technique has sensitivity analogous to interferometric methods [23].

The experimental setup for Z-scan is shown schematically in Fig. 2. A Q-switched Nd-YAG laser (Ekspla NL640 model, 532 nm, 10 ns, 200 Hz) was used as the light source. The sample was traveled in the direction of the light propagation near the focal spot of the lens. The radius of the beam waist ω_0 was calculated to be 37 μm at the focal point. The Rayleigh length, $z_0 = \pi\omega_0^2/\lambda$ was estimated to be 8.08 mm, much greater than the thickness of the sample, which is an indispensable requisite for Z-scan experiments.

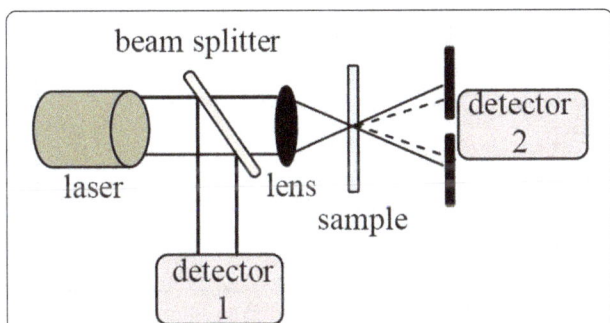

Fig. 2 Schematic diagram of the experimental setup for the measurement of nonlinear optical parameters of Mg/MgO structures prepared in acetone

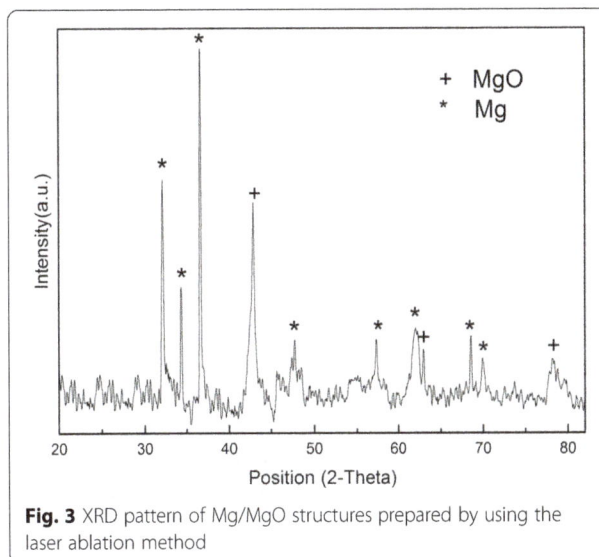

Fig. 3 XRD pattern of Mg/MgO structures prepared by using the laser ablation method

The intensity of the laser beam is 1.1×10^8 Wcm^{-2} at the focal point. This setup had been already applied to investigate NLO properties of $Mg(OH)_2$ nanostructures [24].

Results

Structural investigations

The typical XRD pattern of the structures prepared by the pulsed Nd-YAG laser ablation of magnesium target in acetone is given in Fig. 3. The XRD pattern indicates that both MgO and Mg are presented in the sample with a higher percentage of Mg. It is clear from the XRD pattern that the prepared powders are polycrystalline structures inclusive Mg and MgO. The weight percentage of each Mg and MgO structures can be calculated by Rietveld method [25]. The Mg/MgO ratio in the final product synthesized by laser ablation of magnesium target in acetone is 65/35 percentage. The formed phases in the XRD pattern of Mg/MgO microstructures are the same of those are reported by Phuoc et al. and Abrinaei et al [20, 21].

During the ablation process, each laser pulse with quite high energy passes through the acetone above the magnesium target and increases its temperature. When the front part of the laser pulse interacts with the target, it induces a plasma plume on the surface, which is quite massive. Hence, the heat that is transferred between hot confined plasma and the surrounding liquid increases the temperature of the liquid.

The Mg/MgO microstructures could be formed in three stages. In the first stage, after the interplay among magnesium target and laser beam, the high-temperature and high-pressure plasma is generated in the magnesium target and acetone interface. In the next step, the Mg clusters are produced because of successive ultrasonic and adiabatic expansion of the high-temperature and high-pressure magnesium plasma that makes a cold zone of magnesium plume [26, 27]. In this experiment, the interval between two consecutive pulses of Nd-YAG laser is 0.005 s (repetition rate is 200 Hz) and it is much longer than the lifetime of the magnesium plasma plume. Hence, the next laser pulse does not interact with the previous plasma plume. In the third step, the plasma quenches and the produced Mg clusters envisage the acetone and occur some chemical reactions and lead to the formation of Mg/MgO microstructures. Outwardly, Mg clusters generated by laser ablation of Mg in acetone were oxidized in this solvent, possibly under the high-energy conditions via reaction with the oxygenated solvent. The observed color changes in acetone suggested that the solvent was altered or decomposed due to laser heating or reaction with the Mg particles [28].

Optical properties

Figure 4 shows the UV-VIS optical absorption spectrum of the colloidal suspension of the sample in acetone. Obviously, there is a characteristic absorption band at about 410 nm.

The measurement of the energy band gap, E_g, is important in the micro- and nano-materials. There are various methods for calculation of E_g. In this work, the energy band gap was determined using UV-Vis absorption spectrum of Mg/MgO microstructures. The derivative method was applied for measurement of E_g. In this method, the first derivative of the absorbance was evaluated near the fundamental absorption edge, leading to E_g [29]. The energy band gap value is obtained equal to 2.3 eV for Mg/MgO microstructures.

The electron configuration of magnesium is $1s^2 2s^2 2p^6 3s^2$ that 3 s shell is full. Then, the 3 s shell of Mg would not allow for electrons to increase energy on its own if the 3p band was separated by a gap from the 3 s band. But, in magnesium the energies of 3 s and 3p bands overlap and the 3p band can incorporate six electrons per atom $(2 (2 l + 1) = 6)$, overall 3 s and 3p orbitals from a band can incorporate eight electrons. Consequently, the quarter of conduction band in magnesium is full merely. Thus, magnesium is a good conductor while magnesium oxide is close to an ideal insulating ionic solid with a valence band structure dominated by the strong potential of the ionic cores [19].

The calculated band gap of Mg/MgO microstructures synthesized by the laser ablation method in acetone is significantly smaller than the wide band gap energy 7.8 eV expected for the bulk, pure, crystalline MgO. Therefore, it is clear that Mg/MgO microstructures have metallic conduction behavior [18].

FTIR transmittance spectrum in the wave number range of 4000–500 cm^{-1} for the Mg/MgO microstructures produced by laser ablation of Mg in acetone is presented in Fig. 5.

In the FTIR spectrum, the stretching bands of the superficial OH groups are seen in the region between 4000 and 3500 cm^{-1}. The absorption band around 3448 cm^{-1} is broad and corresponds to the –OH-group-stretching vibration that is dedicated to –OH-stretching mode of residue water and absorbed acetone on the surface of Mg/MgO microstructures. This peak shows the presence of hydroxyl groups at low coordination sites or defects [30]. The absorption bands around 2926 and 2851 cm^{-1} are due to surface OH stretch appearing from hydroxyl groups in dissociated state or C–H stretch of organic residue [31], where these peaks are due to surface OH stretch in-phase and out of phase, respectively. The peak at 1672 cm^{-1} was attributed to the bending

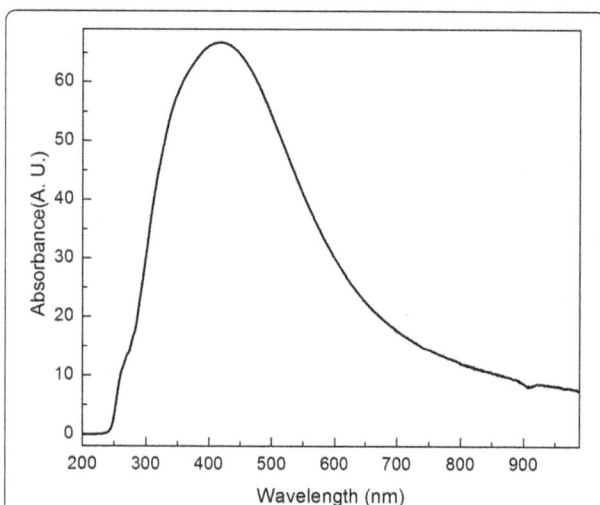

Fig. 4 UV-VIS spectrum of Mg/MgO structures prepared in acetone

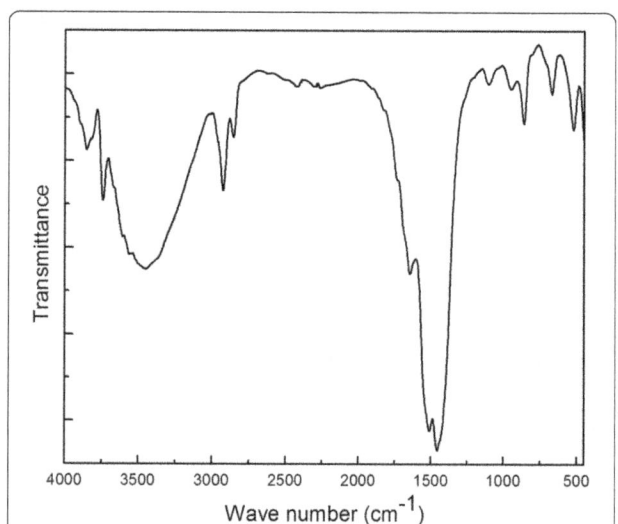

Fig. 5 FTIR spectrum of powder prepared by laser ablation of Mg in acetone

vibration of the water molecule. An absorption peak at 1457 cm^{-1} is assigned to ν_a (C–O) + δ (OC = O) modes. Thus, these contain signatures of adsorption and chemisorption of water and acetone. Tow sharp absorption peaks at 1107 and 970 cm^{-1} are devoted to C–O/C–O stretching modes. While the band at 861 cm^{-1} corresponds to ν (Mg–O) + δ (O–C = O), a peak at 673 cm^{-1} is ascribed to bending mode of O = C = O or vibration of water. The two peaks at 450 cm^{-1} and 515 cm^{-1} affirmed the presence of Mg-O vibrations [32].

Morphological properties

The morphology of the structures examined by scanning electron microscopy (SEM) is shown in Fig. 6.

The SEM image shows that the Mg/MgO microstructures are constructed of particles in the nearly square shape. The obvious SEM picture of Mg/MgO structures was obtained after about 10,000 times grandiosity. The lines appeared in this image related to aluminum foil on which colloidal solution was dried.

The corresponding particle size distribution of the mean sizes is shown in Fig. 7. As shown in the figure, the size distribution width becomes narrow from 1 to 1.5 µm.

Phuoc et al. prepared Mg/MgO nanoparticles in acetone and isopropanol by Nd: YAG (λ = 1064 nm) laser ablation of a magnesium target with distribution sizes were ranged from 15 to 20 nm up to 50 – 100 nm [20]. Abrinaei et al. applied the Nd-YAG (λ = 1064 nm) and cooper vapor laser beam and formed the Mg/MgO spherical and plate-like nanostructures and cubic microstructures with distribution sizes were ranged from 80 to 100 nm and 1–1.1 µm, respectively [21]. As a comparing the results, it approved the results of previous work's author that showed the parameters of laser and liquid

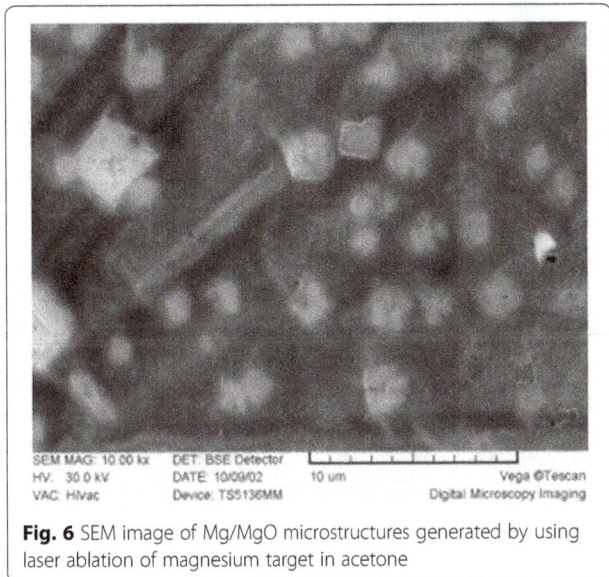

Fig. 7 Particle size distribution of structures generated in acetone

medium in laser ablation experiment significantly alter the shape and size of resultant products [21].

Nonlinear optical properties

NLO parameters, the nonlinear refractive index (n_2) and nonlinear absorption coefficient (β) of the colloidal Mg/MgO structures were obtained by the following relationships [23]:

$$T_{norm} = Ln(1 + q_0(z,t))/q_0(z,t) \qquad (1)$$

$$q_0(z,t) = \beta I L_{eff}/(1 + z^2/z_0^2) \qquad (2)$$

$$L_{eff} = (1 - e^{-\alpha L})/\alpha \qquad (3)$$

$$\alpha = -(1/L)Ln(I/I_0) \qquad (4)$$

$$|\Delta T_{p-v}| = 0.406(1-S)^{0.25}(2\pi/\lambda)n_2 I L_{eff} \qquad (5)$$

Where T_{norm} is the normalized transmission in the open-aperture Z-scan setup and $q_0(z,t)$ is a dimensionless factor, ΔT_{p-v} is the normalized difference between the peak and the valley in the curve of normalized transmittance (T_{norm}) versus location of the sample (z). In these equations, λ is the wavelength of radiation (532 nm), I is the intensity of radiation, S is the fraction of radiation detected by the detector (the transmittance of the aperture), α is the linear absorption coefficient, L refers to the sample length (1 mm), and L_{eff} is an effective sample thickness, which was measured by OL setup shown in Fig. 2.

Optical limiting

In the recent years, the selection of NLO materials in which increase influence of light leads to significant decreases in transmittance is taken into consideration.

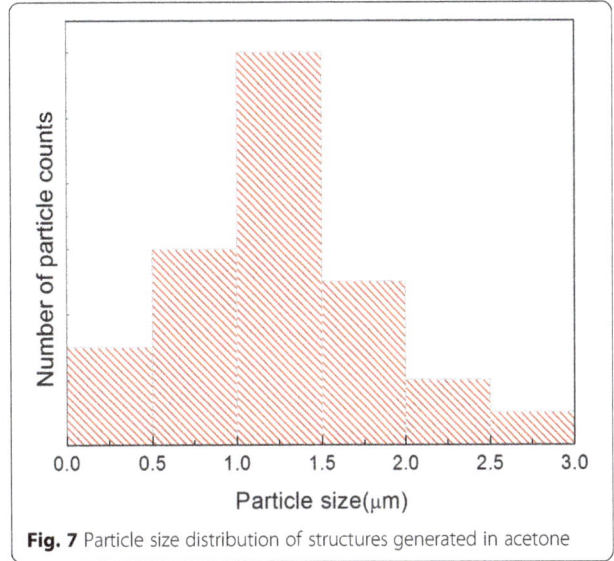

Fig. 6 SEM image of Mg/MgO microstructures generated by using laser ablation of magnesium target in acetone

These materials are used for optical power limiters devices. Investigating of new materials as the optical limiter is important for the protection of a person's eye and optical sensors from laser irradiation. In this work, the OL experiment was carried out by locating the sample at focus location and evaluating the transmitted power thru the aperture for several incident laser powers. By the OL consideration, the critical power of the laser beam at which the nonlinearity starts to affect the transmission can be measured. It is evident that the materials are suitable for OL applications that possess the lower OL threshold.

The experimental setup for OL measurements is shown in Fig. 2. In the OL configuration, the aperture is not used. The sample is put near the focal plane of the lens and the input power is changed after crossing the sample. A 50% beam splitter divides the initial power into the half. The power meter 1 is used to measure the input power. The output power of the transmitted beam through the Mg/MgO solution is measured by power meter 2.

The plot of output power versus input power for Mg/MgO microstructures synthesized by laser ablation of magnesium target in acetone is shown in Fig. 8. A threshold is attained at 20 mW of the input power. After 20 mW, the output power stabilized against the input power. At a low incident power up to 20 mW, the output power alters linearly with a ratio of $I/I_0 = 0.89$. By using the Eq. (4), the linear absorption coefficient for these microstructures is obtained: $\alpha = 1.16$ cm^{-1}.

The plot of output power versus input power for Mg/MgO nanostructures synthesized by laser ablation of magnesium target in isopropanol is shown in Fig. 9. A threshold is reached at 20 mW of the input intensity with slight variation in the output intensity for larger amounts of the input intensities. At a low incident

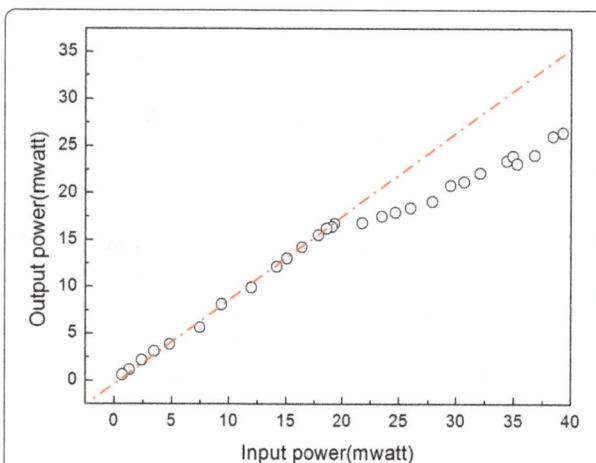

Fig. 9 The plot of output power versus input power for Mg/MgO nanostructures prepared by the laser ablation of magnesium target in isopropanol

power up to 20 mW, the output power alters linearly with a ratio of $I/I_0 = 0.88$. By using the Eq. (4), the linear absorption coefficient for nanostructures prepared by laser ablation of magnesium target in isopropanol is obtained: $\alpha = 1.27$ cm^{-1}.

The OL results confirm that structures prepared by laser ablation of Mg target in acetone and isopropanol are good candidates for OL at 532 nm pulsed lasers.

Closed- aperture Z-scan

The closed-aperture normalized transmittance curve for microstructures synthesized in acetone is shown in Fig. 10. For a sample that exhibits both nonlinear absorption and refraction, their contribution to the far-field beam profile and Z-scan transmittance is coupled. However, it is

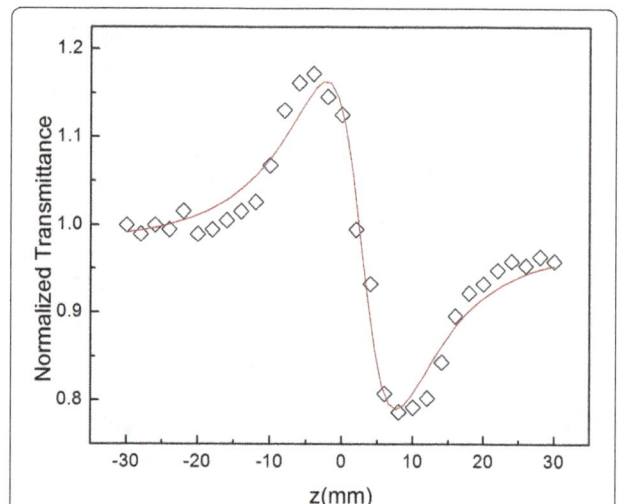

Fig. 8 The plot of output power versus input power for Mg/MgO structures prepared by the laser ablation method in acetone

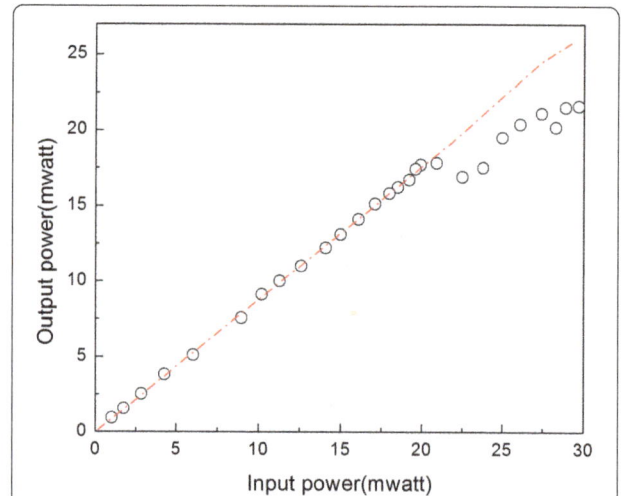

Fig. 10 Closed-aperture Z-scan curve of colloidal Mg/MgO microstructures synthesized in acetone

simple to eliminate the nonlinear absorption contribution to the closed-aperture data. To determine the correct value, the normalized closed-aperture Z-scan data should be divided by the open-aperture Z-scan data to retrieve n_2.

Self-focusing and self-defocusing of laser radiation in nonlinear media is a well-known effect that occurs due to the nonlinear index of refraction n_2. Self-focusing is the effect of positive n_2 while self-defocusing occurs when n_2 is negative. As shown in the Fig. 10, the curves exhibited a peak-to-valley shape indicating a negative value of the nonlinear refractive index, n_2, that shows Mg/MgO microstructures act as a self-defocusing material.

In Fig. 10, the solid curve shows the theoretical fit to the experimental data. The nonlinear refractive index can be measured by fitting the experimental data with the Eq. (5). The nonlinear refractive index of the synthesized structures in acetone is obtained equal to 8.2×10^{-13} cm²/W. Here, the value of aperture linear transmission, S, is 0.3.

The closed-aperture normalized transmittance curve for nanostructures formed in isopropanol is shown in Fig. 11. As shown in the Fig. 11, the curves exhibited a peak-to-valley shape indicating a negative value of the nonlinear refractive index, n_2, that shows Mg/MgO nanostructures act as a self-defocusing material.

The solid curve shows the theoretical fit to the experimental data, In Fig. 11. The nonlinear refractive index can be measured by fitting the experimental data with the Eq. (5). The nonlinear refractive index of the synthesized nanostructures in isopropanol media is obtained equal 2.2×10^{-12} cm²/W.

Open-aperture Z-scan

The open-aperture Z-scan allows measuring the nonlinear absorption coefficient, β. When the sample is located

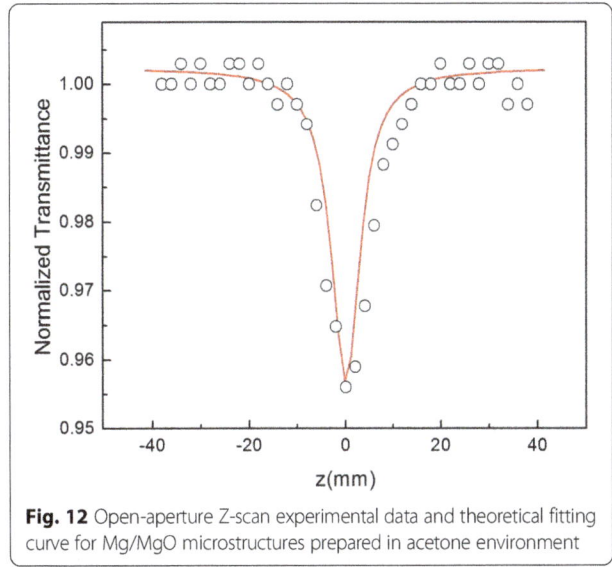

Fig. 12 Open-aperture Z-scan experimental data and theoretical fitting curve for Mg/MgO microstructures prepared in acetone environment

far from the focal point, the laser radiation intensity is low and T(z) is close to 1. The intensity becomes higher as the sample moves closer to the lens focal point. As the result of the positive nonlinear absorption, T(z) becomes smaller and reaches the minimum at the focal point. In the case of negative nonlinear absorption, the reverse picture occurs.

Figure 12 illustrates the open-aperture Z-scan curve of colloidal Mg/MgO microstructures measured by Z-scan setup. It is seen that the open-aperture transmittance has a minimum transmittance. The minimum transmittance confirms the presence of reverse saturation absorption in Mg/MgO microstructures. A fit of the Eq. (1) to the experimental data (solid curve) is depicted in Fig. 12, and yields the value of the nonlinear absorption coefficient $\beta = 1.15 \times 10^{-8}$ cm/W. The open-aperture z-scan data

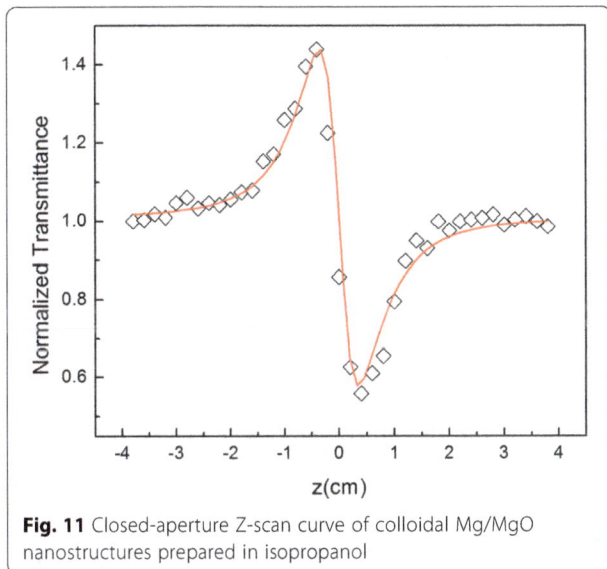

Fig. 11 Closed-aperture Z-scan curve of colloidal Mg/MgO nanostructures prepared in isopropanol

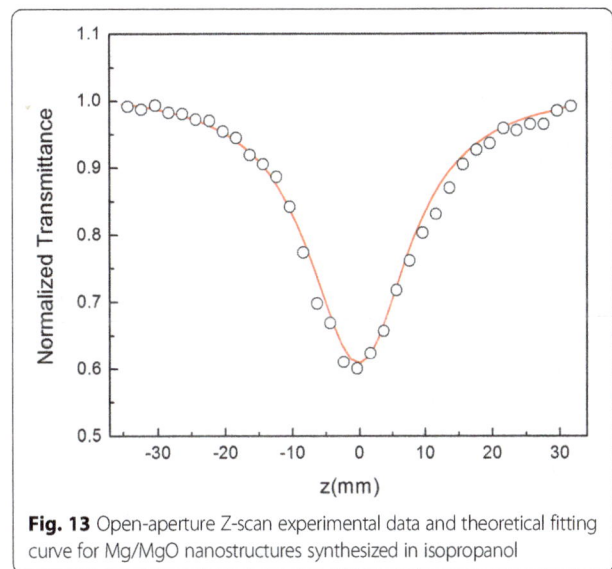

Fig. 13 Open-aperture Z-scan experimental data and theoretical fitting curve for Mg/MgO nanostructures synthesized in isopropanol

for Mg/MgO microstructures synthesized in acetone was fitted with two-photon absorption (2PA) theoretical curve.

The open-aperture normalized transmittance curve for nanostructures formed in isopropanol is shown in Fig. 13. As shown in the Fig. 13, the open-aperture transmittance has a minimum transmittance. The minimum transmittance confirms the presence of reverse saturation absorption in Mg/MgO nanostructures. A fit of the Eq. (1) to the experimental data (solid curve) is depicted in Fig. 13, and yields the value of nonlinear absorption coefficient $\beta = 1.03 \times 10^{-7}$ cm/W. The open-aperture z-scan data for Mg/MgO nanostructures synthesized in isopropanol was fitted to 2PA theoretical curve, properly.

Discussion

The products of ablation simultaneously depend on the properties of the input laser pulses and the surrounding liquid environment [33]. The polarity and the linear refractive index and viscosity of the liquid media, as well as the thickness of the liquid layer on the target surface, are very important in laser ablation process. On the other hands, the parameters of laser that can be effective in products of laser ablation include beam waist, fluence, wavelength, repetition rate, the number of pulses incident per unit area, and pulse duration. Therefore, a minor change in one of these parameters can have a critical impact on products of ablation. The variations of size of products in two cases mentioned in this research are functions of both laser parameters and liquid media.

The calculated values from Z-scan experiments for Mg/MgO microstructures, Mg/MgO nanostructures and Mg(OH)$_2$ structures are shown in Table 1. Furthermore, the real and imaginary parts of $\chi^{(3)}$ are listed in Table 1. The imaginary part of $\chi^{(3)}$ is related to β as [23]:

$$\text{Im}\chi^3(esu) = \left(10^{-2}\varepsilon_0 c^2 n_0^2 \lambda / 4\pi^2\right)\beta(cm/W) \qquad (6)$$

The real part of $\chi^{(3)}$ is related to n_2 as [23]:

$$\text{Re}\chi^3(esu) = \left(10^{-4}\varepsilon_0 c^2 n_0^2 / \pi\right)n_2\left(cm^2/W\right) \qquad (7)$$

Where in Eq. (6) and Eq. (7), n_0 is refractive index, ε_0 is the vacuum permittivity and c is the light velocity in vacuum.

From the Table 1 data, comparing the values of nonlinear refractive indices (n_2) for two samples synthesized in two different liquid environments indicates that n_2 of Mg/MgO microstructures prepared in acetone is an order of magnitude smaller than the value for the n_2 of Mg/MgO nanostructures prepared in isopropanol.

As can be seen in Table 1, comparing the values of the nonlinear absorption coefficient (β) for two samples synthesized by laser ablation of magnesium target in acetone and isopropanol indicates that β for Mg/MgO microstructures is an order of magnitude smaller than the value for the β of Mg/MgO nanostructures. These results show that the Mg/MgO nanostructures are stronger asdsorbent than the Mg/MgO microstructures. Then, the Mg/MgO nanostructures are more appropriate for the protection of eyes and sensors from harmful radiations than the Mg/MgO microstructures.

Therefore, it is clear that a contributing factor in the value of NLO parameters is the particle size of the material.

Furthermore, comparing the results of this study with those of the laser ablation of magnesium in water [24] shows outstanding points. The results of the latter one in the previous study led to the formation of Mg(OH)$_2$ nanostructures and investigation of their NLO properties. The sign of the nonlinearity for Mg(OH)$_2$ nanostructures is opposite to that of Mg/MgO structures.

The low cost of these magnesium-based materials is another advantage that is worth mentioning comparing to other NLO materials. Therefore, each of Mg(OH)$_2$ nanostructure, Mg/MgO microstructure and Mg/MgO nanostructure have its own capability to be used in NLO devices.

A review of the investigation of NLO parameters of semiconductors, fullerenes, dyes, metals, and crystals in various spectral regions was previously reported [34]. Compared to reported amounts in Ref. [34], the Mg/MgO nanostructures and microstructures have satisfactory values despite their low cost.

Conclusion

The major purpose of the present paper is to open a new way to the further development to find suitable materials for producing the highly efficient nonlinear devices that can be used in optical switching and OL.

The pulse laser ablation of magnesium target in acetone led to nearly square Mg/MgO microstructures in the range of 1–1.5 μm with a narrow size distribution. These structures possessed dramatic linear and NLO properties. XRD pattern confirmed the formation of Mg/MgO microstructures. UV-VIS spectrum indicated a

Table 1 Nonlinear optical parameters of the structures prepared by laser ablation of Mg in acetone, isopropanol and water

Nonlinear optical parameters	Mg/MgO microstructures	Mg/MgO nanostructures	Mg(OH)$_2$ structures [24]
n_2 (cm^2/W)	-8.2×10^{-13}	-2.2×10^{-12}	$+7.76 \times 10^{-13}$
β (cm/W)	1.15×10^{-8}	1.03×10^{-7}	1.01×10^{-7}
Im $\chi^{(3)}$ (esu)	2.75×10^{-6}	2.74×10^{-5}	4.58×10^{-5}
Re $\chi^{(3)}$ (esu)	2.47×10^{-11}	7.43×10^{-11}	4.42×10^{-11}
$|\chi^{(3)}|$ (esu)	2.75×10^{-6}	2.74×10^{-5}	4.58×10^{-5}

characteristic absorption band at about 410 nm. The energy band gap of Mg/MgO microstructures calculated equal to 2.3 eV.

Laser ablation of a magnesium target in acetone and isopropanol is shown to be a convenient approach by which to fabricate magnesium-based optical limiters devices.

The nonlinear refraction index, n_2, for Mg/MgO microstructures and nanostructures is negative and of the order of 10^{-13} cm^2/W and 10^{-12} cm^2/W, respectively. The negative sign of n_2 indicates that the self-defocusing phenomenon has taken place and these structures can be considered as a thin negative lens. A nonlinear absorption was detected and is mainly associated with the reverse saturation absorption. The nonlinear absorption coefficient (β) was of the order of 10^{-8} cm/W and 10^{-7} cm/W for Mg/MgO structures synthesized in acetone and isopropanol, respectively.

The third-order optical nonlinearities, $\chi^{(3)}$, are calculated using $|\chi^3| = [(Re(\chi^3))^2 + (Im(\chi^3))^2]^{1/2}$ equation for Mg/MgO microstructures and nanostructures 2.75×10^{-6} and 2.74×10^{-5} esu, respectively.

These results show that the size of the Mg/MgO structures is an effective factor in the order of magnitude of NLO parameters. As well, Mg/MgO microstructures and nanostructures synthesized by the Ng-YAG laser ablation of Mg target in acetone and isopropanol are promising materials for applications in NLO devices.

Abbreviations
2PA: Two-photon absorption; FTIR: Fourier transform infrared spectroscopy; NLO: Nonlinear optical; OL: Optical limiting

Acknowledgements
Not applicable.

Funding
Not applicable.

Competing interests
The author declares that she has no competing interests.

References
1. Boyd, R.W.: Nonlinear optics. Academic, New York (2003)
2. Jeong, H., Mark, A.G., Fischer, P.: Magnesium plasmonics for UV applications and chiral sensing. Chem Commun **52**, 12179–12182 (2016)
3. Sterl, F., Strohfeldt, N., Walter, R., Griessen, R., Tittl, A., Giessen, H.: Magnesium as novel material for active plasmonics in the visible wavelength range. Nano Lett **15**(12), 7949–7955 (2015)
4. Kooi, B.J., Palasantzas, G., Hosson, J.Th.M.De: Gas-phase synthesis of magnesium nanoparticles: A high-resolution transmission electron microscopy study. Appl. Phys. Lett. 89, 161914-1-3 (2006)
5. Zaluska, A., Zaluski, L., Strom-olsen, J.O.S.: Structure, catalysis and atomic reactions on the nano-scale: a systematic approach to metal hydrides for hydrogen storage. Appl Phys A Mater Sci Process **72**(2), 157–165 (2001)
6. Gao, T., Han, F., Zhu, Y., Suo, L., Luo, C., Xu, K., Wang, C.: Hybrid Mg^{2+}/Li$^+$ battery with long cycle life and high rate capability. Adv Energy Mater **1401507**, 1–5 (2014)
7. Habibi, M.K., Joshi, S.P., Gupta, M.: Hierarchical magnesium nano-composites for enhanced mechanical response. Acta Mater **58**, 6104–6114 (2010)
8. Locatelli, E., Matteini, P., Sasdelli, F., Pucci, A., Chiariello, M., Molinari, V., Pini, R., Comes Franchini, M.: Surface chemistry and entrapment of magnesium nanoparticles into polymeric micelles: a highly biocompatible tool for photothermal therapy. Chem Commun **50**, 7783–7786 (2014)
9. Hassan, S.F., Gupta, M.: Development of high strength magnesium copper based hybrid composites with enhanced tensile properties. Mater Sci Technol **19**, 253–259 (2003)
10. Zhang, X., Yang, R., Yang, J., Zhao, W., Zheng, J., Tian, W., Li, X.: Synthesis of magnesium nanoparticles with superior hydrogen storage properties by acetylene plasma metal reaction. Int J Hydrogen Energ **36**, 4967–4975 (2011)
11. Al-Gaashani, R., Radiman, S., Al-Douri, Y., Tabet, N., Daud, A.R.: Investigation of the optical properties of Mg(OH)$_2$ and MgO nanostructures obtained by microwave-assisted methods. J Alloy Compd **52**, 71–76 (2012)
12. Ouraipryvan, P., Sreethawong, T., Chavadej, S.: Synthesis crystalline MgO nanoparticle with mesoporous-assembled structure via a surfactant-modified sol-gel process. Mater Lett **63**(21), 1862–1865 (2009)
13. Mirzaei, H., Davoodnia, A.: Microwave assisted sol-gel synthesis of MgO nanoparticles and their catalytic activity in the synthesis of hantzsch 1,4-Dihydropyridines. Chinese J Catal **33**(9), 1502–1507 (2012)
14. Bertinetti, L., Drouet, C., Combes, C., Rey, C., Tampieri, A., Coluccia, S., Martra, G.: Surface characteristics of nanocrystalline apatites: effect of mg surface enrichment on morphology, surface hydration species, and cationic environments. Langmuir **25**, 5647–5654 (2009)
15. Martinez-Boubeta, C., Bacells, L., Cristofol, R., Sanfeliu, C., Rodriguez, E., Weissleder, R., Lope-Piedrafita, S., Simeonidis, K., Angelakeris, M., Sandiumenge, F., Calleja, A., Casas, L., Monty, C., Martinez, B.: Self-assembled multifunctional Fe/MgO nanospheres for magnetic resonance imaging and hyperthermia. Nanomedicine **6**(2), 362–370 (2010)
16. Di, D.R., He, Z.Z., Sun, Z.Q., Liu, J.: A new nano-cryosurgical modality for tumor treatment using biodegradable MgO nanoparticles. Nanomedicine **8**(8), 1233–1241 (2012)
17. Tang, Z.X., Lv, B.F.: MgO nanoparticles as antibacterial agent: preparation and activity. Braz J Chem Eng **31**(3), 591–601 (2014)
18. Kurth, M., Graat, P.C.J., Mittemeijer, E.J.: The oxidation kinetics of magnesium at low temperatures and low oxygen partial pressures. Thin Solid Films **500**, 61–69 (2006)
19. Canney, S.A., Sashin, V.A., Ford, M.J., Kheifets, A.S.: Electronic band structure of magnesium and magnesium oxide: experiment and theory. J. Phys. Condens Matter **11**, 7507–7522 (1999)
20. Phuoc, T.X., Howard, B.H., Martello, D.V., Soong, Y., Chyu, M.K.: Synthesis of Mg(OH)$_2$, MgO, and Mg nanoparticles using laser ablation of magnesium in water and solvents. Opt Laser Eng **46**, 829–834 (2008)
21. Abrinaei, F., Torkamany, M.J., Hantezadeh, M.R., Sabbaghzadeh, J.: Formation of Mg and MgO nanocrystals by laser ablation in liquid: effects of laser sources. Sci Adv Mater **4**, 501–506 (2012)
22. Gutierrez, Y., Ortiz, D., Sanz, J.M., Saiz, J.M., Gonzalez, F., Everitt, H.O., Moreno, F.: How an oxide shell affects the ultraviolet plasmonic behavior of Ga, Mg, and Al nanostructures. Opt Express **24**(18), 20621–20631 (2016)
23. Sheik-bahae, M., Said, A.A., Wei, T.H., Hagan, D.J., Van Stryland, E.W.: Sensitive measurement of optical nonlinearities using a single beam. IEEE J Quantum Elect **26**(4), 760–769 (1990)
24. Abrinaei, F.: Laser ablation of magnesium in water and investigation of optical nonlinearity by Z-scan technique. J Opt Soc Am B **33**(5), 864–870 (2016)
25. Snellings, R., Machiels, L., Mertens, G., Elsen, J.: Rietveld refinement strategy for quantitative phase analysis of partially amorphous zeolitised tuffaceous rocks. Geol Belg **13**, 183–196 (2010)
26. Zhu, S., Lu, Y.F., Hong, M.H.: Laser ablation of solid substrates in a water-confined environment. Appl Phys Lett **79**(9), 1396–1398 (2001)
27. Zeng, H.B., Cai, W.P., Li, Y., Hu, J.L., Liu, P.S.: Composition/structural evolution and optical properties of ZnO/Zn nanoparticles by laser ablation in liquid media. J Phys Chem B **109**(39), 18260–18266 (2005)
28. Foster, M., D'Agostino, M., Passno, D.: Water on MgO (100)—An infrared study at ambient temperatures. Surf Sci **590**(1), 31–41 (2005)
29. Ferreirada Silva, A., Veissid, N., An, C.Y., Pepe, I., Barrosde Oliveira, N., da Silva AV, B.: Optical determination of the direct bandgap energy of lead iodide crystals. Appl Phys Lett **69**, 1930–1932 (1996)

30. Site, L.D., Alavi, A.: Lynden-Bell RM. Structure and spectroscopy of a monolayer of water on MgO (100). J Chem Phys **113**(8), 3344–3350 (2000)

31. Rezaei, M., Khajenoori, M.: Nematollahi B. Synthesis of high surface area nanocrystalline MgO by pluronic P123 triblock copolymer surfactant. Powder Technol **205**, 112–116 (2011)

32. Alavi, M.A., Morsali, A.: Syntheses and characterization of Mg(OH)2 and MgO nanostructures by ultrasonic method. Ultrason Sonochem **17**, 441–446 (2010)

33. Rao, S.V., Podagatlapalli, G.K., Hamad, S.: Ultrafast laser ablation in liquids for nanomaterials and applications. J Nanosci Nanotechnol **14**, 1364–1388 (2014)

34. Ganeev, RA, Usmanov, T: Nonlinear-optical parameters of various media. Quantum Electron + 37 (7), 605-622 (2007).

13

One–dimensional TiO$_2$/SiO$_2$ photonic crystal filter for thermophotovoltaic applications

Fabrice Kwefeu Mbakop[1], Noël Djongyang[1*] and Danwé Raïdandi[2]

Abstract

Background: The efficiency of a thermal conversion system requires adequate radiative properties. In a thermophotovoltaic system, the optical filter plays a key role into the overall performance of the system. The purpose of this paper is to study a one-dimensional TiO$_2$/SiO$_2$ photonic crystal for application as thermophotovoltaic optical filter.

Methods: The influence of the layers' thicknesses, incidence angle, and number of periods on the spectral reflectance has been investigated through the transfer matrix method.

Results: It was found that, when one varies the number of layers from 6 to 12, an improvement of the optical properties of the spectral filter is observed. The surface wave through a film of photonic crystal generates a resonant transmission for a truncated structure.

Conclusions: These results are in conformity with those found in the literature.

Keywords: Thermophotovoltaic, Spectral filter, Transfer matrix method, Photonic crystal, Multilayered structures

Background

Thermophotovoltaics (TPVs) convert thermal radiation from a man-made high temperature source into electricity through photovoltaic (PV) cells [1, 2]. The concept dates back to 1960s [3]. A TPV system consists of a thermal radiator, PV cells, and selective filter. Low band-gap III–IV semiconductors have led to high performance of modern TPV systems. Advantages of TPV include high power density, good reliability, fuel versatility, low maintenance, and possibility of cogeneration (electricity and heat).

As the temperature of the radiator is typically in the range of 1200–1800 K, the fraction of the power above the electronic band-gap (Eg) in the total radiated power is quite small even for GaSb cells (with a low band-gap of 0.7 eV), leading to an extremely poor overall system efficiency and power density. Fortunately, the system performance can be significantly improved using a

selective filter, which reflects below band-gap photons back to the radiator for re-radiation and transmits above band-gap photons to the cells. One dimensional (1D) Si/SiO$_2$ photonic crystals (PhCs) [4–6] were reported to be promising candidates for these kind of filters.

Most of the investigations on TPV systems focus on increasing conversion's efficiency and power density. For such purpose, the spectral control of thermal radiation from an emitter to a photovoltaic cell is important. One of the effective approaches for the spectral control is to use a filter that control photons selectively from an emitter to a photovoltaic cell. The filter in a TPV system plays its role in the following two aspects: (1) recycling thermal radiation, i.e. reflecting photons whose wavelengths are above the band gap wavelength of a photovoltaic cell back to the emitter. This action helps to maintain a high temperature level of the emitter and reduce the cooling load of the photovoltaic cell. (2) transmitting photons selectively, i.e. selecting the photons from an emitter whose wavelengths are below the band gap wavelength of a photovoltaic cell to

---
* Correspondence: noeldjongyang@gmail.com
[1]Department of Renewable Energy, The Higher Institute of the Sahel, University of Maroua, PO Box 46 Maroua, Cameroon
Full list of author information is available at the end of the article

penetrate the filter and reach the cell. These photons can be absorbed and stimulate electrons out of the photovoltaic cell. The need to develop suitable filter stimulated investigation to improve the conversion's efficiency of TPV systems. Several research groups have been devoting their work to filter's performance, and typical filter structures have been proposed including plasma filters [7, 8], one-dimensional Photonic Crystal [4, 9] filter, and frequency selective surface [6, 10].

One dimensional-photonic crystals (1D-PhCs) were used as selective filters in TPV systems. They present the advantages of being structurally simple and easy to be manufactured. A cascaded inhomogeneous dielectric substrate with different refractive indexes was designed as a frequency-selective structure (FSS), and used as selective filter for TPV system [11]. A 1D photonic crystals consisting of a dielectric-dielectric multilayer (Si/SiO$_2$) mounted on the top part of a TPV cell was used in both thermophotovoltaic (TPV) and micro thermophotovoltaic (MTPV) systems. They presented high efficiency and power throughput [12]. 1D metallic-dielectric photonic crystals (1D-MDPCs) of (Ag/SiO$_2$) were also designed [5] and studied [3]. O'Sullivan et al. [4] proposed and designed a 10-layer quarter-wave periodic structure (Si/SiO2) with thicknesses of 170 and 390 nm and suggested to reduce the first layer thickness to one half of its original thickness for better performances. In this paper, the use of TiO$_2$/SiO$_2$ (1D-PhCs) as selective filter for TPV systems with GaSb PV cell is proposed and studied using the transfer matrix method.

Methods
The assessment of the electric and magnetic fields in a periodic multilayered structure include several calculations' techniques. Banerjee et al. [13], successfully used Transfer Matrix Method (TMM) to evaluate the electric and magnetic fields in a metamaterial structure consisting of positive index materials (PIM) and negative index materials (NIM) alternating structure in both TE and TM modes. The results were compared to those using standard finite element methods (FEM). It was found that the TMM calculations are less computationally demanding, not limited by the thickness of the structures and can be performed for arbitrary angular plane wave spectra. The TMM approach can also be readily applied to a wide variety of cases, such as beam propagation through induced reflection gratings in nonlinear media.

Presentation of the transfer matrix method
The transfer matrix method (TMM) is widely used for the description of the properties of stacked layers [14, 15]. It provides an analytical means for calculation of wave propagation in multilayered media. The method permits exact and efficient evaluation of electromagnetic fields in layered media through multiplications of 2×2 matrices. The solution of the coupled modes (TE and TM) equations is represented by a 2×2 transfer matrix which relates the forward and backward propagating field amplitudes. The grating structure is divided into a number of uniform grating sections, each with its own analytic transfer matrix. The transfer matrix of the entire structure is obtained by multiplying the individual transfer matrices together [16]. The boundary conditions for the vectors of the electric field E in each side of an unspecified interface allow a simple description by a 2×2 matrix. For the kth interface, the relationship between the components of the field is given by [13]:

$$\begin{pmatrix} E_{k-1}^+ \\ E_{k-1}^- \end{pmatrix} = D_{k-1,k} \begin{pmatrix} E_k'^+ \\ E_k'^- \end{pmatrix}, \tag{1}$$

where

$$D_{k-1,k} = \frac{1}{t_{k-1,k}} \begin{pmatrix} 1 & r_{k-1,k} \\ r_{k-1,k} & 1 \end{pmatrix}, \tag{2}$$

The field components at the left and right sides of the kth layer are defined by the propagation matrix P_k [13]:

$$\begin{pmatrix} E_{k^+} \\ E_{k^-} \end{pmatrix} = P_k \begin{pmatrix} E_k^+ \\ E_k^- \end{pmatrix}, \tag{3}$$

where

$$P_k = \begin{pmatrix} \exp(i.\phi_k) & 0 \\ 0 & \exp(-i.\phi_k) \end{pmatrix}, \tag{4}$$

$\phi_k = \frac{2.\pi.n_k \cos(\theta_k)}{\lambda}.d_k$ represents the dephasing for $\theta \neq 0$
The transfer matrix of the system is:

$$M_k = \begin{pmatrix} M_{11} & M_{12} \\ M_{21} & M_{22} \end{pmatrix} = D_k.P_k.D_{k, k+1}, \tag{5}$$

For N layers, the transfer matrix is given by the following relation [13]:

$$M = M_1.M_2 ....M_N, \tag{6}$$

One can directly determine the coefficients of transmission and reflection as [13]:

$$r = \frac{M_{21}}{M_{11}} \text{ and } t = \frac{1}{M_{11}}, \tag{7}$$

Fresnel's coefficients are written [13]:

• For a parallel polarization (S):

$$r_k^s = \frac{E_{k-1,r}^s}{E_{k-1,i}^s} \text{ and } t_k^s = \frac{E_{k,i}^s}{E_{k-1,i}^s}, \tag{8}$$

- For a perpendicular polarization (P):

$$r_k^p = \frac{E_{k-1,r}^p}{E_{k-1,i}^p} \text{ and } t_k^p = \frac{E_{k,i}^p}{E_{k-1,i}^p}, \tag{9}$$

These coefficients are expressed as functions of the refractive indexes n_{k-1} and n_k of the layers using Fresnel laws as follows [13]:

$$r_k^s = \frac{n_{k-1}\cos(\theta_k) - n_k\cos(\theta_{k-1})}{n_{k-1}\cos(\theta_k) + n_k\cos(\theta_{k-1})} \quad t_k^s = \frac{2n_{k-1}\cos(\theta_{k-1})}{n_{k-1}\cos(\theta_k) + n_k\cos(\theta_{k-1})}, \tag{10}$$

$$r_k^p = \frac{n_{k-1}\cos(\theta_{k-1}) - n_k\cos(\theta_k)}{n_{k-1}\cos(\theta_{k-1}) + n_k\cos(\theta_k)} \quad t_k^p = \frac{2n_{k-1}\cos(\theta_{k-1})}{n_{k-1}\cos(\theta_{k-1}) + n_k\cos(\theta_k)}, \tag{11}$$

In case of normal incidence ($\theta_k = 0$), the values of the Fresnel coefficients are independent to the polarization condition. We therefore have:

$$r_k^s = r_k^p = \frac{n_{k-1} - n_k}{n_{k-1} - n_k}, \tag{12}$$

$$t_k^s = t_k^p = \frac{2}{n_{k-1} - n_k}, \tag{13}$$

For $\theta = 0$, the dephasing is:

$$\phi_k = \frac{2.\pi.n_k}{\lambda} . d_k, \tag{14}$$

The energy conservation equation gives [13]:

$$|r|^2 + |t|^2 = 1, \tag{15}$$

$$R = |r|^2 \text{ and } T = |t|^2, \tag{16}$$

Modeling and evaluation of the system

In this paper, we study a TPV system with a spectral filter emitting light toward a GaSb cell. A 1D-PhCs is deposited on NaCl substrate which also serves as front encapsulation glass to a GaSb cell, separated from the emitter by 1 cm. Transfer matrix method is used to assess the performance of the system through the evaluation of the spectral efficiency. We used GaSb as PV cell, which has a low-direct band gap energy of 0.7 eV, corresponding to a wavelength of 1.78 μm. The performance of the 1D-PhCs TiO$_2$/SiO$_2$ filter is characterized by spectral efficiency which can be defined as the ratio of the above-band gap power transmitted through the filter to the PV cell to the net power of the filter got from the emitter [4]. In this paper we will show that the use of a 1D-PhCs TiO$_2$/SiO$_2$ structure as a selective filter can provide better spectral efficiency and system performance.

Spectral distribution of the thermal transmitter

Thermophotovoltaic (TPV) cells convert thermal radiation emitted from a thermal source directly into electricity (Fig. 1). The thermal transmitter behaves like a black body. It entirely absorbs radiations received from the heat source. To improve the reflectance and the spectral transmittance of the one-dimensional filter, the structure (TiO$_2$/SiO$_2$ PhC) must correspond to the spectral distribution of the transmitter at high temperature in the corresponding band of the reflectance and transmittance [7, 16]. According to Planck's law, the spectral power of the radiation of the source at high temperature is [7]:

$$E_{(\lambda, T)} = \frac{2\pi hc^2}{\lambda^5\left(\exp\left(\frac{hc}{\lambda K T}\right) - 1\right)}, \tag{17}$$

where h is the Planck's constant, K the Boltzmann constant and c the speed of the light in the vacuum.

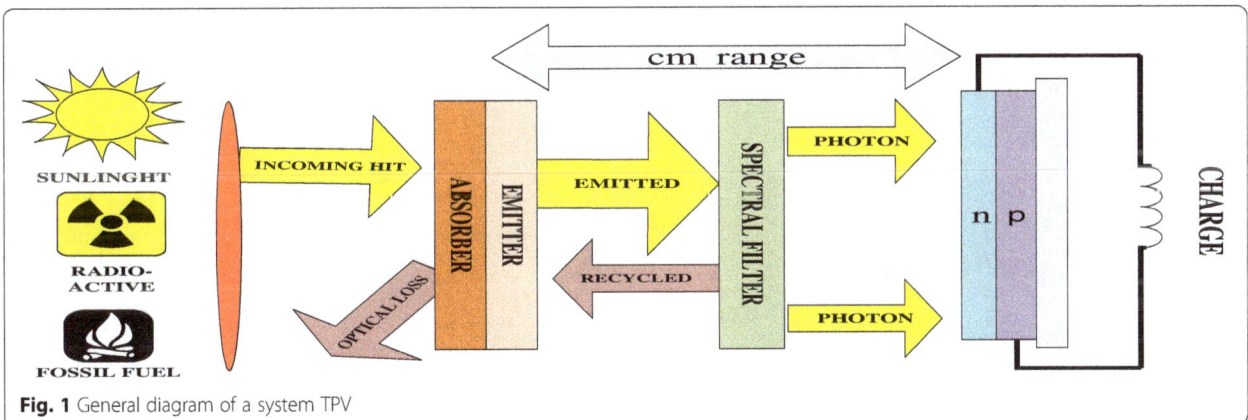

Fig. 1 General diagram of a system TPV

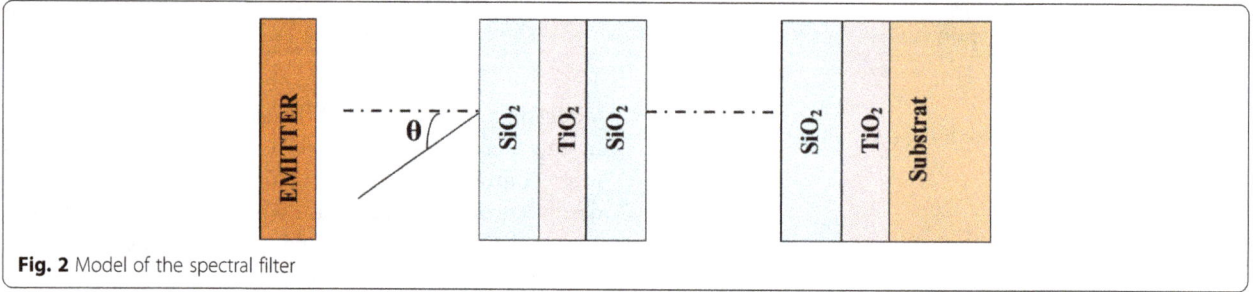

Fig. 2 Model of the spectral filter

The total power radiated by a body is given by Stefan's law [7]:

$$P = \sigma T^4 \qquad (18)$$

The wavelength (λ_m) of the maximum of radiated energy depends on the temperature T following the expression known as Wien's displacement law [7]:

$$\lambda_m . T = 2898 \ \mu m . K, \qquad (19)$$

Studied model

We propose a structure made up of one-dimensional photonic crystals $(TiO_2/SiO_2)^6$ with P periods of Titanium Dioxide (TiO_2) and Silicon Dioxide (SiO_2) deposited on a substrate (NaCl). 1D-PhCs does not exhibit a complete photonic band gap. However, when coupled to free space it exhibits total omnidirectional reflectance, which can be used in TPV systems to improve its performance [17–19].

The central wavelength of the normal-incidence stopband can be expressed as [19]:

$$\lambda_0 = \frac{1}{1 - \frac{2}{\pi} \sin^{-1}\left(\frac{n_H - n_L}{n_H + n_L}\right)} \lambda_g, \qquad (20)$$

$$d_k = \frac{\lambda_0}{4 n_k}, \qquad (21)$$

For thermal energy transformation into electric power, the photovoltaic cell which operates perfectly in the infrared is GaSb. Its energy of gap is Eg = 0.7 eV and its wavelength of gap is $\lambda g = 1.78 \ \mu m$. The TiO_2 and SiO_2 used for the filter are respectively high (H) and low (L) refractive index materials. The refractive indexes are n_H = 2.4 for TiO_2 and n_L =1.46 for SiO_2. The substrate material for the filter has a refractive index taken as n_b = 1.5 in the entire range of wavelengths. The central wavelengths are located in the spectral band around 1300 nm. Figure 2 presents the model of the studied structure.

Results and discussion

The parameters that play an important role in achieving the best performance of the filter are the number of periods P, the thickness of each TiO_2 and SiO_2 layer, and the incidence angle.

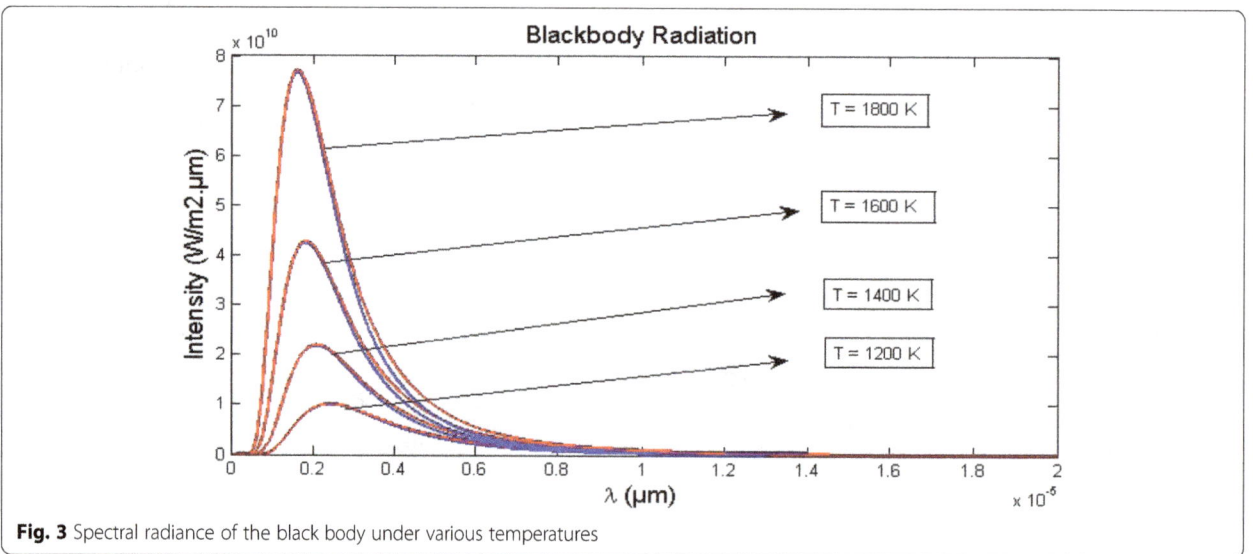

Fig. 3 Spectral radiance of the black body under various temperatures

Behavior of the temperature of the transmitter

One of the important issues for improving the spectral transmission performance of the one-dimensional TiO_2/SiO_2 PhCs filter is to make the filter structure match well with spectral distribution of high temperature emitter within the corresponding transmission band. For the sake of analysis simplicity, the emitter is set as a blackbody [7, 16, 20]. The results corresponding to the wavelength range below 2000 nm are given in Fig. 3. It could be seen that, most of the radiation power from an emitter was limited in the spectral range of 0–2000 nm. Thus, the design of the filter should ensure this wavelength band to be in its high transmission band. Figure 3 presents an overview of the amount of light available for the GaSb cell. The red lines in the figure indicate the fraction of photons from the spectrum that is greater than the energy band gap.

Effect of the number of periods on the reflectance and transmittance

Table 1 presents the values of the transmission and the reflection when the period varies.

The reflectance and transmittance of the structure are obtained by the TMM polarized in TM mode. Figure 4 shows the behavior of the multilayered structure at different periods. In Fig. 4(a), the reflection becomes flatter in the bandwidth $\Delta\lambda$ and reaches 100% when the number of period P increases from $P = 3$ to $P = 6$. In Fig. 4(b), the transmittance rapidly vanishes on bandwidth $\Delta\lambda$ when the period $P = 6$. These results are in conformity with those obtained by Li et al. [21, 22]. In fact, when the wave is generated inside the periodically nonlinear photonic crystal, its spatial evolution inside the 1D-PhCs is coupled to the spatial distribution. However, the structure encounters significant multiple reflections and transmission caused by the variation of the refractive index. Therefore, large oscillations appear in the pass band due to mismatch between the PhCs and the quartz substrate and these oscillations will lead to reduce the amount of power transmitted to GaSb cell. A large enhancement of the conversion efficiency can be obtained by the use of a one-dimensional photonic band gap structure such as a Bragg mirror, where the enhancement comes from the simultaneous availability of a high density of states. The combination of the quasi-phase matching and the photonic band edge effects in a nonlinear

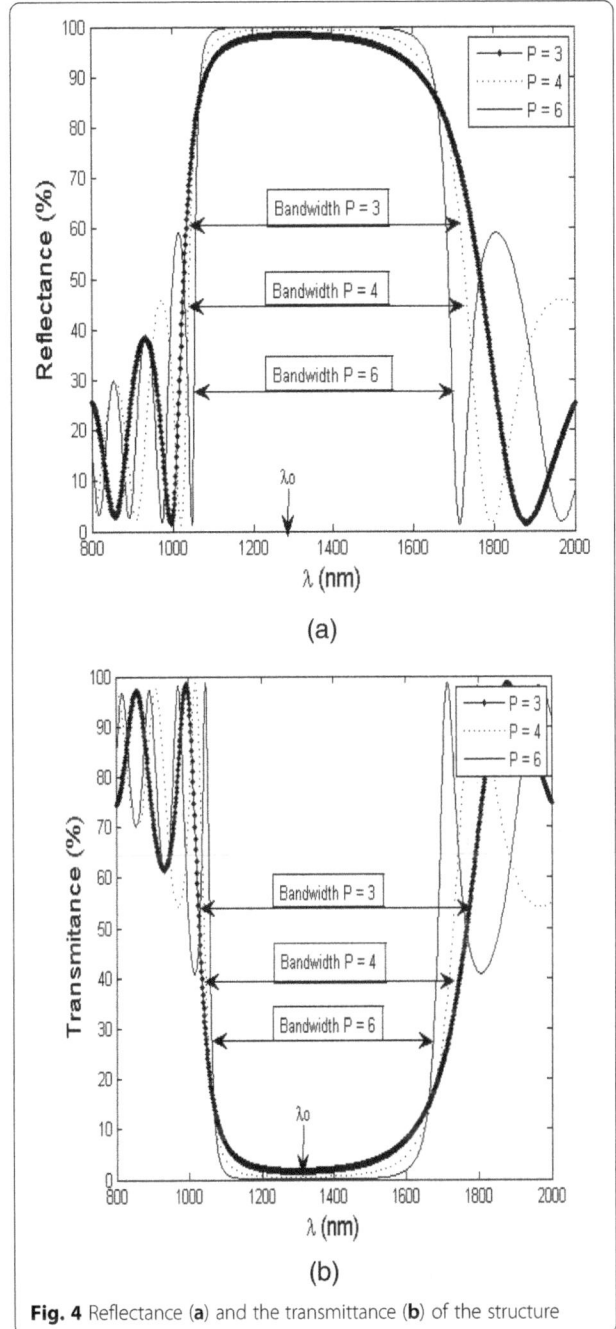

(a)

(b)

Fig. 4 Reflectance (**a**) and the transmittance (**b**) of the structure

structure with periodically poled crystals can significantly increase conversion efficiency, often three-four orders of magnitude compared to using quasi-phase matching only. The period of a nonlinear structure enables substantially to increase the conversion efficiency [21, 22].

Effect of the thickness of the layers on the reflectance and transmittance

Figure 5 presents the reflectance and transmittance of a 1D (TiO2/SiO2) photonic crystal deposited on a sodium

Table 1 Values of the reflection and the transmission when P varies

	λ_1 (nm)	λ_2 (nm)	$\Delta\lambda$ (nm)
$P = 3$	1160	1570	410
$P = 4$	1180	1150	270
$P = 6$	1200	1400	200

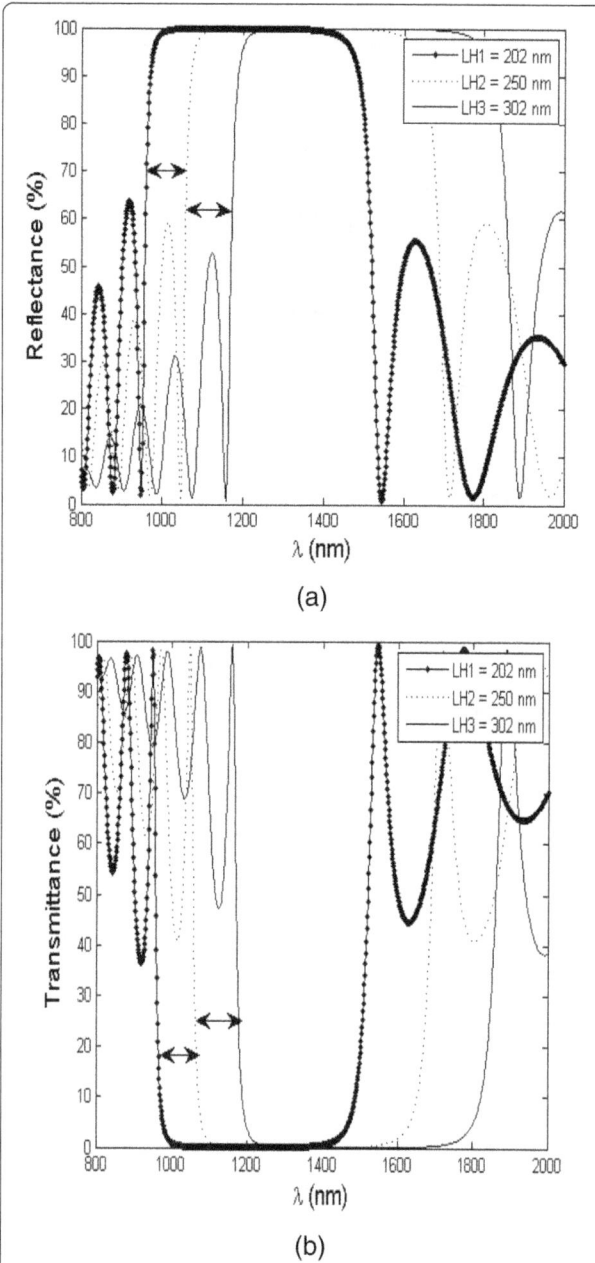

Fig. 5 a and **b** Effect of the variation the thickness LH (thickness of TiO2) on the reflection and the transmission for a structure with $P = 6$

Table 2 Value of the reflection and the transmission when LH varies

LH (nm)	λ_1 (nm)	λ_2 (nm)	$\Delta\lambda$ (nm)
202 nm	1000	1400	400
250 nm	1100	1480	380
302 nm	1220	1700	480

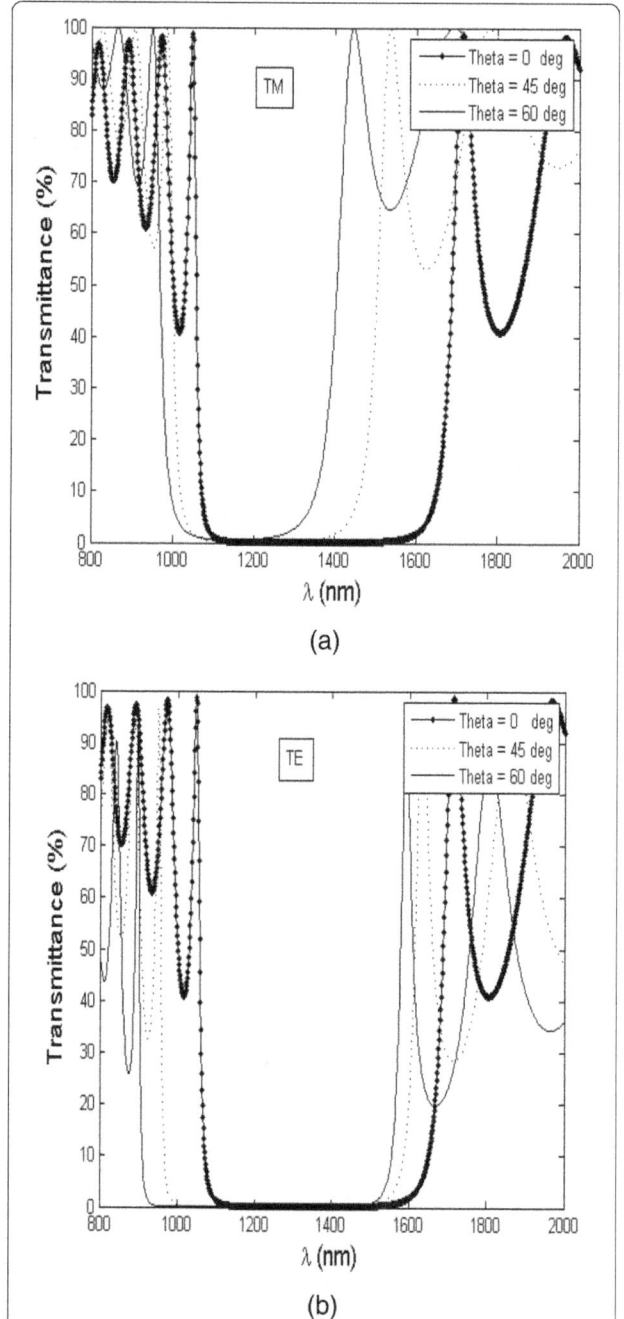

chloride substrate (NaCl) for TM polarization. Table 2 shows the values of the transmission and the reflection when the thickness varies.

Effect of the incidence angle on the transmittance

The transmittance of the proposed structure for TM and TE modes is shown in Fig. 6. Table 3 presents the effect of incidence angle on the transmittance of the structure for both TM and TE modes.

Fig. 6 Effect of incidence angle on the transmittance of the structure (**a**) TM mode, (**b**) TE mode

Table 3 Values of the bandwidth when the angle of incidence varies

	TM			TE		
	λ_1 (nm)	λ_2 (nm)	λ_1 (nm)	λ_2 (nm)	λ_1 (nm)	λ_2 (nm)
Theta = 0 degree	1100	1480	380	1100	1480	380
Theta = 45 degree	1090	1400	310	1000	1580	580
Theta = 60 degree	1080	1250	170	950	1500	550

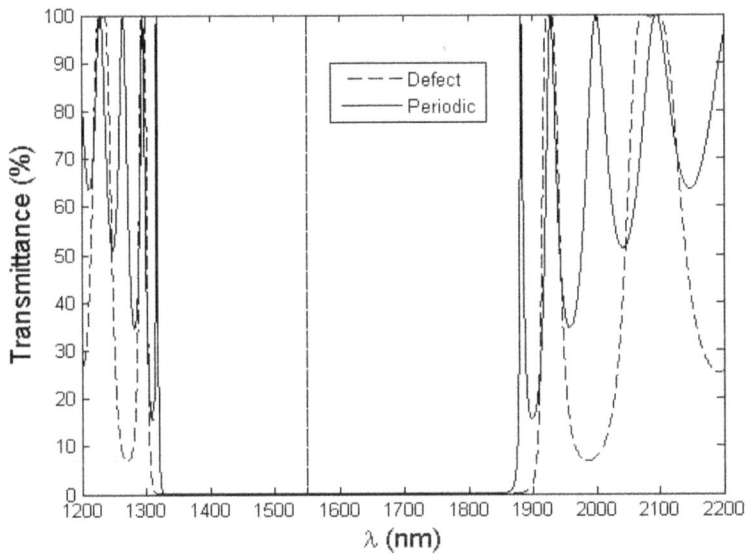

(a)

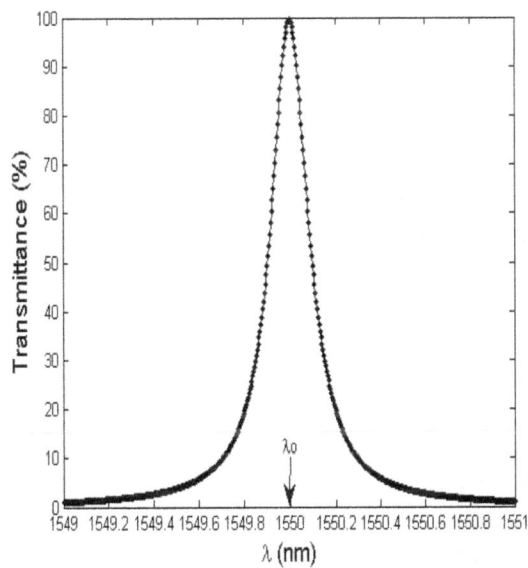

(b)

Fig. 7 a Transmission spectrum of the defect structure at incidence normal polarization TM. **b** Represents of the peak of transmission of the defect layer

When the wave is polarized at normal incidence, both TE and TM modes present similar band gap. However, when the angle of incidence increases to 45°, there is a difference between the TE and the TM modes. The forbidden band becomes narrower when the incidence angle of the light is 60°; the TE polarization transmission spectrum moves rapidly to smaller wavelengths, while in the TM polarization, the movement is gradual. It could also be seen that when the incidence angle increases, the spectral transmittance decreases in passband region. Local field enhancement in a 1D-PhCs structure by introducing an oblique incidence angles may be larger than normal incidence. Oblique incidence provides a simple technique to tune the phase matching and local field enhancements for the sample. It is shown that widely different conversion efficiencies can be obtained for various incidence angles and various thicknesses of the nonlinear material [21, 23]. However, the variation of the angle of incidence allows obtaining different values of the efficiencies of the device; this is in line with the results obtained by Li et al. [22]. They calculated the spatial distributions of the fundamental frequency (FF) and second harmonic (SH) waves. Furthermore, they showed that widely different conversion efficiencies can be obtained for various incident angles of the FF of the nonlinear material.

Transmission through truncated film

The transmission spectrum is obtained when the film is illuminated by a monochromatic wave at normal incidence. The calculation is done in TM polarization as presented in Fig. 7. The considered wavelength range is between 1100 and 2200 nm. The designed wavelength at which the layers are quarter wavelength is taken to be the standard source $\lambda o = 1550$ nm. Figure 7(a) presents the factor of transmission depending of the length of the wave in the case of truncation and no truncation. A peak of transmission of 100% appears in the gap about 1550 nm in the case of truncated photonic crystal. Figure 7(b) shows the resonance peak corresponding to the transmission of the truncated structure. The resonance at the transmission is provided by the excitation of surface waves at the interface of the film of the photonic crystal.

Conclusion

In this paper, we studied a one-dimensional TiO_2/SiO_2 photonic crystal as a spectrally selective filter for TPV applications. The transfer matrix method was used and the influence of the radiative properties (reflectance and transmittance) of the structure was assessed. The processor used in our work is an AMD Dual Core E-450 APU 1.65 GHz with 4 GB of memory for 12-layer structure. The solutions obtained by this method showed that for a 12-layer structure, the solution takes about 1.1 s; this is in

line with the results in obtained by Banerjee et al. [13]. The variation of period and layer's optical thicknesses LH permitted to evaluate the behavior of the structure. The effect of the angle of incidence on the spectral transmittance was also studied in both TM and TE polarization modes. It was found that, a suitable choice of wavelength's value in a well-defined range could lead to 100% resonant transmission.

Abbreviations

μ_o: Permeability; n_b: Refractive index of the substrate; n_H: The high refractive index material; n_L: The low refractive index material; d_k: Thickness of the layer k; θ: Angle of incidence; ε_o: Dielectric constant; λ_0: Central wavelength (nm); λ: Wavelength (nm); λ_1: Opening wavelength (nm); λ_2: Closing wavelength (nm); 1D: One-dimension; a: Dielectric layer; b: Dielectric layer; c: Velocity of light $c = 3.10^8$ m s^{-1}; FF: Fundamental frequency; N: Period; n_a: The index of refraction of air; PhCs: photonic crystal; r: Fresnel reflection coefficient (%); R: Reflectance (%); SH: Second harmonic; SiO_2: Silicon dioxide; STPV: Solar thermophotovoltaics; t: Fresnel transmission coefficient (%); T: Transmittance (%); TE: Electric polarization; TiO_2: Titanium dioxide; TM: Magnetic polarization; TPV: Thermophovoltaic; $\Delta\lambda$: Bandwidth (nm)

Acknowledgements

Authors would like to express their deepest and sincere thanks to all the staff of the Renewable Energy Lab of the University of Maroua for their continuous guidance and support during this work.

Authors' contributions

FKM conceived the work, carried out the thermo-optical simulations and contributed to writing; ND did thermal analysis, offered the conditions for the project and contributed to writing; DR coordinated the work. All authors read and approved the final manuscript.

Competing interests

The authors declare that they have no competing interests.

Author details

^{1}Department of Renewable Energy, The Higher Institute of the Sahel, University of Maroua, PO Box 46 Maroua, Cameroon. ^{2}Department of Mechanical Engineering, National Advanced Polytechnic School, University of Yaounde I, PO Box 8390 Yaounde, Cameroon.

References

1. Coutts, T.J.: An overview of thermophotovoltaic generation of electricity. Sol. Energy Mater. Sol. Cells 66, 443–452 (2001)
2. Mao, L., Ye, H.: New development of one-dimensional Si/SiO2 photonic crystals filter for thermophotovoltaic applications. Renew. Energy 35, 249–256 (2010)
3. Fraas, L.M., Avery, J.E., Huang, H.X., Martinelli, R.U.: Thermophotovoltaic system configurations and spectral control. Semicond. Sci. Technol. 18, 165–173 (2003)
4. O'Sullivan, F., Celanovic, I., Jovanovic, N., Kassakian, J.: Optical characteristic of one dimensional Si/SiO2 photonic crystals for thermophotovoltaic applications. J. Appl. Phys. 97, 033529 (2005)
5. Celanovic, I., O'Sullivan, F., Ilak, M., Kassakian, J., Perreault, D.: Design and optimization of one-dimensional photonic crystals for thermophotovoltaic applications. Opt. Lett. 29, 863–865 (2004)
6. Celanovic, I., O'Sullivan, F., Jovanovic, N., Qi, M., Kassakian, J.: 1D and 2D photonic crystals for thermophotovoltaic applications. In: Proc. SPIE, 5450, pp. 416–422. SPIE, Bellingham (2004)
7. Guang, P.L., Min, X.Y., Yu, G.H., Qian, L.: Investigation of one-dimensional Si/SiO2 photonic crystals for thermophotovoltaic filter. Science in China Series E. Technol. Sci. 51(11), 2031–2039 (2008)
8. Ehsani, H., Bhat, I., Borrego, J.: Optical properties of degenerately doped silicon films for application in thermophotovoltaic systems. J. Appl. Phys. 81(1), 432–439 (1997)

9. Poy, D.M., Fourspring, P.M., Baldasaro, P.F.: Thermophovoltaic spectral control. 2nd International Energy Conversion, pp. 5762–5776. Engineering Conference, Rhode Island (2004)
10. Kritensen, R.T., Beausang, J.F., Depoy, D.M.: Frequency selective surfaces as near infrared electromagnetic filters for thermophotovoltaic spectral control. J. Appl. Phys. **95**(9), 4845–4851 (2004)
11. Lee, H.Y., Yao, T.: Design and evaluation of omnidirectional one-dimensional photonic crystals. J. Appl. Phys. **93**(2), 819–830 (2003)
12. Kiziltas, G., Volakis, L.J., Kikuchi, N.: Design of a frequency selective structure with inhomogeneous substrates as a thermophotovoltaic filter. IEEE Trans. Antennas Propag. **53**(7), 2282–2289 (2005)
13. Banerjee, P.P., Han, L., Aylo, R., Nehmetallah, G.: Transfer matrix to propagation of angular plane wave spectra through metamaterial multilayer structures. Proc. SPIE 8093, Metamaterials: Fundamentals and Applications IV. 80930P1-7 (2011)
14. Chen, S., Wang, Y., Duan, Z.Y., Song, Z.: Absorption enhancement in 1D Ag/SiO2 metallic-dielectric photonic crystals. Opt. Appl. **39**, 473–479 (2009)
15. Born, M., Wolf, E.: Principles of optics, 7th edn, pp. 54–74. Cambridge U. Press, Cambridge (1999). Sect. 1.6
16. Petcu, A.C.: The optical transmission of one-dimensional photonic crystals containing double-negative materials. National Research and Development Institute for Gas Turbines Bucharest, Romania (2012)
17. Holman, J.P.: Heat transfer. China Machine Press, Beijing (2005)
18. Samah, G.B., Yong, S., Mohamed, O.S.-A., Ming, X.: One–dimensional Si/SiO2 photonic crystals filter for thermophotovoltaic applications. WSEAS Trans. Appl. Theor. Mechan. **9**, 97–103 (2014)
19. Fink, Y., Winn, J., Fan, S., Chen, C., Michael, J., Joannopoulos, J., Thomas, E.: A dielectric omnidirectional reflector. Science **282**, 1679 (1998)
20. Southwell, W.H.: Omnidirectional mirror design with quarter-wave dielectric stacks. Appl. Opt. **38**(25), 5464–5467 (1999)
21. Li, H., Haus, J.W., Banerjee, P.P.: Application of transfer matrix method to second-harmonic generation in nonlinear photonic bandgap structures: oblique incidence. JOSA B **32**(7), 1456–1462 (2015)
22. Li, H., Haus, J.W., Banerjee, P.P.: Second harmonic generation at oblique angles in photonic bandgap structures. JOSA B. **32**(7), 1456–1462 (2015)
23. Li, H., Haus, J.W., Banerjee, P.P.: Third harmonic generation in multilayer structures: oblique incidence. In: Laser science. JTu4A. **41**, (2015)

14

Competition between sub-bandgap linear detection and degenerate two-photon absorption in gallium arsenide photodiodes

Benjamin Vest[1], Baptiste Fix[2*], Julien Jaeck[2] and Riad Haïdar[2]

Abstract

This letter is on the response of gallium arsenide p-i-n diodes to sub-bandgap photons (1.55 μm). We investigate the various regimes of sub-bandgap operation by using different light sources delivering pulses ranging from nanosecond to microsecond durations. We evidence two regimes : a regime of degenerate two-photon absorption, with a clear quadratic dependence with respect to the incident flux, and a sub-bandgap, temperature dependant linear regime, that drives photocurrent generation at lower power densities. Both processes are associated to a very low quantum efficiency, around 10–8. We then determine absorption coefficients as well as trap densities, thanks to a model involving a photo-assisted Shockley Read Hall effect.

Keywords: Gallium arsenide, Two-photon absorption, Non-linear optics, Detection, Defects

Background

Semiconductors materials with high non-linear coefficients, such as GaAs or InP are good materials for numerous optical devices, such as optical switches [1], logic gates [2], and quantum detectors [3]. Two-photon absorption is a third order optical process describing the quasi simultaneous absorption of a pair of photons in a material, with a quadratic dependence relatively to light intensity.

Inside semiconductors, two photon absorption using sub-bandgap photons can be described by a two-step process: those photons of energy $E < E_g$ can promote an electron from the valence band to a virtual state (meaning a non stationary state) in the gap. During the lifetime of this state, given by the second Heisenberg principle $\Delta \tau \geq \frac{\hbar}{2(E_g-E)}$, a second photon with enough energy can complete the transition, hence creating an electron-hole pair and generating a photocurrent, named two-photon current (TPC). As the lifetime of this intermediate state is very short in visible gap material (in the fs range), this process is particularly well suited for ultra fast correlation measurements [3–6]. However, the very short lifetime of the virtual state is also responsible for the intrinsically low

efficiency of this process, as it requires the quasi simultaneous occurrence of two photons in this time interval. As a consequence, common two-photon absorption applications in semi-conductors are, so far, essentially limited to the study of very fast processes, more generally to high peak intensity regimes, only reached by pulsed light sources.

Yet, TPC can also be considered as an interesting solution to detect infrared light in wide gap semiconductor. Those materials are less sensitive to thermally generated carriers and, if associated with TPC, they would open new applications in the field of room-temperature infrared detection. However, such configuration implies the detection of low optical power delivered by CW light sources which signifies lower levels of generated TPC.

The quadratic behavior of TPC with the optical power has already been checked over several decades by many authors in highly crystalline detectors such as GaAs photocathodes and PMTs [7]. It has also been observed in GaN photodiodes at high peak- power using pico- and femto-second pulsed lasers [8]. In complement to these previous results, our study of sub-band gap absorption in GaAs photodiodes at lower temporal regimes (from nano- to micro-second pulses) reveals a linear detection process, which appears in competition with TPC.

*Correspondence: baptiste.fix@onera.fr
[2]ONERA, Chemin de la Hunière, 91761 Palaiseau Cedex, France
Full list of author information is available at the end of the article

Such a process might be attributed either to photo-assisted tunneling (PAT) or to photo-assisted Shockley-Read-Hall mechanism (PASRH). Indeed, these two mechanisms involve traps in the energy bandgap and are thus dependant to the quality of the detector in terms of defects.

In this Letter, we investigate the detection of sub-bandgap photons in a series of commercial GaAs p-i-n photodiode at 1.55 μm at very low flux. Our study evidences the existence of sub-bandgap absorptions composed of two competitive processes : a linear, temperature dependant, sub-bandgap contribution attributed to PASRH and a quadratic two-photon absorption regime. We also investigate the competition between the two processes in various temporal regimes from ns to μs allowing to determine the two-photon absorption coefficient and the trap densities in the photodiode material.

Methods

Competition between two transition processes

Our first experimental setup aims at investigating the processes responsible for the sub-bandgap response of several Optowell PP85-B1T0N GaAs PIN photodiode. The ball lens covering the detector has been removed, in order to avoid any effects on the transmission. We focus light emitted by a pulsed laser source, a M-Squared Firefly-IR optical parametric oscillator (OPO 1), delivering pulses of duration $\Delta t = 10$ns under a repetition rate of $f = 150$ kHz (see Table 1). The light is linearly polarised in the direction maximising the TPC by a polariser. The position of the photodiode is controlled with a 3-axis motorized stage. Hence, it can be precisely placed at the focal point of the focused beam, thanks to a Z-scan technique [9], realized under a mean incident power of 40 mW. At this focal point, an I-P characteristic of the photodiode is acquired with a Keithley 6430 source-meter, applying a bias voltage from -1V to 0V. The OPO light is split in two beams using a ZnSe wedge. The weaker one is used as reference, and its intensity is measured with a fast InGaAs detector. We used optical densities to progressively lower the incident flux, thus giving access to a 4-order magnitude dynamic range of optical power. On the I-P characteristic of Fig. 1, the dark current (around 60 pA) has been substracted, so that the graph only shows the evolution of

the photogenerated current. One can clearly identify two distinct regimes: the quadratic regime above 1 mW and a linear behaviour below 100 μW.

Such a linear regime could be explained by two mechanisms, either the PAT or the PASRH. The PAT is based on the curvature of the energy bands and hence depends on the applied bias. However, our experiments did not show such a dependence for biases ranging from -1V to 0V, and we thus neglect this contribution. On the other hand, as it has been observed and described in silicon [10], PASRH is based on the ionization of traps from deep levels in the bandgap to the conduction band. This generation-recombination process is driven by the thermal agitation, meaning that PASRH is strongly temperature dependent.

So, in order to check the SRH nature of the transition, the photodiode was placed inside a cryostat and was illuminated by a pulsed light at 1.55 μm delivered by a A.P.E nanoLevante OPO (OPO 2) with a repetition rate of 15 kHz (see Table 1). As previously, we achieved two I-P characteristics of the diode while measuring the temperature by a PT100 thermo-resistor glued to the photodiode (see Fig. 2). At room-temperature, a clear transition is observed at 20 pA between a linear and a non-linear regime. In comparison with the first experiment (see Fig. 1), we expect a shift of the TPC due to the difference of duty cycles (factor of 10 between OPO1 and OPO2): indeed, the production of charge carriers by two-photon absorption only occurs when the pulse is illuminating the diode, so that TPC is actually dependant of the peak optical power. One can deduce this factor of ten on TPC from the measurements, corresponding to the ratio of the duty cycles.

At 80 K, the I-P characteristic is quadratic over the two decades of measurement. We can notice that the TPC regime is independent of the temperature. But more importantly, no parasitic linear regime is observable at cryogenic temperature due to the freezing of defects and trap levels contribution to this process.

We now modelize the total photocurrent as the sum of both contributions in order to derive their respective parameters. Photocurrent levels are low compared to the incident optical power, so we will consider that the pump is not depleted leading to a simple expression :

$$\langle J_{phot} \rangle = \langle J_{1ph} \rangle + \langle J_{2ph} \rangle = K_{1ph}(T) \langle P_{opt} \rangle + K_{2ph} \langle P_{opt} \rangle^2$$

(1)

Table 1 Summary of the different parameters associated to each experiment

Setup	OPO1	OPO2	AOM
Pulse duration Δt	10 ns	10 ns	1 μs
Repetition rate f	150 kHz	15 kHz	150 kHz
Wavelength	1.55 μm	1.55 μm	1.55 μm
Accessible average power (W)	$10^{-7} - 10^{-2}$	$10^{-4} - 10^{-1}$	$10^{-4} - 10^{-2}$

where J_{1ph} and J_{2ph} are the respective contributions of PASRH and TPC, T is the temperature, P_{opt} is the incident optical power and $\langle \cdot \rangle$ represents the integration over one duty cycle. We can now derive η, the quantum efficiency of the linear process, describing directly the conversion of photons into electrons, and β, the degenerate two-photon

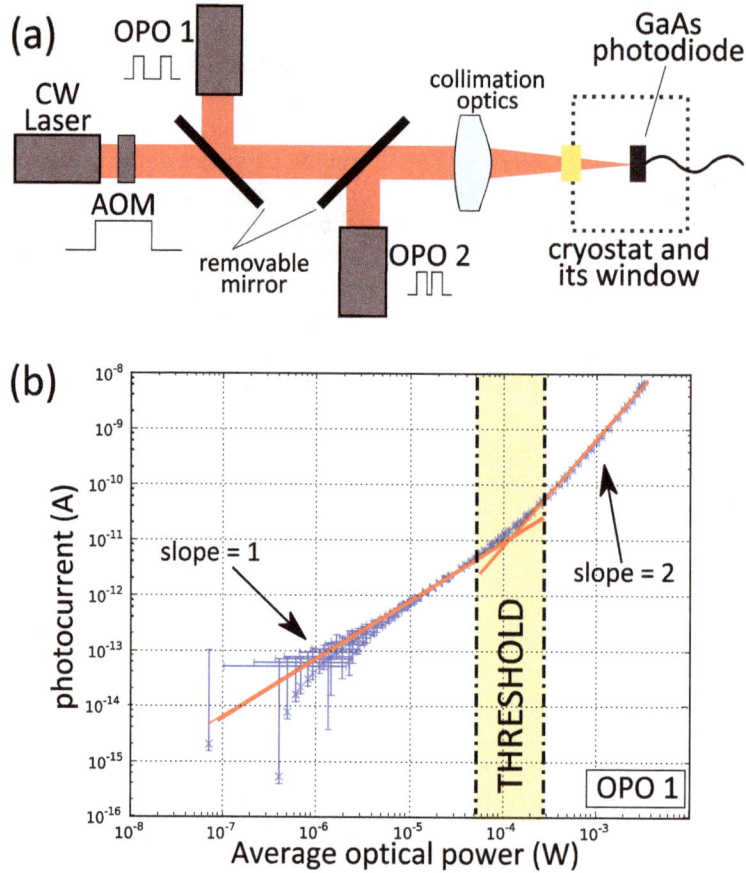

Fig. 1 a Schematic drawing of the experimental set-up. **b** I-P characteristic of the detector under study at -0.5 V using OPO1 (see Table 1). Two regimes are visible : a linear behavior at low flux regime (slope=1, preponderance of PASRH), a quadratic behavior at higher flux (slope=2, preponderance of TPC)

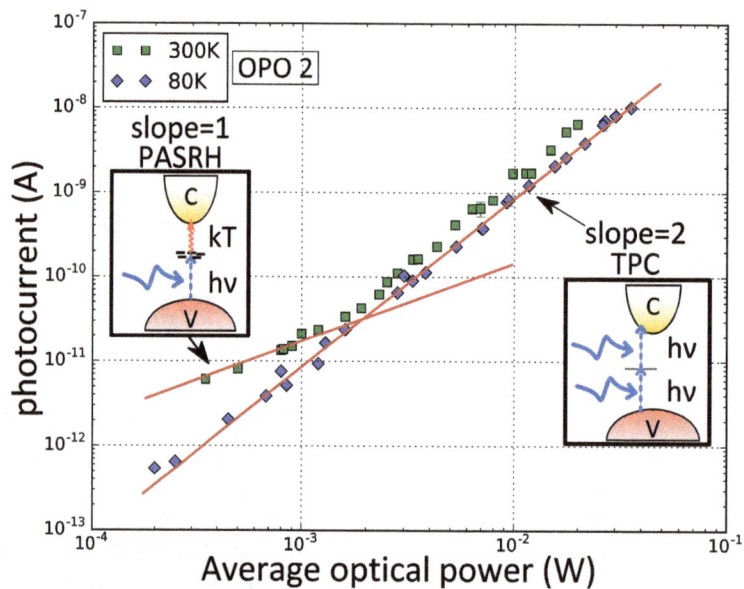

Fig. 2 I-P characteristics of the detector under study mounted in a cryostat using OPO2 (see Table 1). At ambient temperature (*green squares*), the transition between the linear and the non-linear regime is observed by increasing the incident optical power. At cryogenic temperature (80 K), there is no linear regime observable due to the freezing of SRH and PASRH processes, thus allowing a pure non-linear detection regime

absorption coefficient of gallium arsenide at $1.55\,\mu$m from these coefficients.

$$\beta = \frac{2h\nu}{e}\frac{S}{L}\frac{J_{2ph}}{P_{opt}^2} \qquad (2)$$

Since our experiment gives access to mean values, we can rewrite this equation as :

$$\beta = \frac{2h\nu}{e}\frac{S}{L}\frac{\langle J_{2ph}\rangle/R_c}{\left(\langle P_{opt}\rangle/R_c\right)^2} = \frac{2h\nu}{e}\frac{S}{L}K_{2ph}R_c \qquad (3)$$

$$\eta = \frac{h\nu}{e}\frac{\langle J_{1ph}\rangle}{\langle P_{opt}\rangle} = \frac{h\nu}{e}K_{1ph} \qquad (4)$$

where e is the electron charge, $h\nu$ the photon energy, Δt the pulse duration, f the repetition rate, $R_c = \Delta t f$ the duty cycle and assuming a focal spot of area $S = \pi * (40\,\mu\mathrm{m})^2$ and an active medium thickness of $L = 10\,\mu\mathrm{m}$ (determined from capacitance-voltage data).

It is well known that beta depends on the polarisation of the light source with respect to the semiconductor crystallographic orientation [11]. Indeed, the two-photon absorption coefficient in GaAs varies with the direction of polarisation as $\sin^2(2\theta)$ for [100] crystal, as $\left(1+3\cos^2(\theta)\right)\sin^2(\theta)$ for [110] crystal or with the ellipticity of the polarisation for a [111] crystal. We achieved a set of complementary measurements and found that the GaAs is oriented along its [100] axis in our commercial photodiode, for which $\beta^{max}_{[100]} = 45.4 \pm 7.5\,\mathrm{cm/GW}$.

On the other hand, estimation of the quantum yield of the linear process is made difficult because of important uncertainties on the photocurrent due to the very weak signal to noise ratio associated to the determination of the incident power at low flux, as well as predominance of TPC at higher fluxes.

Results and discussion

Improving the measurement precision on the quantum efficiency of the PASRH by changing the temporal regime

In order to improve the precision of measurements on the linear process, we change the temporal regime of excitation of the photodiode and we now provide microsecond optical pulses. Indeed, if the two-photon contribution is sensitive to the squared power of the light (meaning that the relevant parameter to quantify photocurrent generation is the squared peak optical power), we emphasize that the linear contribution should be sensitive only to the average number of photons, assuming that the change of temporal regime has no incidence on the defects in terms of response time. By delivering the same number of photons but on a much longer pulse duration, we reduce significantly the non-linear processes as well as we maintain comparable level of linear photocurrent generation (see inset of Fig. 3).

The setup uses now a $1.55\,\mu$m CW fiber laser source, and an acousto-optic modulator (AOM). This component chop the continuous wave to define a $1\,\mu$s wide laser pulse, with a similar 150 kHz repetition rate. The light is then

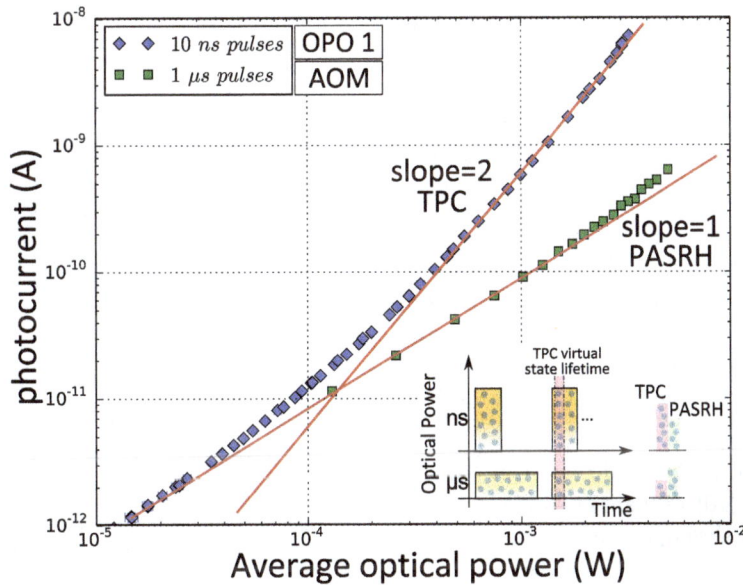

Fig. 3 I-P characteristic of the detector under study in microsecond regime compared with previous results on nanosecond regime (with OPO1) for indentical repetition rate and similar output powers. I-P curve displays a predomining linear behavior, with a smaller quadratic contribution slightly arising. The inset shows the difference in term of photons temporal concentration between the two configuration leading to either a PASRH domination or a TPC domination

collimated at the output of the fiber and refocused onto the detector (see Table 1).

Figure 3 shows the I-P characteristic obtained with this setup. The preliminary Z-scan alignment step does not display a lorentzian shape anymore, meaning that the non-linear contribution is not predominant. The I-P characteristic is coherent with this first observation, as the linear behaviour of the detector is mainly observed in the intensity range under study. Principles leading the modelization step are nevertheless the same, and we must take into account both contributions. Our modelization leads to values of $\eta = [6.1 \pm 0.6].10^{-8}$. Now that the incident power is known with a much better precision than previously, the derived value for η can be considered as reliable. Yet, due to the microsecond regim, this experiment is not adapted to measure the two-photon absorption coefficient with precision.

The PASRH requires the existence of real electronic states inside the bandgap which are mostly due to defects in the semiconductor [10]. Thus, it is possible to retrieve the density of defects from our experiment. By assuming an optical cross section typically in the 10^{-15}cm^{-2} range we find a volumic density of defects in the 10^{14} cm^{-3} range.

Finally, TPC can be compared to PASRH through the introduction of an effective quantum yield. This quantum yield generally describes a probability for a photon to be absorbed in the medium. The quadratic dependance of TPC leads to an intensity-dependent yield, as this probability increases with the number of available photons in the active medium. The inset of Fig. 3 shows a representation of the TPC with respect to the average incident power. It depicts the predominance of absorption of photon pairs in the nanosecond regime, whereas PASRH remains predominant in the microsecond regime. Indeed, the PASRH is proportional to the number of available photons, whereas TPC is sensitive to the photon concentration both in space and time.

Conclusion

In conclusion, we have observed that a real competition appears in terms of photocurrent generation in our device, limiting possibilities to detect subbandgap photon pairs at low optical powers. However, several strategies are accessible to allow a clear predominance of the TPC signal:

- A significant decrease of the PASRH level, by either cooling the detector [10], or thanks to fewer interactions with trap levels in more controlled active medium and smaller interaction volumes.
- A significant increase of the TPC by enhancing the non-linear process in adequate devices, such as resonant nanostructures [12], quantum wells [13] or photonic crystals [14].

Acknowledgements
This work is partially supported by a public grant overseen by the French National Research Agency (ANR) as part of the "Investissements d'Avenir" program (Labex NanoSaclay, reference: ANR-10-LABX-0035) and by a DGA-MRIS scholarship. The authors wish to thank Pr. Jacob Khurgin from Johns Hopkins University for fruitful discussions and Pr. Emmanuel Rosencher for his considerable contribution to this work.

Authors' contributions
BV and BF have led the experiments and have equally contributed to the work. JJ has established the experimental protocol. RH has supervised the work. All have contributed to data analysis. All authors read and approved the final manuscript.

Competing interests
The authors declare that they have no competing interests.

Author details
[1] Laboratoire Charles Fabry, 2 Avenue Augustin Fresnel, 91127 Palaiseau, France. [2] ONERA, Chemin de la Hunière, 91761 Palaiseau Cedex, France.

References
1. Bristow, AD, Rotenberg, N, Van Driel, HM: Two-photon absorption and Kerr coefficients of silicon for 850–2200 nm. Appl. Phys. Lett. **90**, 191104–191104 (2007)
2. Liang, T, Nunes, L, Tsuchiya, M, Abedin, K, Miyazaki, T, Van Thourhout, D, Bogaerts, W, Dumon, P, Baets, R, Tsang, H: High speed logic gate using two-photon absorption in silicon waveguides. Opt. Commun. **265**, 171–174 (2006)
3. Boitier, F, Dherbecourt, J-B, Godard, A, Rosencher, E: Infrared quantum counting by nondegenerate two photon conductivity in GaAs. Appl. Phys. Lett. **94**, 081112 (2009)
4. Lee, CH, Jayaraman, S: Measurement of ultrashort optical pulses by two-photon photoconductivity techniques. Opto-electron. **6**, 115–120 (1974)
5. Boitier, F, Godard, A, Rosencher, E, Fabre, C: Measuring photon bunching at ultrashort timescale by two-photon absorption in semiconductors. Nat. Phys. **5**, 267–270 (2009)
6. Kikuchi, K: Optical sampling system at 1.5 mu;m using two photon absorption in Si avalanche photodiode. Electron. Lett. **34**, 1354–1355 (1998)
7. Roth, JM, Murphy, T, Xu, C: Ultrasensitive and high-dynamic-range two-photon absorption in a GaAs photomultiplier tube. Opt. Lett. **27**, 2076–2078 (2002)
8. Fishman, DA, Cirloganu, CM, Webster, S, Padilha, LA, Monroe, M, David, JH, Van Stryland, EW: Sensitive mid-infrared detection in wide-bandgap semiconductors using extreme non-degenerate two-photon absorption. Nat. Photon. **5**, 561–565 (2011)
9. Sheik-Bahae, M, Said, AA, Wei, T-H, Hagan, DJ, Van Stryland, EW: Sensitive measurement of optical nonlinearities using a single beam. Quantum Electron. IEEE J. **26**, 760–769 (1990)
10. Vest, B, Lucas, E, Jaeck, J, Haidar, R, Rosencher, E: Silicon sub-bandgap photon linear detection in two-photon experiments: A photo-assisted Shockley-Read-Hall mechanism, Vol. 102 (2013)
11. Dvorak, M, Schroeder, W, Andersen, D, Smirl, A, Wherrett, B: Measurement of the anisotropy of two-photon absorption coefficients in zincblende semiconductors. Quantum Electron. **30**, 256–268 (1994)
12. Portier, B, Vest, B, Pardo, F, Péré-Laperne, N, Steveler, E, Jaeck, J, Dupuis, C, Bardou, N, Lemaitre, A, Rosencher, E, et al.: Resonant metallic nanostructure for enhanced two-photon absorption in a thin GaAs pin diode. Appl. Phys. Lett. **105**, 011108 (2014)
13. Capasso, F, Sirtori, C, Cho, AY: Coupled quantum well semiconductors with giant electric field tunable nonlinear optical properties in the infrared. Quantum Electron. IEEE J. **30**, 1313–1326 (1994)
14. Soljačić, M, Joannopoulos, JD: Enhancement of nonlinear effects using photonic crystals. Nat. Mater. **3**, 211–219 (2004)

15

A microscope using Zernike's phase contrast method and a hard x-ray Gabor hologram

Kiyofumi Matsuda[1,2], Juan C. Aguilar[3], Masaki Misawa[1]* ⓘ, Masato Yasumoto[4,5], Shakil Rehman[6], Yoshio Suzuki[5], Akihisa Takeuchi[5] and Ilpo Niskanen[7]

Abstract

Background: In hard X-ray phase imaging using interferometry, the spatial resolution is limited by the pixel size of digital sensors, inhibiting its use in magnifying observation of a sample.

Methods: To solve this problem, we describe a digital phase contrast microscope that uses Zernike's phase contrast method with a hard X-ray Gabor holography associated with numerical processing and spatial frequency domain filtering techniques. The hologram is reconstructed by a collimated beam in a computer. The hologram intensity distributions itself become the reconstructed wavefronts. For this transformation, the Rayleigh-Sommerfeld diffraction formula is used.

Results: The hard X-ray wavelength 0.1259 nm (an energy of 9.85 keV) was employed at the SPring-8 facility. We succeeded in obtaining high-resolution images by a CCD sensor with a pixel size of 3.14 μm, even while bound by the need to satisfy the sampling theorem and by the CCD pixel size. The test samples used here were polystyrene beads of 8 μm, and human HeLa cells.

Conclusions: We thus proved that the resolution 0.951 μm smaller than the pixel size of CCD (3.14 μm) was achieved by the proposed reconstruction techniques and coherent image processing in the computer, suggesting even higher resolutions by adopting greater numerical apertures.

Keywords: X-ray microscopy, Distributed-feedback, Digital holography, X-ray imaging, Interference microscopy, X-ray interferometry

Background

Hard X-rays allow visualization of objects at high resolutions because of their inherent smaller wavelengths [1]; they are therefore highly desirable in various application areas, including biological studies [2, 3]. Even though optical imaging methods are non-destructive (and therefore useful), they are typically limited in spatial resolution because of the long wavelengths used [4] and transmission limitations. Imaging with electron beams depends on fixing the sample under vacuum conditions, and this is detrimental to the morphology of the biological specimens under observation. On the other hand, hard X-rays may suffer the effect of external vibrations, because of their short wavelengths.

Phase imaging using hard X-ray makes it possible to visualize phase objects using interferometry or holography.

Computed tomography using an X-ray interferometer [5] and observation of biological soft tissues using interferometry [6] have been used for phase imaging. Methods using differential interference contrast (DIC) to form Talbot images with hard X-rays have been reported [6]. Given that samples are imaged on a digital sensor, the spatial resolution in such methods is limited, because of the pixel size.

Many reports of digital holography [7] and holographic lateral shear for DIC [8] using laser beam holography have been made. High resolution is possible in holography, because the images are obtained at a position apart from the digital sensor. The two-point resolution in coherent illumination hard X-rays can be obtained by $0.82\lambda/NA$ [9]—if the relative phase between the two points can be considered as zero—where λ is the wavelength and NA is the numerical aperture. It should be pointed out that the sensor pixel size does not appear in the equation for resolution. Accordingly, hard X-ray digital holographic systems may provide high resolution. However, the conjugate images

---

* Correspondence: m.misawa@aist.go.jp
[1]Theranostic Device Research Group, Health Research Institute, National Institute of Advanced Industrial Science and Technology (AIST), 1-2-1 Namiki, Tsukuba, Ibaraki 305-8564, Japan
Full list of author information is available at the end of the article

and reconstructed images overlap, a problem that must be solved. Hard X-rays have been used in Gabor holography [10], to obtain speckle-free coherent illumination [1–11], and in phase contrast microscopy [12]. Hard X-rays have also been used in coherent diffractive imaging [13] with high sensitivity, lens-less imaging with an extended source [14], and phase contrast imaging [15]—using polychromatic hard X-rays [16].

This paper describes a digital holographic microscope using Zernike's phase contrast observation method [4] using a hard X-ray Gabor hologram, which is recorded in a computer. In Gabor holograms, the most important aspect is to reduce the effect of conjugate images. Good quality high-resolution images were obtained after some numerical processing.

Methods

Description of the used hard X-ray Gabor hologram setup

Figure 1 shows the Gabor hologram process using a common path interferometer. In our optical arrangement, the maximum optical path difference between the reference and object beams in the hologram plane is very short. The synchrotron source (the beamline 20XU of SPring-8, Japan) used in this study produces quasi-monochromatic[4] X-rays with a monochromator, as shown in Fig. 1. The central part of the expanded X-ray beam is filtered by aperture CS_1, with a size of 50 μm, and is focused by a Fresnel zone plate (FZP) with a diameter of 104 μm and a focal length of 498 mm. Another cross slit aperture CS_2, with a size of 2 μm, is used to filter the beam and acts as the point source for hologram recording. The beam illuminates an object placed at a distance of Z_0 from CS_2. The size of the hologram is determined by the size of the reference beam at the hologram plane (diameter D = 1.483 mm at a distance of Z_1 from CS_2). The distance between the object and the hologram is Z. A 16-bit Hamamatsu charge-coupled device (CCD) sensor (C4742-98-24) with a 3.14 μm pixel size which consists of 1344 × 1024 pixels was used to record the holograms. Interference fringes with a narrowest spacing of 15.3 μm

were produced at the edges of the sensor. The minimum width of the fringe was about 4.9 times the pixel size, thus satisfying the sampling theorem. The X-ray energy was 9.85 keV, with a corresponding wavelength of 0.1259 nm. The test samples used in this study were 8 μm polystyrene sphere beads and human HeLa cells. The spatial resolution of the reconstructed image in our setup was calculated at 0.951 μm [1], which is less than the sensor pixel size.

The Gabor hologram is reconstructed by numerical processing in the computer. In the conventional reconstruction method, the object image and its twin image appear close to each other, as shown in Fig. 2, and removal of this twin image is therefore crucial to obtain the final image [17]. To reduce the effects of both the dc noise and the twin image, defocus is introduced in the image plane. For reconstruction, a collimated beam parallel to the optical axis is used to illuminate the hologram, so that the complex amplitude of the reconstructed image is obtained by simply carrying out Fresnel back propagation of the hologram to the image plane. In this way, the twin image is produced far from the object image, as shown in Fig. 3. The convolution of the twin image with an impulse response having a diverging curvature produces a defocus, thereby reducing the twin image effects.

Mathematical description of Zernike's method

The Zernike's phase contrast method can be mathematically derived using the complex amplitudes reconstructed from the Gabor hologram, as will be shown. If the complex amplitudes of the reference beam and the phase object beam in the hologram plane at coordinates (x_1, y_1) are denoted by $r(x_1, y_1)$ and $g(x_1, y_1)$ respectively, the intensity distribution $I_h(x_1, y_1)$ of hologram is given by

$$
\begin{aligned}
I_h(x_1, y_1) &= |r(x_1, y_1) + g(x_1, y_1)|^2 \\
&= \left[|r(x_1, y_1)|^2 + |g(x_1, y_1)|^2 \right] + r^*(x_1, y_1)g(x_1, y_1) \\
&\quad + r(x_1, y_1)g^*(x_1, y_1)
\end{aligned}
$$

$$(1)$$

The complex amplitude of $r(x_1, y_1)$ is given by

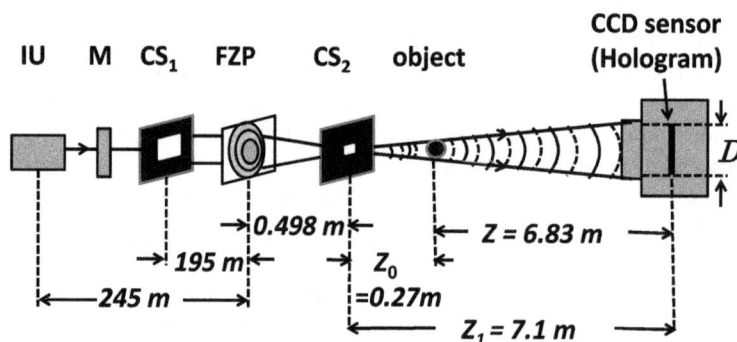

Fig. 1 Optical arrangement for the recording of a hard X-ray Gabor hologram. IU is an in-vacuum undulator, M a monochromatic meter, CS_1 a cross-slit aperture for a pseudo-point source of size 50 × 50 μm, CS_2 a cross-slit aperture for spatial filtering of size 2 × 2 μm, and FZP a Fresnel zone plate

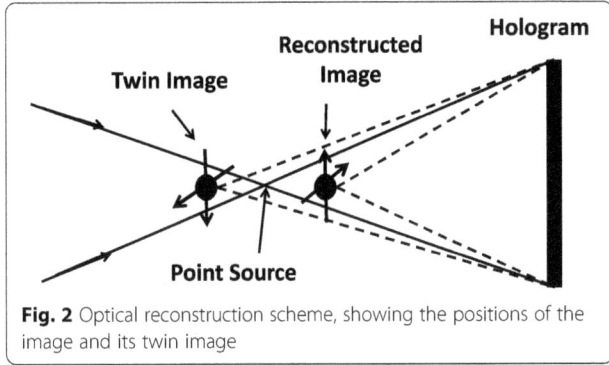

Fig. 2 Optical reconstruction scheme, showing the positions of the image and its twin image

$$r(x_1, y_1) = \left(\frac{1}{j\lambda Z_1}\right)e^{\frac{j2\pi Z_1}{\lambda}}e^{\frac{j\pi}{\lambda Z_1}(x_1^2 + y_1^2)}, \qquad (2)$$

where Z_1 is the distance between the point source and the hologram, and λ is the X-ray wavelength, and j is an imaginary unit, $j = \sqrt{-1}$. If the phase object is illuminated by the hard X-rays originating from a point source, the complex amplitude $g(x_1, y_1)$ of the phase object in the hologram plane is given by

$$g(x_1, y_1) = -\frac{1}{\lambda^2 Z_0 Z}e^{\frac{j2\pi(Z_0 + Z)}{\lambda}}e^{\frac{j\pi}{\lambda Z}(x_1^2 + y_1^2)}$$

$$\times \iint e^{\frac{j\pi}{\lambda}\left(\frac{1}{Z_0} + \frac{1}{Z}\right)(x^2 + y^2) - j\Phi(x,y)}e^{-j\frac{2\pi}{\lambda Z}(x_1 x + y_1 y)}dxdy, \qquad (3)$$

where Z_0 is the distance between the point source and the object, Z is the distance between the object and the hologram, and $\Phi(x, y)$ is the phase distribution of the object. It is noted that if the phase object is thicker, the wavefront is further advanced. The intensity distribution $I_h(x_1, y_1)$ of the hologram is obtained by substituting Eq. (2) and Eq. (3) into Eq. (1).

For reconstruction the Gabor hologram is illuminated by a collimated beam of unity amplitude to separate the

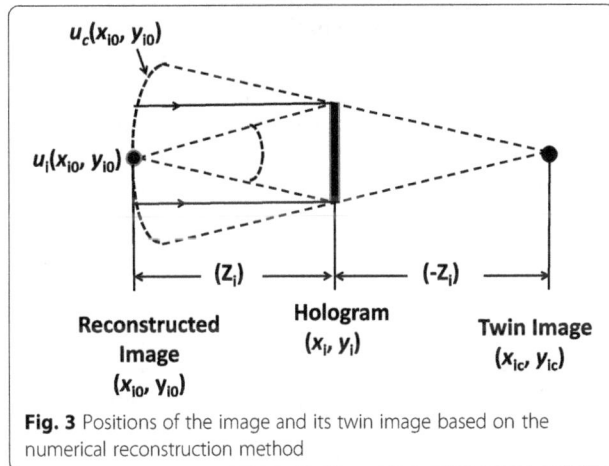

Fig. 3 Positions of the image and its twin image based on the numerical reconstruction method

object image and its conjugate image. The complex amplitudes reconstructed in the image plane are obtained by carrying out a Fresnel Transform of $I_h(x_1, y_1)$ from the hologram plane to the image plane. The complex amplitude of image $u_i(x_i, y_i) = Fr[r^*(x_1, y_1)g(x_1, y_1)]$ is given by

$$u_i(x, y)$$

$$= \frac{1}{j\lambda Z_i}e^{\frac{j2\pi Z_i}{\lambda}}\iint r^*(x_1, y_1)g(x_1, y_1)e^{\frac{j\pi}{\lambda Z_i}[(x_i - x_1)^2 + (y_i - y_1)^2]}dx_1 dy_1$$

$$= -\frac{1}{\lambda^4 Z_0 Z\, Z_1 Z_i}e^{\frac{j2\pi Z_i}{\lambda}}e^{\frac{j\pi}{\lambda Z_i}(x_i^2 + y_i^2)}\iint e^{\frac{j\pi}{\lambda}\left(\frac{1}{Z_0} + \frac{1}{Z}\right)(x^2 + y^2) - j\Phi(x,y)}$$

$$\times \left\{ \begin{matrix} \iint e^{\frac{j\pi}{\lambda}\left(\frac{1}{Z} - \frac{1}{Z_1} + \frac{1}{Z_i}\right)(x_1^2 + y_1^2)} \\ e^{-\frac{j2\pi}{\lambda}\left[\left(\frac{x}{Z} + \frac{x_i}{Z_i}\right)x_1 + \left(\frac{x}{Z} + \frac{x_i}{Z_i}\right)y_1\right]}dx_1 dy_1 \end{matrix} \right\}dxdy. \qquad (4)$$

The image is reconstructed at position Z_{i0} where the condition $1/Z - 1/Z_1 + 1/Z_{i0} = 0$ is satisfied; the image plane position is therefore given by $Z_{i0} = -ZZ_1/Z_0$ (using $Z_1 = Z_0 + Z$). The complex amplitude of the reconstructed image is given by

$$u_i(x_{i0}, y_{i0}) = \frac{1}{\lambda^2 Z_1^2}e^{\frac{j2\pi}{\lambda}Z_{i0}}e^{\frac{j\pi}{\lambda}\left(\frac{1}{Z_{i0}} + \frac{1}{M^2 Z_0} + \frac{1}{M^2 Z}\right)(x_{i0}^2 + y_{i0}^2)}e^{-j\Phi\left(\frac{x_{i0}}{M}, \frac{y_{i0}}{M}\right)}, \qquad (5)$$

where $\Phi(x_{i0}/M, y_{i0}/M)$ is a real periodic function (with period d) and $M = |Z_{i0}/Z|$ is the system magnification. We assume that the magnitude of $\Phi(x_{i0}/M, y_{i0}/M)$ is small compared to unity, so that we may write [4]

$$u_i(x_{i0}, y_{i0})$$

$$= \frac{1}{\lambda^2 Z_1^2}e^{\frac{-j2\pi Z_{i0}}{\lambda}}e^{\frac{j\pi}{\lambda}\left(\frac{1}{2Z_{i0}}\right)(x_{i0}^2 + y_{i0}^2)}e^{\frac{j\pi}{\lambda}\left(\frac{1}{2Z_{i0}} + \frac{1}{M^2 Z_0} + \frac{1}{M^2 Z}\right)(x_{i0}^2 + y_{i0}^2)}$$

$$\times \left[1 - j\Phi\left(\frac{x_{i0}}{M}, \frac{y_{i0}}{M}\right)\right] \qquad (6)$$

where the amplitude $k_i = 1/(\lambda^2 Z_1^2)$ is constant, and $Fr[]$ stands for the Fresnel transform. $Z_{i0} = -ZZ_1/Z_0$ is thus obtained. It is also noted that $\exp[j\Phi] = 1 + j\Phi - \Phi^2/2! + \cdots$.

Similarly, we can derive the complex amplitude of the conjugate image $u_c(x_i, y_i)$ in the plane where the conjugate image is reconstructed; $u_c(x_i, y_i)$ is then given by

If the conjugate image is reconstructed at position $Z_i = Z_{ic}$, the condition $1/Z - 1/Z_1 - 1/Z_{ic} = 0$ must be satisfied. The position in the conjugate image plane (x_{ic}, y_{ic}) is $Z_{ic} = ZZ_1/Z_0$, and the complex amplitude of conjugate image is given by

$$u_c(x_i, y_i) = \frac{1}{j\lambda Z_i} e^{\frac{j2\pi Z}{\lambda}} \iint r(x_1, y_1) g^*(x_1, y_1) e^{\frac{j\pi}{\lambda Z_i}\left[(x_i-x_1)^2 + (y_i-y_1)^2\right]} dx_1 dy_1 \tag{7}$$

$$= \left(\frac{1}{\lambda^4 Z_0 Z Z_1 Z_i}\right) e^{\frac{j2\pi Z_i}{\lambda}} e^{\frac{j\pi}{\lambda Z_i}(x_i^2 + y_i^2)}$$

$$\times \iint e^{-\frac{j\pi}{\lambda}\left(\frac{1}{Z_0}+\frac{1}{Z}\right)(x^2 + y^2) + j\Phi(x,y)} \left[\begin{array}{c} \iint e^{-\frac{j\pi}{\lambda}\left(\frac{1}{Z}-\frac{1}{Z_1}-\frac{1}{Z_i}\right)(x_1^2 + y_1^2)} \\ \times \\ e^{j\frac{2\pi}{\lambda Z_i}\left[\left(x\frac{Z_i}{Z}-x_i\right)x_1 + \left(y\frac{Z_i}{Z}-y_i\right)y_1\right]} dx_1 dy_1 \end{array} \right] dxdy$$

$$u_c(x_{ic}, y_{ic})$$

$$= \frac{Z}{\lambda^2 Z_0 Z_1 Z_{ic}} e^{\frac{j2\pi}{\lambda}Z_{ic}} e^{\frac{-j\pi}{\lambda Z_{ic}}(x_{ic}^2 + y_{ic}^2)}$$

$$e^{-\frac{j\pi}{\lambda}\left(\frac{1}{Z_0}+\frac{1}{Z}\right)\left(\frac{Z}{Z_{ic}}\right)^2(x_{ic}^2+y_{ic}^2)+j\Phi\left(\frac{Z}{Z_{ic}}x_{ic}, \frac{Z}{Z_{ic}}y_{ic}\right)} . \tag{8}$$

If the distance between the reconstructed image and its conjugate image is very long, a wavefront coming from the conjugate image plane may be considered as being emanated from a point source. This condition will be met if the radius $MD/2$ of the aperture in the image plane is smaller than the Airy disc caused by its aperture at the conjugate image plane; that is, the coherent illumination condition is given by

$$D^2 < \frac{4(0.82\lambda Z_{ic})}{M} \tag{9}$$

It is noted that the magnitudes at the image plane and conjugate image plane are equal. We shall now calculate the complex amplitude of the conjugate image in the image plane caused by a point in the conjugate image plane. In Eq. (8), $x_{ic} = y_{ic} = 0$ is substituted and given by

$$u_c(x_{ic}, y_{ic}) = \frac{Z}{\lambda^2 Z_0 Z_1 Z_{ic}} e^{\frac{j2\pi}{\lambda}Z_{ic}} e^{j\Phi(0,0)} \tag{10}$$

In the image plane (x_{i0}, y_{i0}), the complex amplitude of conjugate image is given by

$$u_c(x_{i0}, y_{i0}) = -\frac{j}{2\lambda^3 Z_1^2 Z_{ic}} e^{-\frac{j2\pi}{\lambda}Z_{ic}} e^{-\frac{j\pi}{2\lambda Z_{ic}}(x_{i0}^2 + y_{i0}^2)} e^{j\Phi(0,0)}. \tag{11}$$

When $\exp[j\Phi(0, 0)] \doteq 1 + j\Phi(0, 0)$ and the magnitude of $\Phi(0, 0)$ is small compared to unity,

$$u_c(x_{i0}, y_{i0}) \sim -\frac{j}{2\lambda^3 Z_1^2 Z_{ic}} e^{-\frac{j2\pi}{\lambda}Z_{ic}} e^{-\frac{j\pi}{2\lambda Z_{ic}}(x_{i0}^2 + y_{i0}^2)}[1 + j\Phi(0,0)] \tag{12}$$

The magnification of $(j\pi/2\lambda Z_{ic})(x_{i0}^2 + y_{i0}^2)$ is small compared to unity and we may also write

$$u_c(x_{i0}, y_{i0}) \sim \frac{1}{\lambda^2 Z_1^2} \frac{-j}{2\lambda Z_{ic}} e^{-\frac{j2\pi}{\lambda}Z_{ic}} e^{-\frac{j\pi}{2\lambda Z_{ic}}(x_{i0}^2 + y_{i0}^2)}[1 + j\Phi(0,0)] \tag{13}$$

Moreover, the complex amplitude of the zero-order term is expressed by $u_0(x_i, y_i) = \text{Fr}[|r(x_1, y_1)|^2 + |g(x_1, y_1)|^2]$ and is given by

$$u_0(x_i, y_i) = \frac{-1}{j\lambda Z_i} e^{\frac{-j2\pi}{\lambda}Z_i} \iint[|r(x_1, y_1)|^2 + |g(x_1, y_1)|^2] e^{\frac{j\pi}{\lambda Z_i}[(x_i-x_1)^2 + (y_i-y_1)^2]} dx_1 dy_1 \tag{14}$$

$$u_0(x_{i0}, y_{i0}) = \left(\frac{1}{\lambda^2 Z_1^2} + \frac{1}{\lambda^2 Z_0^2}\right) e^{\frac{-j2\pi Z_{i0}}{\lambda}}$$

$$= k_0 e^{\frac{-j2\pi Z_{i0}}{\lambda}} \tag{15}$$

In these equations, $k_0 = (1/\lambda^2 Z_1^2 + 1/\lambda^2 Z_0^2)$ is a constant. Three wavefronts of $u_i(x_{i0}, y_{i0})$, $u_c(x_{i0}, y_{i0})$, and $u_0(x_{i0}, y_{i0})$ exist in the image plane. To remove the zero-order term, all complex amplitudes in the image plane are Fourier transformed, and the zero frequency is excluded; this means that $u_0(x_i, y_i)$ expressed by Eq. (14) is also removed. These complex amplitudes are then inverse Fourier transformed to the image plane. If we expand $\exp[-j\Phi(x_{i0}/M, y_{i0}/M)]$ with a Fourier series

$$e^{-j\Phi\left(\frac{x_{i0}}{M}, \frac{y_{i0}}{M}\right)} = \sum_{m=-\infty}^{m=\infty} c_m e^{\frac{j2\pi m}{d}x_{i0}}, \tag{16}$$

then $c_0 = 1$ and $c_{-m} = -c_m^*$, $(m \neq 0)$. We Fourier transform the complex amplitudes of $u_i(x_i, y_i)$ and $u_c(x_i, y_i)$ again, and we add $\exp(\pm j\pi/2) = \pm j$ at the zero frequency. Eq. (7) and Eq. (14) then become:

$$u_i(x_i, y_i) \sim \frac{1}{\lambda^2 Z_1^2} e^{-\frac{j2\pi}{\lambda}Z_{i0}} e^{\frac{j\pi}{2\lambda Z_{i0}}}\left(x_i^2 + y_i^2\right) e^{\frac{j\pi}{\lambda}\left(\frac{1}{2Z_{i0}} + \frac{1}{M^2 Z_0} + \frac{1}{M^2 Z}\right)(x_i^2 + y_i^2)}$$
$$\times \left[\pm j - j\Phi\left(\frac{x_i}{M}, \frac{y_i}{M}\right)\right] ,$$

$$(17)$$

$$u_c(x_i, y_i) \sim \frac{1}{\lambda^2 Z_1^2} \frac{-j}{2\lambda Z_{ic}} e^{-\frac{j2\pi}{\lambda}Z_{ic}} e^{-\frac{j\pi}{2\lambda Z_{ic}}}\left(x_i^2 + y_i^2\right)[\pm j + j\Phi(0,0)]$$

$$(18)$$

Using $Z_{ic} = -Z_{i0} = ZZ_1/Z_0$, the intensity distribution $I_i(x_{i0}, y_{i0})$ in the image plane is obtained as

$$I_i(x_{i0}, y_{i0})$$
$$= |u_i(x_{i0}, y_{i0}) + u_c(x_{i0}, y_{i0})|^2$$
$$= \left(\frac{1}{\lambda^2 Z_1^2}\right)^2 \left| \left[\pm j - j\Phi\left(\frac{x_i}{M}, \frac{y_i}{M}\right)\right] e^{\frac{j\pi}{\lambda}\left\{\left(\frac{1}{2Z_{i0}} + \frac{1}{M^2 Z_0} + \frac{1}{M^2 Z}\right)(x_{i0}^2 + y_{i0}^2) + 2Z\frac{Z_1}{Z_0}\right\}} \right.$$
$$\left. - \frac{1}{2\lambda Z_{ic}}[\pm j + j\Phi(0,0)] e^{-\frac{j\pi}{\lambda}\left(2Z\frac{Z_1}{Z_0}\right)} \right|^2$$

$$(19)$$

In Eq. (19), if $\lambda Z_{ic} \gg 1 + \Phi(0,0)$, the complex amplitude of the conjugate image can be neglected. The phase term of the complex amplitude of the reconstructed image is cancelled by calculation of the intensity, so that Eq. (19) becomes

$$I_i(x_{i0}, y_{i0}) = k\left\{1 \pm 2\Phi\left(\frac{x_{i0}}{M}, \frac{y_{i0}}{M}\right)\right\},$$

$$(20)$$

where $k = \{1/(\lambda^2 Z_1^2)\}^2$ is a constant, and the term Φ^2 has been neglected because of its smallness. We should note that Eq. (20) shows that if a Gabor hologram is recorded in the computer, it is possible to use Zernike's phase contrast observation method.

Results

The hard X-ray Gabor hologram is reconstructed by a collimated beam parallel to the axis, so the hologram intensity distribution is multiplied by unity. This means that the intensity distribution itself can be regarded as the complex amplitude and transformed from the hologram plane to the object plane. For this transformation, the Rayleigh-Sommerfeld diffraction formula was used. The image wavefront is produced at a long distance from the hologram.

The correct position determination in the image plane is important. A method to obtain an autofocused image has been proposed [18]. However, we will present a method to elegantly find the position in the image plane of simple objects recorded in the hologram, such as the polystyrene sphere beads. In our method, we use the diffraction effect that results from the distance between the image plane and the hologram being very long. The principle of the method is that if the object is located at the image plane, no diffraction pattern appears around the image; however, if the image is located at a position away from the image plane, a ring-shaped diffraction pattern appears around the image. Therefore, the position of the image plane can be elegantly determined by

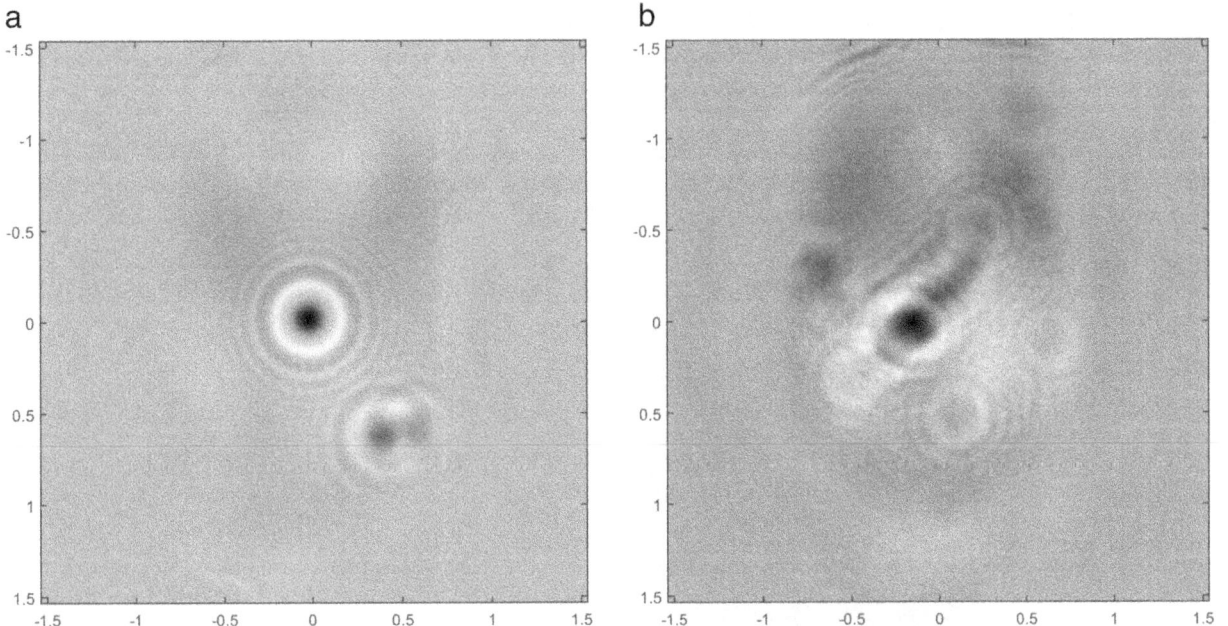

Fig. 4 Holograms constructed by hard X-ray, with $\lambda = 0.1259$ nm (X-ray energy of 9.85 keV). **a** Polystyrene sphere beads with 8 µm diameter. **b** HeLa cells (human cells)

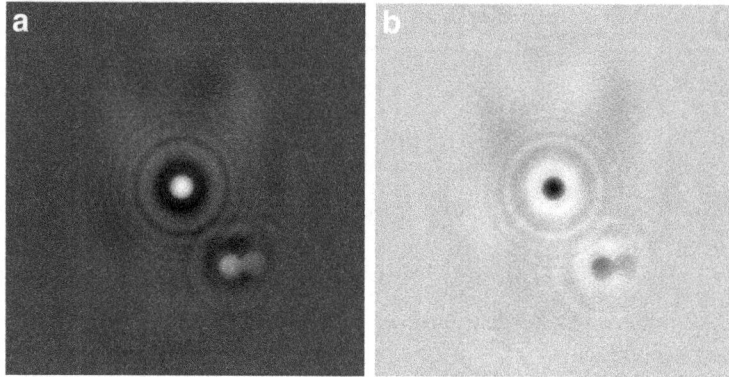

Fig. 5 Experimental results of Zernike's phase contrast method using polystyrene sphere beads with *8* μm diameter. **a** $I_i(x_i, y_i) = k\{1 - 2\Phi(x_{i0}/M, y_{i0}/M)\}$. **b** $k\{1 + 2\Phi(x_{i0}/M, y_{i0}/M)\}$

observing the diffraction patterns (the image plane Z_{i0} was 174.5 m). The magnification M of the optical system [19] was calculated to be $M = |Z_{i0}/Z| = 25.55$, given that the distance between object and hologram was $Z = 6.830 \, m$.

Figure 4a and b show the hard X-ray Gabor holograms of the polystyrene beads and the HeLa cells, respectively. Since the wavefront produced by the X-rays transmitted through the phase object is advanced in comparison with the wavefront traveling in the air, the hologram becomes dark. Figures 5 and 6 show the results of Zernike's phase contrast method. Figure 5 shows the results obtained with the 8-μm polystyrene sphere bead samples. Figure 5a shows $I_i(x_i, y_i) = k\{1 - 2\Phi(x_{i0}/M, y_{i0}/M)\}$, and Fig. 5b shows $k\{1 + 2\Phi(x_{i0}/M, y_{i0}/M)\}$ as noted in Eq. (20). The reconstructed phase object is the bright sphere in Fig. 5a and the dark sphere in Fig. 5b. Figure 6 shows the results obtained for the dried HeLa cell samples. Figure 6a shows $I_i(x_i, y_i) = k\{1 - 2\Phi(x_{i0}/M, y_{i0}/M)\}$, and Fig. 6b shows $k\{1 + 2\Phi(x_{i0}/M, y_{i0}/M)\}$. It should be pointed out that $\Phi(x_{i0}/M, y_{i0}/M)\}$ is negative, because with hard X-rays the object refractive index is smaller than that of air.

Discussion

It should be noted that the refractive index related to the phase difference in the hard X-ray regime is less than unity, contrary to what happens with visible light; the well-known refractive index equation is approximately given by $n = 1 - 1.35 \times 10^{-6}\rho\lambda^2$, where ρ (g/cm^3) is the density, λ (Å) is the wavelength [12], and the terms representing absorption and scattering are ignored for simplicity. The maximum value of the optical path difference in the polystyrene sphere is about $\delta = 0.017$ nm, if the density is taken as $\rho \doteq 1$. The condition needed to derive Eq. (20) is $\Phi(x_i/M, y_i/M) < 1$; the magnitude of $\Phi(x_i/M, y_i/M) = 2\pi\delta/\lambda$ recorded in the hologram is about 0.848. Since this value is smaller than unity, the condition is satisfied [4]. In Eq. (10), a value of $(4(0.82\lambda Z_{ic})/M)^{1/2} = 53$ μm in the object plane was obtained, so that the wavefront of the conjugate image in the image plane can be regarded as a wavefront caused from a point source. Samples of polystyrene sphere beads with 8-μm diameter and dried HeLa cells are used as phase objects; in the used hologram, the aperture diameter magnitude can be considered to produce a wavefront originating from a point source. The

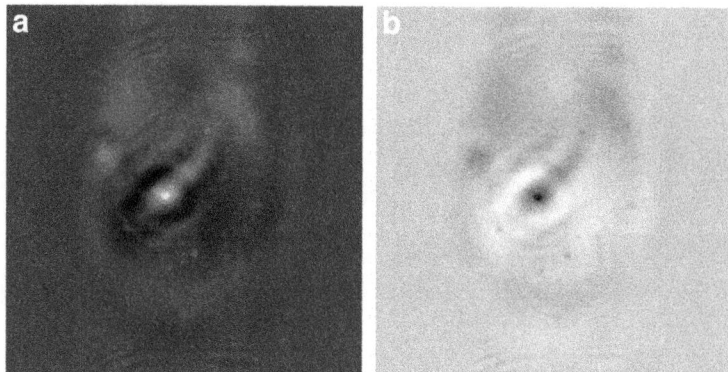

Fig. 6 Experimental results of Zernike's phase contrast method using HeLa cells. **a** $I_i(x_i, y_i) = k\{1 - 2\Phi(x_{i0}/M, y_{i0}/M)\}$. **b** $k\{1 + 2\Phi(x_{i0}/M, y_{i0}/M)\}$

numerical value of $(1 + \Phi(0, 0))/(2\lambda Z_{ic})$ in Eq. (20) is calculated, to check whether this term can be neglected or not, resulting in $\Phi(0, 0) < 1$, $\lambda = 0.1259 \times 10^{-3} \mu m$, and $Z_{ic} = 174.5 \times 10^6 \mu m$; therefore, $(1 + \Phi(0, 0))/(2\lambda Z_{ic}) \sim 4.2 \times 10^{-5} \ll 1$, and we can safely neglect this term.

Conclusions

We proposed a microscope using Zernike's phase contrast observation method and a hard X-ray Gabor hologram recorded in a computer. Two different sample types were used for demonstration purposes: polystyrene sphere beads with an 8-μm diameter, and dried HeLa cells. Recording the hologram in a computer makes it possible to perform computational experiments on coherent X-ray processing. This is very important, because X-ray resources are limited. Moreover, coherent numerical processing becomes easy and simple; creating a perfect phase delay of $\pi/2$, for example, is trivially produced numerically. Even though a small 0.1259 nm wavelength was used, there is no influence of the external vibrations in image reconstruction. It is also pointed out that the use of holography with the exception of an image hologram can produce images with high resolution, since the pixel size of the CCD detector does not have influence on resolution, but its diameter has.

Abbreviations

CCD: Charge coupled device; DIC: Differential interference contrast; FZP: Fresnel zone plate; NA: Numerical aperture; λ: Wavelength

Acknowledgements
The authors would like to thank Dr. Y. Koseki, a group leader, and Dr. K. Chinzei, a sub-director of AIST, Japan, for providing research facilities. Technical advices from Dr. K. Hibino, Dr. T. Eijyu and Dr. M. Yamauchi were highly appreciated.

Funding
Consejo Nacional de Ciencia y Tecnología(CONACYT), Mexico, has awarded to Dr. Juan C. Aguilar with a scholarship to make a postdoctoral research at National Institute of Advanced Industrial Science and Technology (AIST), Japan (250204). A part of this study was supported by 2016 Saga Prefecture Leading Industry Incubation Program, Japan.

Authors' contributions
KM: He engaged in the idea of the method, mathematical analysis, simulation of laser and X-ray holography and design of computer algorithm. JCA: He engaged in a computer code for hard X-ray holography. MM: He engaged in the management of this study, discussion and advice of X-ray measurement method. MY: He engaged in X-ray optical design, production of hard X-ray hologram and advice for simulation of laser holography. SR: He engaged in a part of the idea of the method, simulation of laser holography and computer algorithm. YS: He engaged in production of hard X-ray hologram. AT: He engaged in production of hard X-ray hologram. IN: He engaged in discussion of hard X-ray optics and simulation of laser holography. All authors read and approved the final manuscript.

Competing interests
The authors declare that they have no competing interests.

Author details
[1]Theranostic Device Research Group, Health Research Institute, National Institute of Advanced Industrial Science and Technology (AIST), 1-2-1 Namiki, Tsukuba, Ibaraki 305-8564, Japan. [2]The graduate School for the Creation of New Photonics Industries, 1955-1 Kurematsu, Nishi-ku, Hamamatsu, Shizuoka 431-1202, Japan. [3]Instituto Nacional de Astrofísica,, Óptica y Electrónica, Luis Enrique Erro #1, Tonantzintla, Puebla, Mexico. [4]Research Institute for Measurement and Analytical Instrumentation, NMIJ, National Institute of Advance Industrial Science and Technology, Tsukuba 305-8568, Ibaraki, Japan. [5]Japan Synchrotron Radiation Research Institute, SPring-8, Sayo, Hyogo 679-5198, Japan. [6]Singapore-MIT Alliance for Research and Technology (SMART) Centre1 CREATE Way #09-03, CREATE Tower, Singapore 138602, Singapore. [7]Faculty of Technology, University of Oulu, PO Box 73009014 Oulu, Finland.

References
1. Suzuki, Y, Takeuchi, A: "Reduction of Speckle Noises by Spatial Filter Method in Hard X-ray Region", AIP Conf. Proc. 1234, p. 453 (2010). (http://scitation.aip.org/content/aip/proceeding/aipcp/10.1063/1.3463238. Accessed 15 Nov 2016)
2. Momose, A., Takeda, T., Tani, Y., Hirano, K.: Phase-contrast X-ray composed tomography for observing biological soft tissues. Nat. Med. 2, 473–475 (1996)
3. Xu, W., Jericho, M.H., Meinertzhagen, I.A., Kreuzer, H.J.: Digital in-line holography for biological application, pp. 11301–11305. (2001). PNAS98
4. Born, M, Wolf, E: Principles of Optics 2nd Ed, pp. 424–425, and p. 524. London, Pergamon Press Ltd. (1991)
5. Momose, A.: Demonstration of phase–contrast X-ray computed tomography using an X-ray interferometer. Nucl. Instrum. Methods. A352, 622–628 (1995)
6. Momose, A, Fujii, A, Kadowaki, H, Jinnai, H: Three dimensional observation of polymer blend by X-ray phase tomography. Macromolecules. 38, (2005) pp. 622–628
7. Matsuda, K, Rehman, S, Oohashi, H, Niskanen, I, Yamauchi, M, Nakatani, T, Homma, K, and Peiponen, K : Digital holographic microscopy using a single mode fiber, Proc. The Eighth Finland-Japan Joint Symposium on Optics in Engineering, 3-5 September 2009, Technical Digest, Tokyo, Japan, pp. 17–19 (2009)
8. Matsuda, K., Namiki, M.: Holographic lateral shear interferometer for differential interference contrast method. J. Optics. (Paris) 11, 81–85 (1980)
9. Goodman, JW: Introduction to Fourier Optics, 2nd Ed, p.157 and p.74. New York, McGraw Hill Co. Inc. (1988)
10. Gabor, D.: A new microscopic principle. Nature 161, 777 (1948)
11. Suzuki, Y., Takeuchi, A.: Gabor holography with speckle-free spherical wave in hard X-ray region. Jpn. J. Appl. Phys. 51, 086701–1 (2012)
12. Matsuda, K, Rehman, S, Lopez, JCA, Suzuki, Y, Takeuchi, A, Misawa, M, Niskanen, I: Phase contrast microscope using a hard X-ray Gaborhologram, Proc. The Eleventh Finland-Japan Joint Symposium on Optics in Engineering, 1-3 September 2015, Joensuu, Finland, p. 6–7 (2015)
13. Chapman, H.N., Nugent, K.A.: Coherent lensless X-ray imaging. Nat. Photonics 4, 833–839 (2010)
14. Rodenburg, J.M., Hurst, A.C., Cullis, A.G., Dobson, B.R., Pfeiffer, F., Bunk, O., David, C., Jefimovs, K., Johnson, I.: Hard X-ray lensless imaging of extended objects. Phys. Rev. Letts 98, 034801 (2007). 1–4
15. Snigirev, A., Snigireva, I., Kohn, V., Kuznetsov, S., Schelokov, I.: On the possibilities of X-ray phase contrast microimaging by coherent high-energy Synchrotron radiation. Rev. Sci. Instrum. 66, 5486–5492 (1995)
16. Wilkins, S.W., Gureyev, T.E., Gao, D., Pogany, A., Stevenson, A.W.: Phase contrast imaging using polychromatic hard X-ray. Nature 384, 335–338 (1996)
17. Latychevskaia, T., Fink, H.W.: Solution to the twin image problem in holography. Phy. Rev. Lett 98, 233901 (2007). 1–4
18. Langehanenberg, P., Kemper, B., Dirksen, D., Von Bally, G.: Autofocusing in digital holographic phase contrast microscopy on pure phase objects for live cell imaging. Appl Optics. 47, D176–D182 (2008)
19. Meier, R.W.: Magnification and third-order aberrations in holography. J. Opt. Soc. Am. 55, 46–51 (1965)

16

Optical system for measuring the spectral retardance function in an extended range

Abdelghafour Messaadi[1], María del Mar Sánchez-López[2], Pascuala García-Martínez[3], Asticio Vargas[4] and Ignacio Moreno[1]* (iD)

Abstract

Background: Optical retarders are key elements for the control of the state of polarization of light, and their wavelength dependance is of great importance in a number of applications.

Methods: We apply a well-known technique for determinig the spectral retardance by measuring the transmission spectra between crossed or parallel polarizers. But we we develop an optical system to perform this measurement in a wide spectral range covering the visible (VIS) and near infrared (NIR) spectrum in the range from 400 to 1600 nm.

Results: As a result we can measure the spectral retardance of different retarders and easily identify the kind of reterder (multiple order, zero-order, achromatic). We show results with tunable liquid-crystal retarders as well, where the technique is applied to determine the spectral retardance as a function of the applied voltage. Finally, the accuracy of the technique is verified by the generation of a birefringent spectral filter.

Conclusions: A technique to measure the spectral retardance of a linear retarder in a wide spectral range is applied to identify different types of retarders, and provide an accurate description of the spectral polarization conversion properties of these elements.

Keywords: Optical retarders, Liquid-crystals, Spectroscopy, Filters

Background

Optical linear retarders are very useful components for any optical application requiring control of the state of polarization [1]. High quality retarders are usually fabricated with anisotropic optical materials such as quartz or calcite. Lower cost retarders are fabricated with birefringent polymers, having additionally the advantage of being produced with much larger areas. Tunable retarders can be fabricated with liquid crystal (LC) materials, where the application of a relative low voltage yields a large variation of the effective retardance, due to the tilt of the liquid-crystal director. Liquid crystal retarders (LCR) can be manufactured in the form of a single retarder element, or in the form of one or two-dimensional arrays, as in the liquid-crystal on silicon (LCOS) displays [2]. Other tunable retarders are fabricated with electro-optic materials, such as lithium

niobate (LiNbO3). They require higher voltages and have much smaller areas than LC retarders, but can be switched at much faster rates [3]. Therefore, these tunable retarders are becoming very useful in all kind of applications that require programmable control of the intensity, the phase, or the state of polarization of an input light beam, thus becoming key components in advanced optical instruments for optical microscopy, interferometry, polarimetry or optical communications.

Usually, linear retarders are designed introducing a specific retardance (typically a half-wave or a quarter-wave) for a given operating wavelength. However, characterization of their spectral retardance properties can be very valuable for several reasons: 1) the retarder can be used at wavelengths different to the original design; 2) the retarder can be applied to build spectral birefringent filters, which are based on the wavelength variation of the retardance [4], 3) it allows the simple identification of the ordinary and extraordinary neutral axes of the retarder [5], and 4) the retardance modulation of tunable LC retarders can be characterized [6, 7].

---
* Correspondence: i.moreno@umh.es
[1]Departamento de Ciencia de Materiales, Óptica y Tecnología Electrónica, Universidad Miguel Hernández, 03202 Elche, Spain
Full list of author information is available at the end of the article

In addition, the spectral retardance function can provide very useful information about the fabrication characteristics of the retarder, allowing a simple identification of multiple-order, low-order or zero-order retarders, as well as achromatic retarders.

Several works have demonstrated different techniques for the spectral retardance characterization. A usual technique consists in inserting the retarder in between two linear polarizers, which are oriented at ±45° with respect to the retarder neutral axes. The system is illuminated with a light source with broadband spectrum, and the transmission is analyzed with a spectrometer [6–10]. The transmitted spectrum typically shows an oscillatory dependence with wavelength from which the spectral retardance function can be retrieved. Similar spectral methods sequentially rotate the polarizers to achieve more data [11, 12]. In addition, this kind of spectral measurements provide a simple test to identify whether the retarder presents multiple-reflection Fabry-Perot interferences [13].

Most of the works mentioned above use visible (VIS) light. However, there is an increasing interest in extending the spectral range in the near infra-red (NIR) range, for applications such as optical fiber communications, with its transparency window centered at 1550 nm [14], or biomedical imaging, where the therapeutical windows in the ranges of 650–950 nm (first window) or 1100–1350 nm (second window) are conventionally used for tissue imaging, and deeper IR windows seem to have potential great interest [15].

Here we apply the above mentioned technique of measuring the transmittance spectra between crossed or parallel polarizers to determine the spectral retardance function [6–12], but we use an optical calibration system developed for extending the measurement range to wavelengths from 450 nm to 1600 nm. The system incorporates a thermal broadband light source or a super-continuum laser, two broadband beam-splitter polarizers, and two spectrometers that operate in the VIS and in the NIR band regions respectively. As a result, we can determine the spectral

retardance function of different retarders in a very wide spectral range by fitting the measured and the simulated transmission curves. In some cases, a Cauchy-like dispersion relation can be applied, which has been proved to give good approximations far from the absorption bands of anisotropic materials [6, 7, 10, 12].

We apply the developed system to different types of retarders such as multiple-order, zero-order and achromatic retarders. We show how their spectral characteristics allow a very simple identification of these different types of retarder designs. In all cases we determine the spectral retardance function, and we also include some interesting configurations that can be obtained by simply placing two retarders.

Methods

Figure 1 shows a scheme of the optical system, including a picture in the inset. We use a quartz tungsten halogen lamp from Oriel, model 66882, with a power that can be adjusted from 10 to 250 watts. It provides white light of continuous broadband spectrum that covers the wavelength range from 400 to 1600 nm. The housing includes a fused silica condenser that can be adjusted to provide a collimated output beam with a diameter of 33 mm.

As linear polarizers (P1 and P2) we use two high-quality calcite Glan-Taylor cube polarizers from Edmund Optics, covering a spectral range from 350 to 2200 nm, with a nominal extinction ratio less than 5×10^{-6}. This kind of polarizers is required since common commercial polaroid sheets do not act properly as polarizers in the IR range. They have been mounted on rotatable mounts, so the angle of the transmission axis can be rotated continuously. The retarder to be characterized is placed in between the two polarizers. Then, the transmitted light is divided in two beams by means of a B270 Glass Polka Dot beam-splitter from Thorlabs. Again, this kind of beam-splitter is required since it operates in a wide range of wavelengths from 350 nm to 2.0 μm. These two

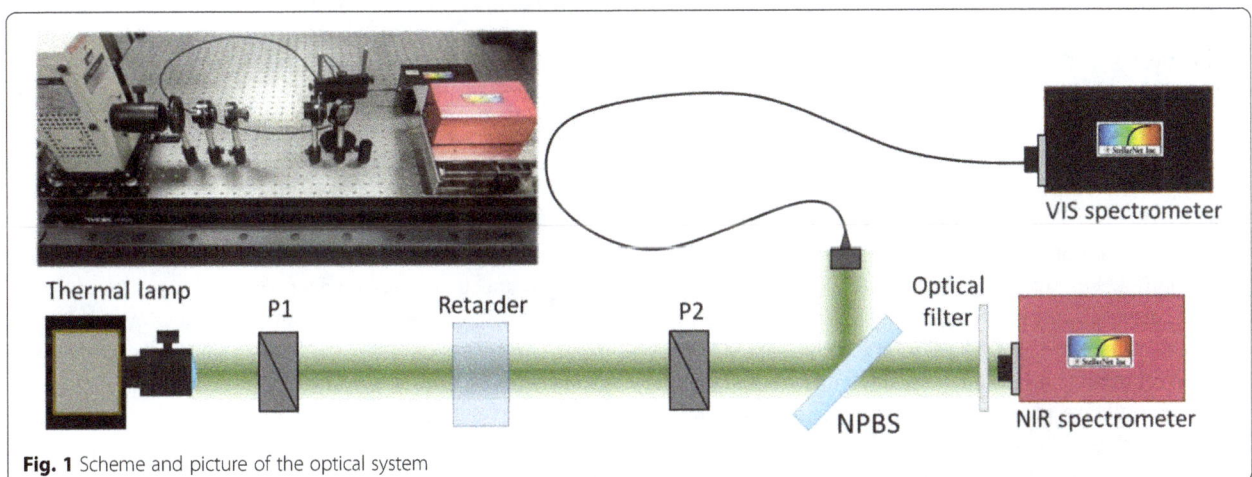

Fig. 1 Scheme and picture of the optical system

beams are analyzed with two different spectrometers. The beam reflected by the beam-splitter is captured with a STN-F600-UVVIS-SR optical fiber that is connected to a VIS spectrometer from Stellar-Net, STN-BLK-C-SR model, which measures the spectrum in the range from 200 nm to 1080 nm with a resolution of 2 nm. The second beam is directly sent to another spectrometer from Stellar-Net, model STE-RED-WAVE-NIR-512-25, which measures the spectrum from 900 nm to 1700 nm, with a resolution of 3 nm. In this case we do not use a fiber to avoid absorption bands in the IR region. Finally, in order to avoid second-order contribution from the visible light that enters this IR spectrometer, we include a filter in front of the slit entrance, which filters the visible spectrum.

The retarder is inserted between two parallel or crossed linear polarizers, with the principal c-axis oriented with a relative angle of 45° with respect the transmission axes of the polarizers. In this situation, the normalized transmission output is given by [4]:

$$T_{par} = \cos^2\left(\frac{\phi}{2}\right), \quad T_{cros} = \sin^2\left(\frac{\phi}{2}\right), \qquad (1)$$

where subindices "par" and "cros" refer to having the two polarizers parallel or crossed respectively. Here, ϕ denotes the wave-plate retardance. These relations assume ideal retarders and polarizers, where no other polarization phenomena different than linear retardance occurs. This is a reasonable approximation for linear retarders, and no additional polarimetric measurements are required.

In order to normalize the experimental spectral data, the intensity of the transmitted light is measured in two ways: one first measurement with parallel polarizers, $I_{par}(\lambda)$, and a second with crossed polarizers, $I_{cros}(\lambda)$. The retarder is inserted in between the polarizers with the principal axis rotated 45° to polarizer P1. Then, data are normalized for each wavelength as:

$$T_{par} = \frac{I_{par}}{I_{par} + I_{cros}}, \quad T_{cros} = \frac{I_{cros}}{I_{par} + I_{cros}}. \qquad (2)$$

This normalization makes the experimental data directly comparable to Eqs. (1). Again, note that this kind of normalization ignores possible spectral variations in the transmission/extinction of the analyzer, and therefore high quality polarizers must be employed. Our goal here is to measure the spectral retardance function, i.e., the function $\phi(\lambda)$ which describes the dependence of the retardance with wavelength λ. For that purpose the function $\phi(\lambda)$ that best fits the curves $T_{par}(\lambda)$ and $T_{cros}(\lambda)$ must be determined.

In fact, different types of retarders show very different spectral retardance functions [16]. Therefore, these spectral measurements are of interest to easily identify the kind of retarder. For instance, a simple retarder composed of a single layer of uniaxial plate, the retardance is given by

$$\phi = (k_e - k_o)\ d = \frac{2\pi}{\lambda}(n_e - n_o)\ d, \qquad (3)$$

where k_e and k_o are the wavenumbers for the extraordinary and ordinary waves, n_e and n_o are the extraordinary and ordinary indices of refraction respectively, and d denotes the thickness of the plate. In multiple-order retarders, the thickness d is large, and the total retardance for the design wavelength is

$$\phi = m2\pi + \delta, \qquad (4)$$

where m is the order of the retarder, and δ denotes the modulo 2π retardance. For these multiple-order retarders, the function $\phi(\lambda)$ change very rapidly and the retarder suffers from larger retardance variations with temperature and wavelength. These variations are reduced significantly with zero-order retarders, where the retardance is directly the design value, $\phi = \delta$, i.e., $m = 0$, and consequently the function $\phi(\lambda)$ changes slowly.

In some applications it is of interest to use retarders with a retardance that does not change with wavelength. Achromatic retarders are made by placing together two retarder layers of different materials with opposite dispersion relations [17, 18]. The difference in thickness and refractive index of these two anisotropic layers can be adjusted to provide the same retardance for two separated wavelengths, and $\phi(\lambda)$ only shows a very small amount of deviation from this value in between. Alternatively, Fresnel rhombs are retarders with almost perfect wavelength independent retardance [19], since they are not based on a material's birefringence, but on the difference in phase-shift for the s and p polarized components in a total internal reflection.

All these different types of retarders exhibit very different spectral retardance functions, that can be easily visualized in the spectrometer, as shown next.

Results and discussion
In this section we show results of the spectral retardance measured for different retarders.

Multiple-order and zero-order retarders
We start by using two different quartz quarter-wave plate (QWP) multiple order retarders, designed for wavelengths of 514 nm and 488 nm respectively. We denote them as QWP$_{514}$ and QWP$_{488}$ respectively. Figure 2(a) and (b) show the normalized data $T_{par}(\lambda)$ for these two retarders. Blue and red points denote the data captured with the VIS and NIR spectrometers respectively.

Two features are clearly visible in these graphs: 1) A rapid oscillation as a function of wavelength is observed in both cases, and 2) a value $T_{par} = 0.5$ is obtained at the design wavelengths. The rapid oscillation observed in Fig. 2(a) and (b) indicates that the retardance is experimenting a very rapid change with wavelength, as

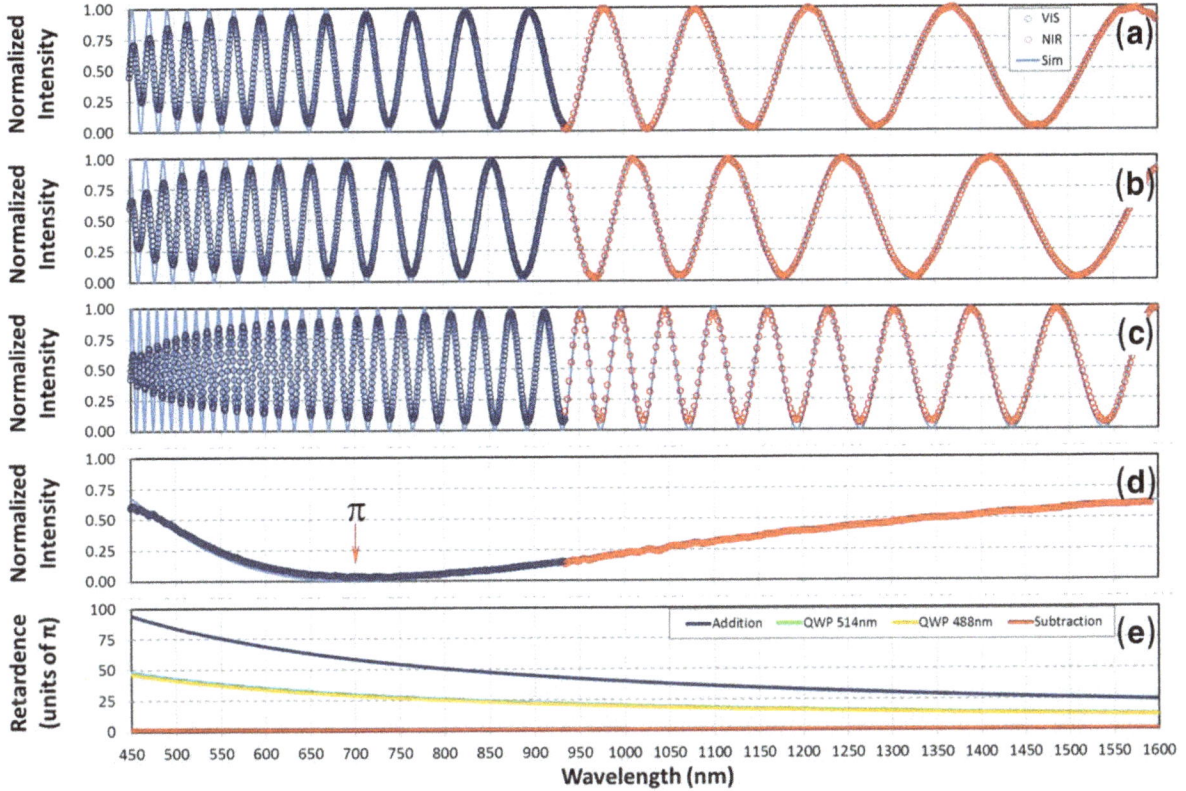

Fig. 2 Normalized spectral intensity transmission $T_{par}(\lambda)$ for **a** Multiple-order QWP for 488 nm; **b** Multiple-order QWP for 514 nm; **c** Addition of the two QWPs; **d** Subtraction of the two QWPs. In all cases the continuous lines correspond to the simulation that best fits the experimental data. **e** Spectral retardance for the four cases derived after fitting the experimental data

expected in a multiple-order retarder. The number of complete oscillations for QWP_{488} is slightly larger than the total oscillation for QWP_{514}, and the total retardance variation is around 36π radians in the covered spectral range for the two retarders. Another interesting aspect to note is that, although the design wavelength is located at the lower extreme of the measured wavelength range, the oscillatory behavior is maintained up to the other extreme at 1600 nm. This denotes that these retarders operate properly in the entire spectral range, although they are normally commercialized for a single specific designed wavelength.

We added two other curves in Fig. 2(c) and (d). Here the two retarders are placed in between the polarizers. In Fig. 2(c) the two retarders are aligned with the fast axis in the same orientation, while in Fig. 2(d) the second one is rotated 90° with respect to the first one. Thus, in Fig. 2(c) the total retardance is the addition $\phi(\lambda) = \phi_1(\lambda) + \phi_2(\lambda)$, while in Fig. 3(d) the total retardance is the subtraction $\phi(\lambda) = \phi_1(\lambda) - \phi_2(\lambda)$. Note that the retardance addition doubles the spectral oscillation. On the contrary, when the retardances are subtracted, a very slow oscillation remains because the two retarders have small thickness difference. A retardance difference of π radians is obtained for the wavelength of

700 nm. Note that this last case mimics a zero-order retarder. And this result shows how the spectral method is a very simple technique to clearly distinguish between multiple-order and zero-order waveplates.

These results show an interesting limiting factor. Note how the amplitude of the oscillations is reduced in the low part of the spectrum. This is due to the resolution limit of the spectrometer. As the retardance varies so fast in this region, so does the spectral transmission. Therefore, the limited size of the pixel detector cannot detect this rapid oscillation, and zero and one transmission are not properly detected.

Nevertheless, these experiments can be used to fit the spectral retardance function which can then be adjusted according to a microscopic physical model, as for instance the Cauchy-type series that are usually a good approximation far from the material absorption bands [20]. The experimental curves in Fig. 2(a) and (b) were thus fitted to a numerical simulation of Eq. (1) assuming a spectral dependence of $\phi(\lambda)$ as:

$$\phi(\lambda) = \frac{A}{\lambda} + \frac{B}{\lambda^3} + \frac{C}{\lambda^5} + D\lambda. \tag{5}$$

The first three terms correspond to a third order Cauchy approximation for the refractive indices in Eq. (2), while the

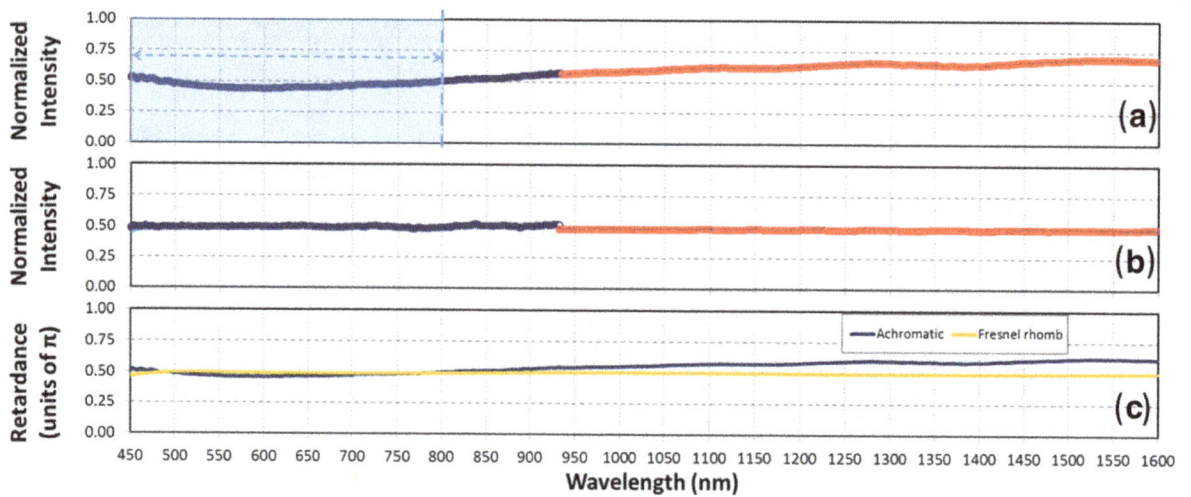

Fig. 3 Normalized spectral intensity transmission $T_{par}(\lambda)$ for **a** An achromatic quarter-wave retarder designed for the indicated spectral range from 450 to 800 nm; **b** A quarter-wave Fresnel rhomb; **c** Spectral retardance for these two retarders

last term provides good results for quartz in the IR region [10]. A numerical search for the constants A, B, C, and D that minimize the difference between simulation and experimental data was performed for the two retarders. This is done by numerically evaluating the mean absolute error (MAE) between the normalized transmission experimental data and the simulated data, and seeking for the values that minimize this difference. This was programmed in Microsoft Excel and was solved with the SOLVER routine, which employs a generalized reduced gradient algorithm (https://support.microsoft.com/en-us/kb/214115).

Since we have used two quartz waveplates from the same supplier, purchased at the same time, we can assume exactly the same retardance dispersion for the two retarders, with a simple multiplicative factor. Therefore, we have considered the retardance for QWP$_{514}$ as $\phi_{514}(\lambda)$ following the relation in Eq. (5), and we have considered the retardance for QWP$_{488}$ follows a relation $\phi_{488}(\lambda) = t\phi_{514}(\lambda)$, where t is a multiplicative factor that takes into account the small amount of thickness difference between the two plates. Thus, the numerical fit consists in a single search of the A, B, C, and D constants for $\phi_{514}(\lambda)$ and the constant t, that simultaneously match for the four curves in Fig. 2. This way we obtain a more confident result than simply fitting the result for a single retarder. Figure 2(a), (b), (c) and (d) show the simulated curves as well, revealing a very good agreement with the experimental data. The corresponding spectral retardance functions are shown in Fig. 2(e). The spectral retardance is very similar for the two waveplates, since the thickness difference parameter is $t = 0.9657$.

Achromatic retarders and Fresnel rhombs

A second interesting example involves using retarders with flat spectral retardance functions. We consider here two examples: an achromatic QWP retarder from Thorlabs, model

AQWP05M-600, designed for the range 400–800 nm, and a quarter-wave Fresnel rhomb also from Thorlabs, model FR600QM, designed for the range 400–1550 nm. Figure 3 shows the corresponding experimental data for the normalized intensity transmission T_{par}. In these cases, the spectral oscillations present in the previous retarders do not appear, and the normalized transmission is approximately constant at the value $T_{par} = 0.5$, as expected for a QWP. But for the achromatic retarder, this is approximately true only in the spectral range of design, while the Fresnel rhomb shows a much better flat transmission in the wide spectral range. The VIS spectral region between 450 and 800 nm, where the achromatic QWP retarder operates, has been marked in Fig. 3(a). The two extremes of this regions shows the exact normalized transmission of 50 %, and it shows only a small variation for wavelengths in between. On the contrary, for wavelengths larger than 800 nm, the normalized transmission is slowly but progressively increasing, thus showing the deviation from the quarter-wave retardation at these wavelengths.

Figure 3(b) displays the corresponding experimental data for the Fresnel rhomb. In this case a perfect flat normalized transmission $T_{par} = 0.5$ is obtained in the complete spectral range, showing the superior behavior of this retarder in providing a wavelength-independent quarter-wave retardance. Finally, Fig. 3(c) shows the derived spectral retardances $\phi_{ACRH}(\lambda)$ and $\phi_{FR}(\lambda)$ for the achromatic retarder and the Fresnel rhomb respectively.

Liquid-crystal tunable retarders

As a final example we consider a liquid crystal retarder (LCR). These are tunable retarders where the retarder layer is made of nematic liquid crystal, showing maximum retardance when the device is off and the liquid crystal director is aligned to the plane of the retarder.

When a voltage is applied to the device electrodes, the liquid-crystal director tilts and the effective retardance is reduced.

In this work we consider a LCR device from ArcOptix [21]. Figure 4 shows the measurement for this retarder. Again, the oscillatory behavior in the normalized intensity as in Fig. 2 is observed. But the number of oscillations is much lower since the LCR is a low-order retarder. Secondary oscillations are observed in the IR range from 1400 to 1600 nm. This is due to a Fabry-Perot interference effect at the LC layer, as studied in Ref. [13]. For simplicity, we ignore here this secondary effect, and we will consider the retarder simpler approximation. A fit of the experimental data to the spectral retardance function in Eq. (5) was performed. The locations of the maxima and minima indicate the wavelengths for which the retardance is an integer multiple of π radians. These points are indicated in Fig. 4(b), being $\phi = 2\pi$ for 1030 nm, $\phi = 3\pi$ for 710 nm, and $\phi = 4\pi$ for 560 nm.

However, in order to make a more precise spectral retardance fit, we combined the LCR with the QWP Fresnel rhomb. The reason for this combination is related to the fact that measurements show the maximum accuracy around quarter-wave retardance values (i.e., where the normalized transmission is 50 %) [10, 12]. The QWP Fresnel rhomb introduces an additional $\pi/2$ retardance that can be added or substracted to the LCR retardance depending on their relative orientation. In Fig. 4(a) the LCR and the Fresnel rhomb are oriented such that

their retardances add, and therefore the oscillations are shifted to lower wavelengths. In Fig. 4(b), the LCR is the only retarder in the system. Finally, in Fig. 4(c) the LCR is rotated by 90°, and therefore the retardances subtract. In this case the oscillations shift to higher wavelengths. In both cases the shift introduced by the Fresnel rhomb transforms the maxima and minima in Fig. 4(b) into points at 50 % transmittance, therefore improving the accuracy at these wavelengths. Note that an equivalent technique has been used to measure the retardance of half-wave retarders with monochromatic light [22].

Figure 4(d) shows the retardance that best simultaneously fits the three curves in Fig. 4(a), (b) and (c). Again, a spectral response given by Eq. (5) is assumed for the LCR spectral retardance $\phi_{LCR}(\lambda)$. The Cauchy dispersion relation assumed in this equation for the refractive indices has been shown to be a good approximation for liquid-crystal materials [23, 24]. Figures 4(b) shows the theoretical curve together with the experimental data, and the agreement is excellent. Figure 4(a) and (c) show the theoretical curves derived using Eq. (1) for $\phi_{LCR}(\lambda) + \phi_{FR}(\lambda)$ and $\phi_{LCR}(\lambda) - \phi_{FR}(\lambda)$ respectively, again with excellent agreement with the experimental data. The simultaneous fit of the three curves in Fig. 4 thus provides a very reliable procedure to accurately determine the LCR retardance function.

In the results in Fig. 4, the LCR is off. But LCR devices are of interest mainly because the retardance

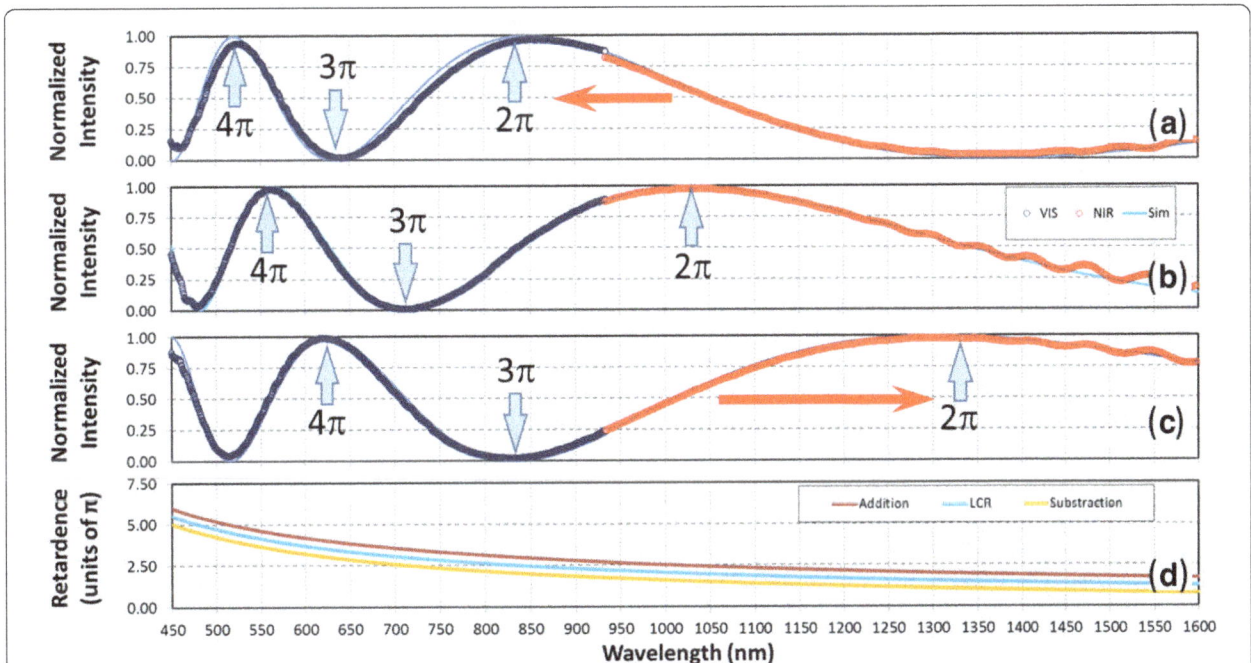

Fig. 4 Normalized spectral intensity transmission $T_{par}(\lambda)$ for the LCR without applied voltage; **a** the LCR plus the Fresnel rhomb; **b** the LCR alone and **c** the LCR minus the Fresnel rhomb. **d** Spectral retardance for the three cases considered

can be controlled via an applied voltage. Normally, in parallel aligned nematic LCR devices, the maximum retardance occurs in the absence of voltage, and the application of voltage reduces the retardance due to the tilt of the liquid-crystal director [7]. Figures 5(a)-(d) show the spectral transmittance and retardance for the LCR without applied voltage, and when a 1.6 KHz square-amplitude signal with polarity inversion is applied, with peak to peak voltages Vpp = 1 V, Vpp = 1.5 V and Vpp = 2 V respectively. The first result that becomes apparent is the shift of the oscillations to the left part of the graphs (lower wavelengths) due to the reduction of the retardance. Because the peak to peak voltage can be tuned continuously, we can follow the shift of the maxima, and therefore identify where integer values of π radians are obtained, as indicated in the figures.

In Fig. 5(b), (c) and (d), the spectral retardance was derived by fitting the experimental data to a function $\phi(V,\lambda) = g(V)\phi(V = 0,\lambda)$, where $\phi(V = 0,\lambda)$ is the LCR spectral retardance without applied voltage (result in Fig. 4(d)), and $g(V)$ is a voltage transfer function that can take values between 1 and 0, and which allows describing the spectral retardance modulation with a single value [7]. For every voltage, the value $g(V)$ that best fits the experimental data is retrieved, obtaining values $g(V_{pp} = 1$ V$) = 0.732$, $g(V_{pp} = 1.5$ V$) = 0.467$ and

$g(V_{pp} = 1$ V$) = 0.294$ respectively. Note that all cases again show a very good agreement between the experimental data and the numerically fitted curve. The evolution of the spectral retardance with voltage is given in Fig. 5(e). Note that the retardance in Fig. 5(c) ($V_{pp} = 2$ V) is slightly less than half of the retardance in Fig. 5(a) ($V_{pp} = 0$). We will use these results in the next section.

Spectral birefringent filter
One of the interesting uses of the spectral properties of retarders is their application to build birefringent filters [25], i.e., spectral filters based on the variations in the state of polarization for different wavelengths caused by the birefringence dispersion. They have become more interesting with the development of liquid-crystal technology since they can be tuned, and nowadays we can find commercial tunable spectral filters based on this technology [26, 27]. The successful realization of such filters depends critically on the correct characterization of the spectral retardance of the retarders used to compose the filter. Therefore, in order to confirm the validity of the previous results, this last section of the paper shows as an example the classical Lyot-Ohmann (LO) birefringent filter [28] made by combining two LCR retarders.

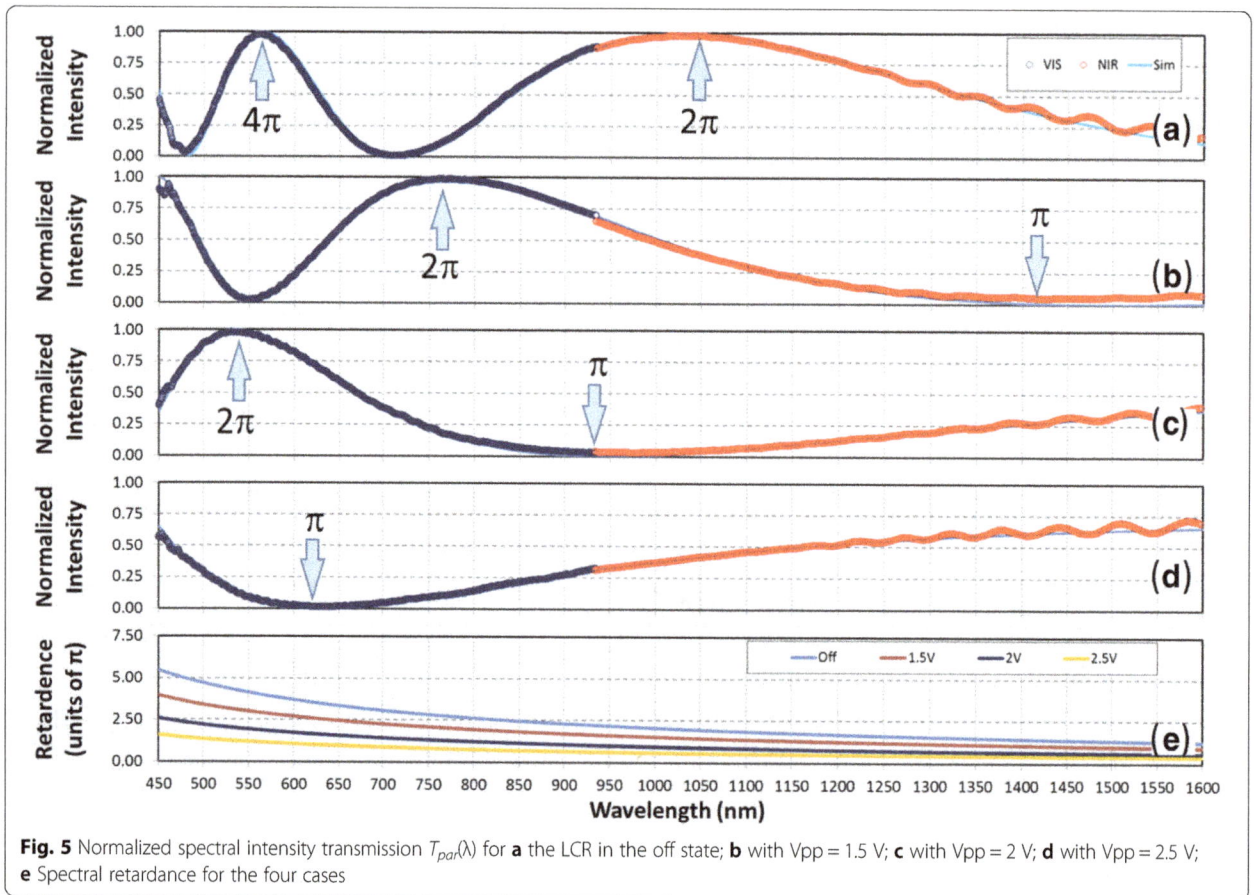

Fig. 5 Normalized spectral intensity transmission $T_{par}(\lambda)$ for **a** the LCR in the off state; **b** with Vpp = 1.5 V; **c** with Vpp = 2 V; **d** with Vpp = 2.5 V; **e** Spectral retardance for the four cases

The LO filter is generated by cascading various polarizer – retarder – polarizer subsystems, where the retarder is oriented at 45° relative to the parallel polarizers, and where the retardance in each consecutive subsystem doubles that of the previous subsystem. Each polarizer – retarder – polarizer subsystem generates an oscillatory spectrum, such as those we have presented in the previous sections. A subsystem with double retardance provides a spectrum with doubled oscillations. Therefore, cascading various subsystems generates a maximum transmission only at the wavelengths where all subsystems coincide to have maximum transmission. In order to properly generate the filter, it is important that the retarders are made with the same material, to ensure that all retarders show the same type of retardance dispersion.

We have generated a LO filter by using two LCR devices as that calibrated in Fig. 5. The system is therefore composed of a first polarizer, LCR1, a second polarizer, LCR2 and a third polarizer. The three polarizers are oriented at 45° to the vertical direction, while the LC director of the LCR devices is vertically oriented. The advantage of using LCR devices is that the LO filter can be tuned to different wavelengths [29, 30]. And if combined with other types of filters is able to provide narrowband multispectral tunable filters [31].

Figure 6 shows our experimental results for the LO filter with two stages (LCR1 and LCR2). In this case, since our LCR devices have (only approximately) the same thickness, we have to play with the applied voltage to reduce the retardance of one of them to become half the retardance of the other. In Fig. 6(a), LCR1 is left without applied voltage ($V_1 = 0$), so its transmission between polarizers is that in Fig. 5(a). The device LCR2 is then tuned to provide half the retardance, i.e., $\phi_{LCR1} = 2\phi_{LCR2}$.

This is achieved by applying a voltage $V_2 = 2.07$ volts, thus yielding a transmission between polarizers as shown in Fig. 6(b). We adjusted the maximum transmittance to be located at the wavelength of 565 nm, the same wavelength where there is a maximum in Fig. 6(a). Therefore, the combination of the two elements in cascade to generate a LO filter provides a single transmission band around 565 nm, as can be seen in Fig. 6(c). Note that the retarder with lower retardance (in this case LCR2) fixes the free spectral range of the filter (wavelength range between consecutive maxima). In this case, since the retardance of LCR2 must be reduced significantly, only one single maximum is observed in the entire wavelength range from 450 to 1600 nm, and the IR has been completely removed. This type of filter might be useful to highly remove the IR content and only transmit the visible range.

Conclusions

In summary, we have applied a classical spectral technique for measuring the retardance of linear retarders, but in a very wide spectral range from 450 to 1600 nm. For that purpose, we developed an optical system that uses two spectrometers, one for the VIS range and another for the NIR range. With this system we measured the spectral retardance function of different types of crystal retarders as well as of LCRs. The measured spectral content allows a very simple identification of the type of retarder according to its order (multiple, low or zero-order retarders). Also, the wavelength shifts of the oscillations observed in the spectral transmittance allows a simple identification of situations where the retardance increases or decreases, that can be useful with fixed retarders, and specially with variable LCRs.

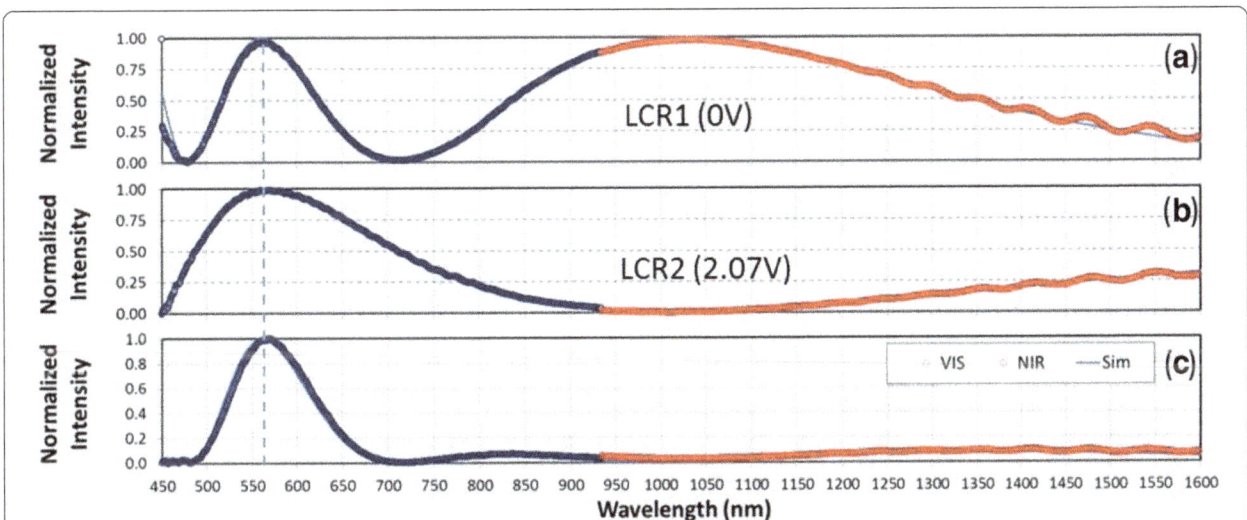

Fig. 6 Normalized spectral intensity transmission for the polarizer – LCR – polarizer system with: a LCR1 without applied voltage b LCR2 tuned to have half retardance than LCR1. c Normalized spectral intensity transmission of the Lyot-Ohmann filter made of the two previous systems showing a maximum transmission 565 nm

We have provided some useful tricks to be applied in the fitting procedure of the experimental data in order to derive an accurate spectral retardance function. For instance, the combination of two retarders made of the same material (thus having exactly the same spectral birefringence dispersion) helps to obtain additional curves by adding or subtracting retardance, as we showed in Fig. 3. Or the combination of a retarder under evaluation with a Fresnel rhomb retarder that adds/subtracts a constant shift of one quarter oscillation, which is useful to achieve an accurate measurement in all the spectral range (in opposition to a single measurement, which shows less accuracy at the maxima and minima of the spectral transmittance curve). Finally, we have confirmed the accuracy of the spectral measurements by demonstrating the realization of a classical birefringent Lyot-Ohmann filter. As a result of the correct calibration of the spectral retardance of the LCRs involved in the filter, the spectral transmittance was predicted with great accuracy.

Funding

This work received financial support from Ministerio de Economía y Competitividad and FEDER funds (grant ref.: FIS2015-66328-C3-3-R). A. Vargas acknowledges financial support from Fondecyt (grant ref.: 1151290).

Authors' contributions

All coauthors contributed to the paper. AM contributed with the realization of the optical system, taking the measurements data, and analyzing them. MMS-L contributed in the design of the experiments, the analysis of the results, and writing the manuscript. PG-M participated in the design of the experiments and in the discussion and analysis of the results. AV participated in the realization of the experimental system, and designed the procedure for taking the experimental data. Finally, IM contributed in the design of the experiments, the analysis of the results, and the preparation of the manuscript. All authors read and approved the final manuscript.

Competing interests

The authors declare that they have no competing interests.

Author details

[1]Departamento de Ciencia de Materiales, Óptica y Tecnología Electrónica, Universidad Miguel Hernández, 03202 Elche, Spain. [2]Departamento de Física y Arquitectura de Computadores, Instituto de Bioingeniería, Universidad Miguel Hernández, 03202 Elche, Spain. [3]Departament d'Òptica, Universitat de València, 46100 Burjassot, Spain. [4]Departamento de Ciencias Físicas, Universidad de La Frontera, Temuco, Chile.

References

1. Collett, E.: Field Guide to Polarization. SPIE Press, Bellingham (2005).
2. Zhang, Z., You, Z., Chu, D.: Fundamentals of phase-only liquid crystal on silicon (LCOS) devices. Light: Sci Appl **3**, e213 (2014)
3. Davis, C.C.: Lasers and Electro-Optics. Cambridge University Press, Cambridge (2002).
4. Velásquez, P., Sánchez-López, M.M., Moreno, I., Puerto, D., Mateos, F.: Interference birefringent filters fabricated with low cost commercial polymers. Am. J. Phys. **73**, 357–361 (2005)
5. Sánchez-López, M.M., Vargas, A., Cofré, A., Moreno, I., Campos, J.: Simple spectral technique to identify the ordinary and extraordinary axes of a liquid crystal retarder. Opt. Commun. **349**, 105–111 (2015)
6. Wu, S.T., Efron, U., Hess, L.D.: Birefringence measurements of liquid crystals. Appl. Opt. **23**, 3911–3915 (1984)
7. Vargas, A., Donoso, R., Ramírez, M., Carrión, J., Sánchez-López, M.M., Moreno, I.: Liquid crystal retarder spectral retardance characterization based on a Cauchy dispersion relation and a voltage transfer function. Opt. Rev. **20**, 378–384 (2013)
8. Nagib, N.N., Khodier, S.A., Sidki, H.M.: Retardation characteristics and birefringence of a multiple-order crystalline quartz plate. Opt. Laser Technol. **35**, 99–103 (2003)
9. Emam-Ismail, M.: Spectral variation of the birefringence, group birefringence and retardance of a gypsum plate measured using the interference of polarized light. Opt. Laser Technol. **41**, 615–621 (2009)
10. Wang, W.: Determining the retardation of a wave plate by using spectroscopic method. Opt. Commun. **285**, 4850–4855 (2012)
11. Safrani, A., Abdulhalim, I.: Spectropolarimetric method for optic axis, retardation, and birefringence dispersion measurement. Opt. Eng. **48**, 053601 (2009)
12. Abuleil, M.J., Abdulhalim, I.: Birefringence measurement using rotating analyzer approach and quadrature cross points. Appl. Opt. **53**, 2097–2104 (2014)
13. Vargas, A., Sánchez-López, M.M., García-Martínez, P., Arias, J., Moreno, I.: Highly accurate spectral retardance characterization of a liquid crystal retarder including Fabry-Perot interference effects. J. Appl. Phys. **115**, 033101 (2014)
14. Hetch, J.: City of Light: the story of Fiber Optics. Oxford University Press, Oxford (1999).
15. Sordillo, L.A., Pu, Y., Pratavieira, S., Budansky, Y., Alfano, R.R.: Deep optical imaging of tissue using the second and third near-infrared spectral windows. J. Biomed. Opt. **19**, 056004 (2014)
16. Barbarow, W.: A wave plate for every application. Photonics Spectra **43**(7), 54–55 (2009)
17. Hariharan, P.: Achromatic and apochromatic half wave and quarter wave retarders. Opt. Eng. **35**, 3335–3337 (1996)
18. Abuleil, M.J., Abdulhalim, I.: Tunable achromatic liquid crystal waveplates. Opt. Lett. **39**, 5487–5490 (2014)
19. Mawet, D., Hanot, C., Leanerts, C., Riaud, P., Defréfre, D., Vandormael, D., Loicq, J., Fleury, K., Plesseria, J.Y., Surdej, J., Habraken, S.: Fresnel rhombs as achromatic phase shifters for infrared nulling interferometry. Opt. Express **15**, 12850–12865 (2007)
20. Abdulhalim, I.: Dispersion relations for liquid crystals using the anisotropic Lorentz model with geometrical effects. Liq. Cryst. **33**, 1027–1041 (2006)
21. ArcOptix, Variable phase retarder: http://www.arcoptix.com/variable_phase_retarder.htm (Visited 2016, June 10th)
22. Wang, Z.P., Li, Q.B., Tan, Q., Huang, Z.J., Shi, J.H.: Novel method for measurement of retardance of a quarter-wave plate. Opt. Laser Technol. **36**, 285–290 (2004)
23. Wu, S.T.: Birefringence dispersions of liquid crystals. Phys. Rev. A **33**, 1270–1274 (1986)
24. Li, J., Wen, C.H., Gauza, S., Lu, R., Wu, S.T.: Refractive indices of liquid crystals for display applications. J. Displ. Technol. **1**, 51–61 (2005)
25. Wu, S.T.: Design of a liquid crystal based tunable electro-optic filter. Appl. Opt. **28**, 48–52 (1989)
26. Meadowlark Optics, Selectable Bandwidth Tunable Optical Filter: http://www.meadowlark.com/(Visited 2016, June 10th)
27. Thorlabs Inc., Liquid Crystal Tunable Filters: http://www.thorlabs.de/(Visited 2016, June 10th)
28. Yeh, P.: Some applications of anisotropic layered media, Ch. 10 in Optical Waves in Layered Media. John Wiley & Sons (2005).
29. Staromlynska, J., Rees, S.M., Gillyon, M.P.: High-performance tunable filter. Appl. Opt. **37**, 1081–1088 (1998). https://www.osapublishing.org/ao/abstract.cfm?uri=ao-37-6-1081.
30. Aharon, O., Abdulhalim, I.: Liquid crystal Lyot tunable filter with extended free spectral range. Opt. Express **17**, 11426–11433 (2009)
31. Abuleil, M., Abdulhalim, I.: Narrowband multispectral liquid crystal tunable filter. Opt. Lett. **41**, 1957–1960 (2016)

Focusing characteristics of a 4π parabolic mirror light-matter interface

Lucas Alber[1,2]* ⓘ[†], Martin Fischer[1,2†], Marianne Bader[1,2], Klaus Mantel[1], Markus Sondermann[1,2] and Gerd Leuchs[1,2,3]

Abstract

Background: Focusing with a 4π parabolic mirror allows for concentrating light from nearly the complete solid angle, whereas focusing with a single microscope objective limits the angle cone used for focusing to half solid angle at maximum. Increasing the solid angle by using deep parabolic mirrors comes at the cost of adding more complexity to the mirror's fabrication process and might introduce errors that reduce the focusing quality.

Methods: To determine these errors, we experimentally examine the focusing properties of a 4π parabolic mirror that was produced by single-point diamond turning. The properties are characterized with a single $^{174}Yb^+$ ion as a mobile point scatterer. The ion is trapped in a vacuum environment with a movable high optical access Paul trap.

Results: We demonstrate an effective focal spot size of 209 nm in lateral and 551 nm in axial direction. Such tight focusing allows us to build an efficient light-matter interface.

Conclusion: Our findings agree with numerical simulations incorporating a finite ion temperature and interferometrically measured wavefront aberrations induced by the parabolic mirror. We point at further technological improvements and discuss the general scope of applications of a 4π parabolic mirror.

Keywords: Atom-photon coupling, Free space, Quantum optics, Ion trapping, 4Pi parabolic mirror, 4Pi microscopy, Confocal microscopy

Background

Free space interaction between light and matter is incorporated as a key technology in many fields in modern science. The efficiency of interaction influences measurements and applications ranging from various kinds of fundamental research to industrial applications. New innovations and new types of high precision measurements can be triggered by improving the tools needed for a light-matter interface. To achieve high interaction probability with a focused light field in free space, an experimental scheme using parabolic mirrors for focusing onto single atoms has been developed in recent years [1–3]. This scheme relies on mode matching of the focused radiation to an electric dipole mode (cf. Ref. [4] and citations therein).

---

*Correspondence: lucas.alber@mpl.mpg.de
[†]Equal contributors
[1]Max-Planck-Institute for the Science of Light, Staudtstr. 2, 91058 Erlangen, Germany
[2]Department of Physics, Friedrich-Alexander University Erlangen-Nürnberg (FAU), Staudtstraße 7/B2, 91058 Erlangen, Germany
Full list of author information is available at the end of the article

Focusing in free-space experiments is usually done with state-of-the-art lens based imaging systems [5–8]. Single lenses, however, suffer from inherent drawbacks like dispersion induced chromatic aberrations, optical aberrations, and auto-fluorescence, respectively. Most of these limitations can be corrected to a high degree by precisely assembling several coated lenses in a lens-system, e.g. in a high numerical aperture (NA) objective. Although solving some problems, multi-lens-systems induce new problems such as short working distances, low transmission for parts of the optical spectrum, the need for immersion fluids, and high costs, respectively. Therefore, multi-lens systems are often application specific providing best performance only for the demands that are most important for the application.

Mirror based objectives are an alternative to lens-based systems and can overcome some of these problems. The improvement is based on a mirror's inherent property of being free from chromatic aberrations. The nearly wavelength independent behavior also leads to a homogeneous reflectivity for a large spectral window. Comparing the reflectivity of mirrors to the transmission of lens based objectives, mirrors can sometimes also surpass lens-based systems. But surprisingly, they are rarely used when high interaction efficiency is required. This lack in application may be due to the fact that reflecting imaging systems, like the Cassegrain reflector, cannot provide a high NA. A high NA is however needed for matching the emission pattern of a dipole, which spans over the entire solid angle. The limitation in NA consequently constitutes a limitation in the maximum achievable light-matter coupling efficiency.

High NA parabolic mirrors ($NA = 0.999$) have meanwhile been successfully applied as objectives in confocal microscopy [9, 10], demonstrating the potential for imaging applications. The parabolic mirror (PM) is a single optical element that, in theory, can cover nearly the complete 4π solid angle for tight focusing [11]. In this article we report on the detailed characterization of such a 4π parabolic mirror (4π-PM), in which we sample the focal intensity distribution with a single ^{174}Yb$^+$ ion, trapped in a stylus like movable Paul trap [12].

In contrast to our previous studies [13], we measure the response of the ion at a wavelength different to the one used for excitation. This approach is standard in fluorescence microscopy and has also been used in experiments with trapped ions [14]. It renders unnecessary a spatial separation of focused light and light scattered by the ion, thus lifting the limitation of focusing only from half solid angle as in Ref. [13]. However, we will find below that by using the solid angle provided by our 4π-PM the measured effective excitation point spread function (PSF) is worse when using the full mirror as compared to focusing from only half solid angle. As outlined below, this is not a general restriction but specific to the aberrations of the mirror used in our experiments. It is a challenge to determine the aberrations of such a deep parabolic mirror [15] and we discovered the full extent of these aberrations only when scanning the 3D field distribution with the single ion, revealing an error in the earlier interferometric measurements. Here, we present a reasonable agreement of the experiments with results of simulations incorporating a finite ion temperature and new interferometrically measured wavefront aberrations of the parabolic mirror itself.

Despite of these aberrations, the efficiency obtained here for coupling the focused light to the linear dipole transition of the ^{174}Yb$^+$ ion is better than reported previously [13], using the full mirror as well as focusing from half solid angle. As a further improvement in comparison to Ref. [13] we keep the excitation of the ion well below saturation making sure that the size of the ion's wave function stays approximately constant as much as possible. All in all, the ion constitutes a nanoscopic probe with well defined properties throughout the measurement range.

In the concluding discussion of this paper, the parabolic mirror is compared to other high NA focusing tools, especially to lens-based 4π microscopes. Its possible field of application is discussed and further improvements to the existing set-up are proposed.

Methods

Our main experimental intention is to focus light to a minimal spot size in all spatial directions simultaneously. The highest electric energy density that can be realized with any focusing optics is created by an electric dipole wave [16]. We therefore choose this type of spatial mode in our experiment. The electric dipole wave is created by first converting a linear polarized Gaussian beam into a radially polarized donut mode via a segmented half-wave plate (B-Halle) [17, 18]. Second, the radially polarized donut mode is focused with a parabolic mirror onto the trapped ion. This, in theory, enables us to convert approximately 91 % of the donut mode into a linear dipole mode [19]. The conversion efficiency is limited since the donut mode is only approximating the ideal spatial mode that is necessary to create a purely linear dipole mode [2, 20] by being focused with the parabolic mirror. The donut mode, however, yields the experimental advantage of being propagation invariant and comparably easy to generate.

Our focusing tool, the parabolic mirror, is made of diamond turned aluminum (Fraunhofer Institute for Applied Optics and Precision Engineering, Jena) with a reflectivity of 64 % for the incident mode at a wavelength of $\lambda_{exc} = 369.5$ nm. Its geometry has a focal length of 2.1 mm and an outer aperture of 20 mm in diameter. In total, the geometry covers 81 % of the complete solid angle. This fraction corresponds to 94 % of the solid angle that is relevant for coupling to a linear dipole oriented along the axis of symmetry [11, 19]. Furthermore, the mirror has three bores near its vertex: two bores with a diameter of 0.5 mm for dispensing neutral atoms and for illuminating the ion with additional laser beams, respectively; and one bore with a diameter of 1.5 mm for the ion trap itself. The ion trap is a Stylus-like Paul trap similar to [12] with high optical access. The trap is mounted on a movable xyz piezo translation stage (PIHera P-622K058, Physik Instrumente) that is used for measuring the effective excitation PSF. The effective excitation PSF is defined as the convolution of the focal intensity distribution with the spatial extent of the ion.

We measure the effective excitation PSF by probing the focal spot at different positions. In order to do so, we use the translation stage to scan the ion through the focal spot with an increment of 25 nm. At each position, the incoming dipole mode excites the $S_{1/2}$-$P_{1/2}$ transition that has a wavelength of $\lambda_{exc} = 369.5$ nm and a transition linewidth of $\Gamma/2\pi = 19.6$ MHz [21]. The relevant energy levels of

^{174}Yb$^+$ are shown in Fig. 1. We weakly drive this transition such that the probability for exciting the ion into the P$_{1/2}$ state is proportional to the electric energy density at any point in the focal area. During these measurements, the ion is Doppler cooled by the focused donut mode. Hence, the detuning of this mode relative to the S$_{1/2}$- P$_{1/2}$ transition and its power determine the temperature of the ion, see Appendix for further details.

Since we excite the ion from the complete solid angle that is covered by the 4π parabolic mirror and since almost all excitation light is reflected into the detection beam path by the parabolic mirror, we cannot directly detect the fluorescent response of the ion at the same wavelength. Instead, we detect photons at a wavelength of $\lambda_{det} = 297.1$ nm allowing us to independently focus and detect from nearly the complete solid angle. Photons at the detection wavelength λ_{det} are emitted during the spontaneous D[3/2]$_{1/2}$- S$_{1/2}$ decay [21]. The D[3/2]$_{1/2}$ level is populated when the ion spontaneously decays from the excited P$_{1/2}$ state into the metastable D$_{3/2}$ state (branching ratio $\beta = 0.5$ %, lifetime of 52 ms [22], see Table 1). From this state, we optically pump the ion into the D[3/2]$_{1/2}$ state by saturating the D$_{3/2}$- D[3/2]$_{1/2}$ transition with a strong laser field at a wavelength of 935.2 nm (DL-100, Toptica Photonics). The upper state of this transition decays to the S$_{1/2}$ ground state with a probability of 98 % [21]. The infrared laser is co-aligned with a second laser at the excitation wavelength λ_{exc} (TA-SHG pro, Toptica Photonics) and both are sent through the focus of the parabolic mirror via one of its backside bores (see Fig. 1). The second laser at the excitation wavelength λ_{exc} is used for ionization.

The emitted fluorescent photons at the detection wavelength λ_{det} are out-coupled from the excitation beam path via a dichroic mirror (FF310-Di01, Semrock) and

Table 1 Branching ratios and decay rates for the relevant transitions of ^{174}Yb$^+$ taken from [21, 22] and citations therein

Transition	Branching ratio	Decay rate [$\Gamma/2\pi$]
^{2}P$_{1/2}$- ^{2}S$_{1/2}$	99.5 %	19.6 MHz
^{2}P$_{1/2}$- ^{2}D$_{3/2}$	0.5 %	
^{3}D[3/2]$_{1/2}$- ^{2}D$_{3/2}$	1.8 %	4.2 MHz
^{3}D[3/2]$_{1/2}$- ^{2}S$_{1/2}$	98.2 %	
^{2}D$_{3/2}$- ^{2}S$_{1/2}$		3 Hz

two clean up filters (FF01-292/27-25, Semrock). Afterwards, we detect them with a photomultiplier tube in Geiger mode operation (MP-942, Perkin Elmer) that has a remaining underground/dark count rate of 10 - 20 cps. The overall detection efficiency η_{det} at the detection wavelength λ_{det} was measured via pulsed excitation and amounts to $\eta_{det} \approx 1.4$ % (see Appendix). The detection efficiency is needed for the determination of the coupling efficiency to the trapped ion.

Based on the atomic decay rate on the detected transition, the total photon emission rate would be approximately R = $\beta \frac{\Gamma}{2} \frac{1}{2}$ = 154 kcps for $S = 1$ (see Eq. 1). Taking into account the finite detection efficiency, we would expect to measure approximately $R_{det} = \eta_{det}$ 154 kcps = 2160 cps.

The coupling efficiency is measured by recording the detection count rate R_{det} as a function of the excitation power P_{exc}. Analyzing the four-level quantum master equation we find that both quantities are proportional to each other in the limit of strong repumping powers and saturation parameters $S \ll 1$. The latter condition is met by keeping $S \leq 0.1$ in our measurements. This also ensures that the spatial extent of the ion is approximately constant throughout the measurement, see Appendix.

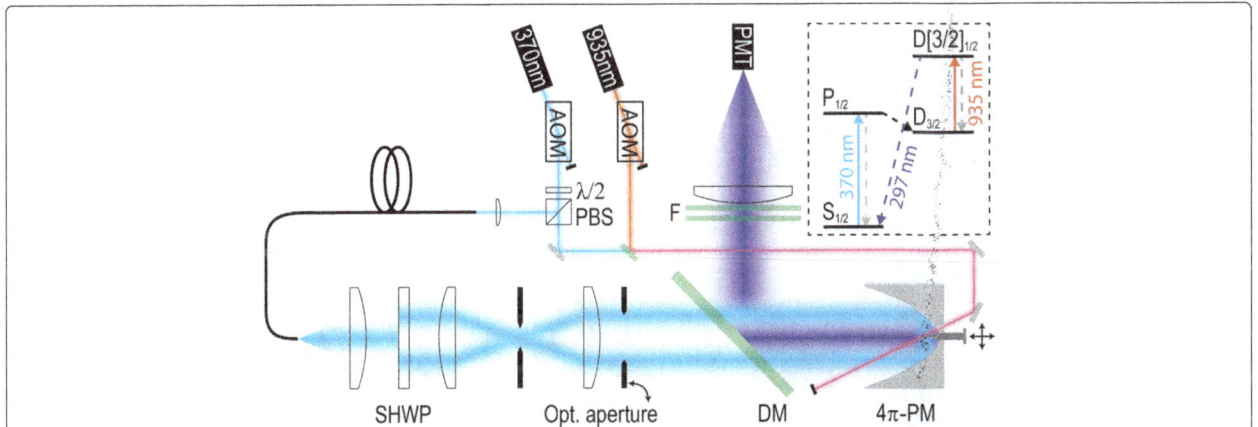

Fig. 1 Optical set-up of the experiment and relevant energy levels of ^{174}Yb$^+$. The cooling laser (*blue*) and the repump laser (*red*) are focused onto the ion through a hole at the backside of the parabolic mirror. AOM - acousto optical modulator, DM - dichroic mirror, F - clean up filter, 4π-PM - 4π parabolic mirror, PMT - photo multiplier tube, SHWP - segmented half-wave plate

The dependence of R_{det} for varying excitation power is given by

$$R_{det} = \eta_{det} \beta \frac{\Gamma}{2} \frac{S}{S+1} = \eta_{det} \beta \frac{\Gamma}{2} \frac{GP_{exc}/P_{sat}}{GP_{exc}/P_{sat}+1} \quad (1)$$

with S denoting the saturation parameter, and G the coupling efficiency, respectively. The saturation power P_{sat} is defined as $P_{sat} = 3 \cdot \frac{hc}{\lambda_{exc}} \frac{\Gamma}{8} (1 + 4(\Delta/\Gamma)^2)$. The factor 3 accounts for the fact that we are not driving a closed linear-dipole transition but a J=1/2 ↔ J=1/2 one. The relation between saturation parameter, saturation power and coupling efficiency is $S = GP_{exc}/P_{sat}$ [13]. Δ is the detuning of the excitation laser from the $S_{1/2}$- $P_{1/2}$ resonance. The formula for the detection count rate enables us to determine the coupling efficiency by curve fitting of our measured data for R_{det} as a function of P_{exc}. For the curve fitting, all parameters except the coupling efficiency are kept constant. During the measurement of the coupling efficiency, we position the ion exactly in the maximum of the excitation PSF, i.e. we measure the maximum coupling efficiency obtainable in the focal region under the current experimental conditions.

Results

The experimental results for the effective excitation PSF are shown in Fig. 2. We measure a spot size of 237 ± 10 nm

(FWHM) in the lateral direction (a, e, f). In the axial direction (b - d), however, the focal peak is broadened due to optical aberrations. The influence of the aberrations is reduced, when we limit the front aperture of the 4π-PM to half solid angle (Fig. 3). The reduced aperture results in a lateral width of 209 ± 20 nm and an axial width of 551 ± 27 nm. These values include the influence of the finite spatial extent of the trapped ion (see Appendix). In the Doppler limit, this extent is approximately 140 nm in lateral and 80 nm in axial direction considering the trap frequencies $\omega_{lateral}/2\pi \cong 490$ kHz and $\omega_{axial}/2\pi \cong 1025$ kHz, respectively, and a detuning from resonance of about 14.1 MHz.

To determine the minimal contribution of the ion-size to the focal broadening when Doppler cooling, we simulate the excitation PSF based on a generalization of the method presented in [23]. Our simulation also includes the aberrations of the parabolic mirror which were measured interferometrically beforehand [15]. The intensity distributions resulting from simulations only accounting for mirror aberrations are subsequently convolved with the spatial extent of the ion to achieve the effective excitation PSF.

The outcomes of our simulations are shown in Table 2. The effective PSF obtained in the simulations exhibits a good qualitative agreement with the PSF obtained in the experiment, cf. Figs. 2 and 3. Based on these results, the

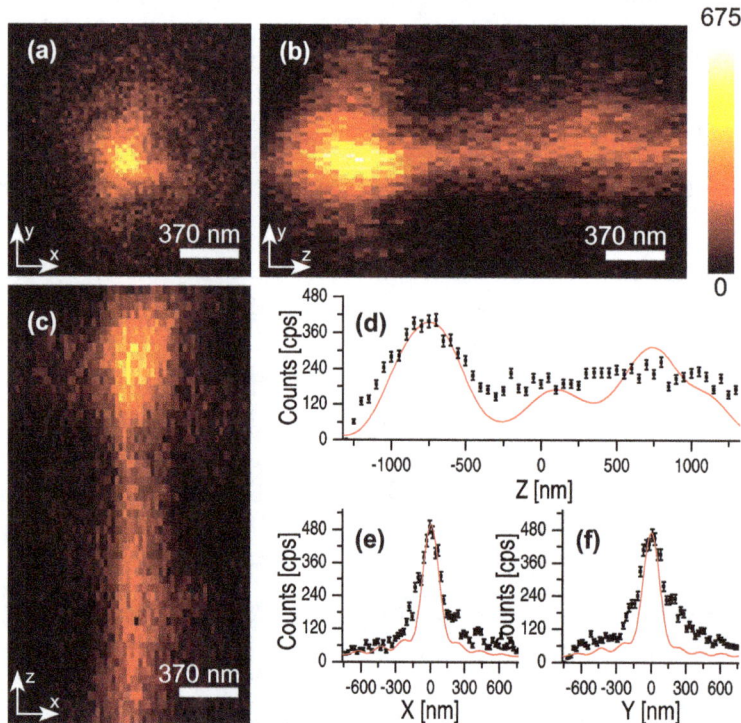

Fig. 2 Effective excitation PSF in the xy (**a**), zy (**b**), and xz (**c**) plane when illuminating the *full solid angle* covered by the 4π-PM. The corresponding line profiles through the center of the focus are shown in (**d–f**). They are overlayed to the line profiles resulting from numerical simulations (*red*)

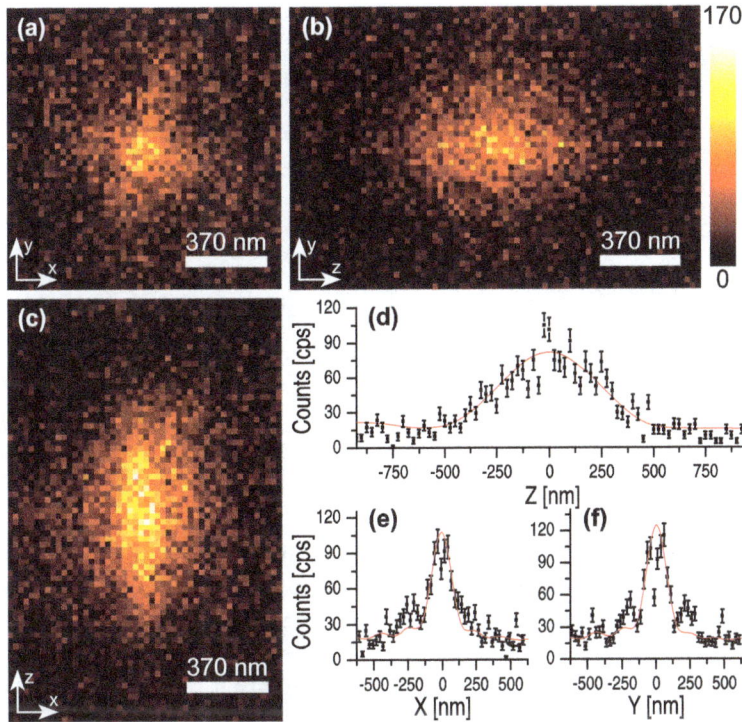

Fig. 3 Effective excitation *PSF* similar to Fig. 2 but for illuminating *half of the solid angle* of the 4π-PM. This is accomplished by using the optional aperture shown in Fig. 1. All panels shown in the figure have the same meaning as in Fig. 2

coupling efficiency G is expected (see Appendix) to be $G = 8.7\%$ for illuminating the full aperture and $G = 14.3\%$ for limiting the aperture of the 4π-PM to half solid angle. From the data shown in Fig. 4, we measure a coupling efficiency of $G = 8.6 \pm 0.9\%$ (full solid angle) and $G = 13.7 \pm 1.4\%$ (half solid angle). These values are close to the expected values from the simulations taking into account all current deficiencies of the set-up.

Discussion

Concentration of light into a narrow three dimensional volume is involved in many scientific applications. The scope ranges from applications that require "classical" light fields, like light microscopy, optical traps and material processing, to applications in quantum information science. In quantum information science, tight focusing of light is the key ingredient for free-space light-matter interfaces with a high coupling efficiency. This kind of free-space set-up may be an alternative to cavity based light-matter interfaces also providing high interaction strength. But in contrast to cavity assisted set-ups, free-space experiments often have a low level of instrument complexity and provide higher bandwidth. This is important considering the scalability and flexibility of an experimental set-up.

Technically, concentration of light is done by using focusing optics. How tight the focusing will be, depends on the numerical aperture of the focusing optics that

Table 2 Influence of experimental factors on the effective excitation *PSF* (eff. exc. *PSF*)

Simulation		Lateral		Axial	
		HSA	FSA	HSA	FSA
Exc. *PSF*	ideal mirror (i.m.)	139 nm	142 nm	412 nm	253 nm
	aberrated mirror (a.m.)	135 nm	135 nm	536 nm	
	i.m. with ion-extent	189 nm	192 nm	418 nm	266 nm
	a.m. with ion-extent	189 nm	189 nm	542 nm	
Eff. exc. *PSF* (measurement)		209 ± 20 nm	237 ± 10 nm	551 ± 27 nm	

The full width at half maximum is given for illuminating full solid angle (FSA) and half solid angle (HSA) of the 4π-PM, respectively. In case of illuminating the full solid angle and considering optical aberrations, no distinct peak can be identified along the axial direction

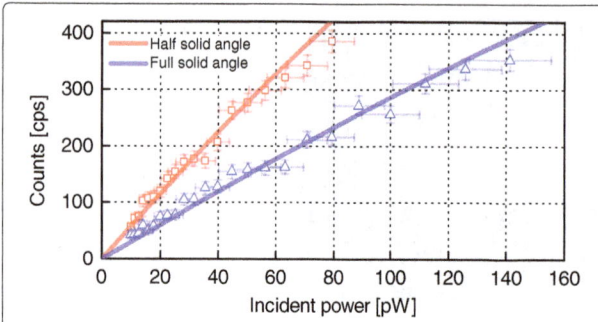

Fig. 4 Measurement of the ion's fluorescence vs. excitation power for determining the coupling efficiency (symbols). *Solid lines* show the results from curve fitting when illuminating half of the solid angle (*red*) and full solid angle (*blue*). To ensure constant coupling for the curve fitting, we only use data points for weak excitation ($R_{det} \leq 392$ cps equals $S < 1/10$, see Appendix) which corresponds to the nearly linear region of a saturation curve

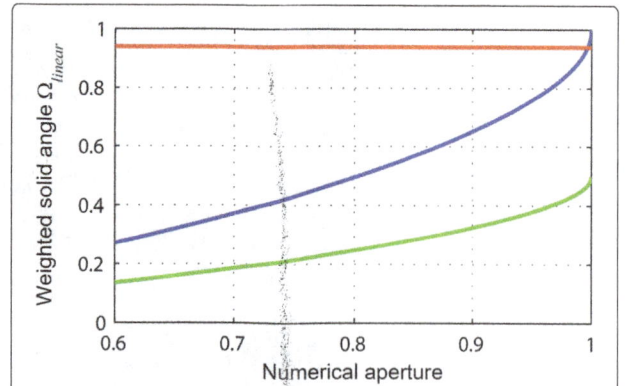

Fig. 5 Weighted solid angle Ω_{linear} covered by a single objective lens (*green*), by a lens based 4π-microscope (*blue*), and by the 4π-PM geometry used in the experiment (*red*)

is given by $NA = n \sin(\alpha)$, $NA \in [0, n]$, where α is the half-opening angle of the optics' aperture and n is the refractive index of the surrounding medium. For high NA objectives, the exact dependence of the focal volume as a function of NA can only be calculated numerically incorporating a vectorial treatment of the electric field. But as the numerical aperture increases, the focal volume will basically decrease.

For a 4π focusing optic, the numerical aperture is no longer defined. Nevertheless, it seems obvious that diffraction limited focusing from more than half of the hemisphere would produce a smaller three-dimensional focal volume. Consequently, a different quantity has to be used for comparing the performance of different 4π focusing optics. One suitable quantity is the weighted solid angle $\Omega \in [0, 1]$ (normalized to $8\pi/3$) [19]. It is the solid angle that is covered by the focusing optics weighted by the dipole's angular irradiance pattern. Ω defines the maximum fraction of incident power that can be coupled into the dipole mode of a single emitter. Consequently, it provides information about the ability to concentrate light since an electric dipole mode achieves the highest possible energy concentration [16]. $\Omega = 1$ therefore means, that all of the light is coupled into the dipole mode, assuming the ideal radiation pattern. The focusing capabilities of such a focusing optics can not be exceeded by any other optics. In Fig. 5, the maximal conversion efficiency into a linear dipole wave Ω_{linear} is compared for different (4π) focusing systems also including the 4π-PM geometry used in the experiment. In case of our 4π-PM, one has $\Omega_{linear} = 0.94$. The same fraction of the weighted solid angle can be covered using two opposing objective lenses each having a NA of 0.997 in vacuum. High quality objectives of such high numerical aperture are, however, not available.

In the experiment, the weighted solid angle covered by the focusing optics is one quantity that determines the

measured light-matter coupling efficiency. Other important experimental factors are the optical aberrations and the spatial extent of the ion, respectively. Under ideal conditions, we expect a coupling efficiency of $G = \eta^2 \cdot 94\% \approx 91\%$ where $\eta \approx 0.98$ is the field overlap with the ideal dipole mode [18]. But in our measurements, we are not able to reach this limit. Our numerical simulations imply, however, that we are currently not primarily limited by the covered solid angle, but by the spatial extent of the ion and the optical aberrations (see Table 2). The optical aberrations reduce the Strehl ratios for focusing from full solid angle and half solid angle to a different degree, see Appendix for details. Since the Strehl ratio is by far worse in the case of full solid angle illumination, the half solid angle focusing yields the better coupling efficiency. The measured coupling efficiency $G = 13.7 \pm 1.4\%$ is, however, approximately twice as high as measured with our setup in half solid angle configuration previously [13]. We conjecture that this is due to the fact that here we do not fit a full saturation curve but restrict the experiment to low saturation parameters, preventing an increase of the ion's spread in position space and the associated stronger averaging over the focal intensity distribution. This effect was not considered in Ref. [13]. Therefore, it is possible that the increase of the ion's extent at higher excitation powers has affected the saturation of the ion there, resulting in a larger saturation power and thus a seemingly smaller coupling efficiency. Furthermore, part of the improvement might be attributed to a better preparation of the incident beam.

Enhancing the optical properties of our focusing system and therefore also the coupling efficiency can be done by correcting for the aberrations over the full aperture. The predominant aberrations in our set-up are due to form deviations of the mirror from the ideal parabolic shape. A higher degree of form accuracy could be provided by including interferometric measurement techniques [15] into the mirror's production process. Alternatively, the aberrations can be corrected by preshaping the incident wavefront before it enters the parabolic mirror. Wavefront shaping techniques may rely on adaptive optical elements (e.g. liquid crystal display, deformable mirror) or a (gray tone) phase plate. The latter technique has already been tested for a 4π-PM of the same geometry as used here [18]. Involving a second optical element for wavefront correction in front of the parabolic mirror would only slightly add complexity to the system. If the corrective element is reflection and refraction based, like a continuous membrane deformable mirror is, the wavelength-independent character of the imaging system is retained.

The wavelength-independent character is the reason why the 4π-PM is intrinsically free of chromatic aberrations. This is beneficial for applications that require tight focusing not only for a monochromatic light source. Typical applications can be found in the field of microscopy: Confocal fluorescence microscopy [24] usually requires correction for the excitation and the detection wavelength; RESOLFT-type far-field Nanoscopy [25] in addition requires correction for the depletion beam. Further examples are two- and three-photon microscopy [26, 27] or Raman microscopy [28]. The variety of reflective materials allows to specialize the 4π-PM for a specific application, e.g. for high power applications, high sensitivity measurements or even for applications where wavelengths from the deep UV to the far IR are used at the same time. This ability may enable new illumination or imaging techniques that are not possible with today's technology.

Conclusion

We experimentally characterized a light-matter interface based on a 4π-PM. By limiting its aperture, we could demonstrate an effective excitation *PSF* having a lateral spot size of 209 ± 20 nm in vacuum. That corresponds to $0.57 \cdot \lambda_{exc}$. Using the full mirror we observed a strong splitting of the focal peak along the axial direction due to form deviations of the parabolic mirror. Measuring the induced aberrations interferometrically and including the results in numerical simulations yields values that are consistent with the experiment. In addition, we measured the light-matter coupling efficiency to be $G = 13.7 \pm 1.4\%$. This value is also in good agreement with our simulations. Our findings allow us to infer that we are currently limited by the aberrations of the parabolic mirror and the spatial extent of the ion.

We can surpass our current technical limitations by correcting the aberrations of the 4π-PM. This would even further reduce the focal spot size and increase the coupling efficiency. Ways for wavefront correction were given in the discussion section of this work. We also may apply

higher trap frequencies or ground state cooling techniques to reduce the spatial extent of the ion. Alternatively, we can trap doubly ionized Ytterbium in our experimental set-up [29]. When trapping a ^{174}Yb^{2+} ion we may not need to change the trapping or cooling techniques because the ion's spatial extent is smaller in the Doppler limit (higher charge, narrower transition linewidth). Furthermore, ^{174}Yb^{2+} provides a closed two-level transition that is desirable in many experiments on the fundamentals of light-matter interaction. This may be a path for realizing a set-up capable of reaching the ultimate limitations of focusing in free space [2, 11].

Appendix
Simulating the expected coupling efficiency G
We simulate the focal intensity distribution along the principal axes of our system (x, y, z) based on a generalization of the method presented in [23] including the aberrations of our parabolic mirror. The electric field, that the ion experiences is an average [30] over the electric field in the focus by the spread of the wavefunction of the ion due to its finite temperature. In order to calculate the expected effective PSF of the PM we convolve the simulated intensity distribution in the focus with the spatial extent of the ion along the corresponding trap axis. In order to deduce the coupling efficiency G, we compare the resulting intensity in the focus to the one of a perfectly focused linear dipole wave [18], with equal ingoing power, in all three directions. The product of these three ratios yields the expected G.

Influence of the spatial extent of the ion's wave function
For determining the spatial extent of the Doppler cooled ion, we assume it to be in a thermal state [31]. Hence the spatial extent in each dimension is described by a Gaussian shaped wave packet with width σ_i, $i \in (x, y, z)$. The width can by calculated from the ground state wave packet $\sigma_{i,0}$ by [31]

$$\sigma_i = \sqrt{2\bar{n}_i + 1}\,\sigma_{i,0},$$

where $\bar{n}_i$ denotes the mean phonon number of the harmonic oscillator in i-th direction and $\sigma_{i,0} = \sqrt{\hbar/(2\,m\,\omega_i)}$. m is the ion's mass and ω_i the trap frequency in direction i. The probability for finding the ion at x_i is hence described by

$$|\Psi_i(x_i)|^2 = \frac{1}{\sigma_i\sqrt{2\pi}}e^{-\frac{1}{2}(\frac{x_i}{\sigma_i})^2}.$$

The mean phonon numbers are calculated using the semiclassical rate equations approach [32, 33]. For the common case of cooling with laser beams impinging from small parts of the solid angle the average number of motional quanta along trap axis i can, in the Lamb-Dicke regime, be approximated to be

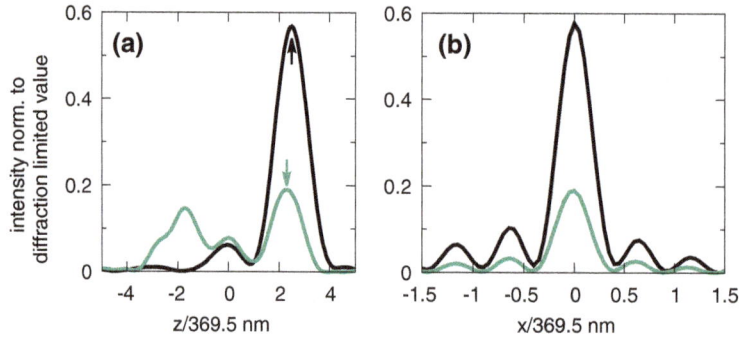

Fig. 6 Axial (**a**) and transverse (**b**) cuts through simulated focal intensity distributions. Green lines denote the case of focusing with the full mirror, i.e. covering 94 % of the solid angle relevant for a linear dipole. Blue lines represent the results for focusing from half solid angle as explained in the text. The transverse cuts in (**b**) are taken at the axial positions with maximum intensity as marked by the arrows in (**a**). The intensities are given relative to the ones obtained without any aberrations but at same solid angle. The spread of the ion's wave function is neglected in all simulations underlying this figure

$$\bar{n}_i = \frac{\alpha \rho_{22}(\Delta, S) + \cos^2 \vartheta \rho_{22}(\Delta - \omega_i, S)}{\cos^2 \vartheta (\rho_{22}(\Delta + \omega_i, S) - \rho_{22}(\Delta - \omega_i, S))}, \quad (2)$$

where $\rho_{22}(\Delta, S) = S/2 (1 + (2\Delta/\Gamma)^2 + S)$ is the upper level population with respect to detuning and saturation parameter, ϑ the angle between the i-th trap axis with the $\vec{k}$-vector of the laser beam, and α a factor depending on the emission pattern of the ion. For ^{174}Yb$^+$ ions the emission pattern is isotropic and therefore $\alpha = 1/3$ [32]. In the case of cooling the ion with a dipole wave, the cooling mode has a continuous spectrum of $\vec{k}$-vectors, hence the factor determining the overlap of the beam and the trap axis $\cos^2 \vartheta$ has to be averaged over the incoming field linear dipole field, that is has the form $\sin^2 \vartheta$. With the geometry of the trap and focusing employed in the setup described here, one of the trap axes is parallel to the optical axis of the mirror, while the other two are perpendicular. The mean overlap for these two cases, the averaged value of

$$\eta_{ax} = \iint d\vartheta \, d\varphi \, \frac{3}{8\pi} \sin^3 \vartheta \cos^2(\vartheta + \pi/2)$$
$$\eta_{rad} = \iint d\vartheta \, d\varphi \, \frac{3}{8\pi} \sin^3 \vartheta \cos^2(\vartheta + 0) \quad (3)$$

with the integration along polar angle φ and azimuthal angle ϑ. This leads to a mean overlap of $\eta_{ax} = 1/5$ of the linear dipole mode with the axial trap axis and $\eta_{rad} = 2/5$ with the radial ones. For negligible excitation, a detuning of $\Delta/2\pi = 14.2$ MHz and the trap frequencies $\omega_x/2\pi = 482.6$ kHz, $\omega_y/2\pi = 491.7$ kHz, and $\omega_z/2\pi = 1025$ kHz, respectively, this yields $\bar{n}_{x,y} \approx 20$, and $\bar{n}_z \approx 14$. The trap frequencies are determined by applying an AC signal to one of the compensation electrodes and scanning the applied frequency over the frequency range supposed to contain the trap frequencies while monitoring the the rate of fluorescence photons.

We only use excitation powers P_{exc} such that $S \leq 1/10$. This equals a detection count rate of $R_{det} \leq 392$ cps. For larger excitation powers, we expect the spatial extent of the ion to be comparable to the size of the focal intensity distribution and the coupling efficiency to be reduced. Since $\bar{n}$ grows linearly with S [34], keeping $S \leq 0.1$ ensures that the spatial extent of the ion is approximately constant over the whole measurement range. A change of 10 % in S corresponds to a change of approximately 5 % in G.

Interplay of mirror aberrations and focusing geometry

To clarify the reasons for the counter-intuitive finding of a larger coupling efficiency when focusing from only half solid angle, we discuss the influence of the mirror's aberrations in more detail. Here, focusing from half solid angle means that no light is incident onto the parabolic mirror for radial distances to the optical axis larger than twice the focal length of the parabola. Figure 6 shows the results of simulations of the focal intensity distributions when only accounting for the aberrations of the parabolic mirror but neglecting the trapped ion's wave function finite extent. For a parabolic mirror free of aberrations, i.e. a Strehl ratio of 1 no matter which portion of the mirror is used for focusing, one would expect coupling efficiencies of 90% when using the full mirror and 48% when focusing from half solid angle, assuming a mode overlap of $\eta \approx 0.98$ in both cases. But as is apparent from the data in Fig. 6, the Strehl ratio drops by a factor of 3 when focusing with the full mirror in comparison to using only half of the solid angle for focusing. Therefore, the expected increase in coupling efficiency by focusing with the full mirror is hindered by the large decrease in Strehl ratio. This particular, device-specific distribution of the aberrations results in a larger coupling efficiency for the half-solid-angle case, as it is found in our experiments.

Determination of the detection efficiency

The main contributions to the detection efficiency η_{det} are the reflectivity of the parabolic mirror for the detection mode, the quantum efficiency of the PMT, the covered solid angle of the parabolic mirror, the reflectivity of an additional beam splitter, the transmission of the dichroic mirror, and the transmission of the clean up filters, respectively. The reflectivity of the parabolic mirror and the quantum efficiency of the PMT are only known for a wavelength of 369.7 nm and amount to 67 % and 13 % [35], respectively. We expect this value to be lower for the detection wavelength of 297.1 nm. Together with the covered solid angle of the parabolic mirror of 81 % [35] and the polarization averaged design parameters for the beam splitter, dichroic mirror and the clean up filters of 43 %, the detection efficiency has to be lower than 3 %.

For a precise value, we additionally measure the detection efficiency via pulsed excitation. We pump the ion from the ground state into the metastable $D_{3/2}$ dark state by focusing a strong 30 μs long laser pulse at a wavelength of 369.5 nm through the backside hole of the parabolic mirror. After that, we drive the $D_{3/2}$- $D[3/2]_{1/2}$ transition with a strong laser pulse for about 30 μs to ensure that the $D[3/2]_{1/2}$- $S_{1/2}$ decay takes place. During this decay, a photon at the detection wavelength of 297.1 nm is emitted. While applying the infrared light, no UV light is driving the ion and only one detection photon can be emitted. For repeating the experiment with a pulse sequence rate of 10 kHz the background corrected count rate amounts to 142 cps. This yields the detection efficiency $\eta_{det} = 142/10000 = 1.4\%$.

Acknowledgements
The authors thank S. Heugel, B. Srivathsan, I. Harder and M. Weber for fruitful discussions and M. Weber for valuable comments on the manuscript.

Funding
G. L. acknowledges financial support from the European Research Council via the Advanced Grant 'PACART'.

Authors' contributions
LA, MF and MB conducted the experiments with a single ion. KM and MS performed the interferometric characterization of the parabolic mirror. LA and MS carried out the simulations. MF performed the calculations on the ion's motional state. MS and GL planned and supervised the experiments. LA and MF wrote the manuscript with contributions from all other authors. All authors read and approved the final manuscript.

Competing interests
The authors declare that they have no competing interests.

Author details
^{1}Max-Planck-Institute for the Science of Light, Staudtstr. 2, 91058 Erlangen, Germany. ^{2}Department of Physics, Friedrich-Alexander University Erlangen-Nürnberg (FAU), Staudtstraße 7/B2, 91058 Erlangen, Germany. ^{3}Department of Physics, University of Ottawa, 75 Laurier Avenue East, ON K1N 6N5 Ottawa, Canada.

References
1. Quabis, S, Dorn, R, Eberler, M, Glöckl, O, Leuchs, G: Focusing light to a tighter spot. Opt. Commun. **179**, 1–7 (2000). doi:10.1016/S0030-4018(99)00729-4
2. Sondermann, M, Maiwald, R, Konermann, H, Lindlein, N, Peschel, U, Leuchs, G: Design of a mode converter for efficient light-atom coupling in free space. Appl. Phys. B. **89**(4), 489–492 (2007). doi:10.1007/s00340-007-2859-4
3. Stobinska, M, Alber, G, Leuchs, G: Perfect excitation of a matter qubit by a single photon in free space. EPL (Europhys. Lett.) **86**(1), 14007 (2009)
4. Leuchs, G, Sondermann, M: Light-matter interaction in free space. J. Mod. Opt. **60**(1), 36–42 (2013). doi:10.1080/09500340.2012.716461. http://dx.doi.org/10.1080/09500340.2012.716461
5. Piro, N, Rohde, F, Schuck, C, Almendros, M, Huwer, J, Ghosh, J, Haase, A, Hennrich, M, Dubin, F, Eschner, J: Heralded single-photon absorption by a single atom. Nat. Phys. **7**(1), 17–20 (2011). doi:10.1038/nphys1805
6. Tey, MK, Maslennikov, G, Liew, TCH, Aljunid, SA, Huber, F, Chng, B, Chen, Z, Scarani, V, Kurtsiefer, C: Interfacing light and single atoms with a lens. New J. Phys. **11**(4), 043011 (2009)
7. Wrigge, G, Gerhardt, I, Hwang, J, Zumofen, G, Sandoghdar, V: Efficient coupling of photons to a single molecule and the observation of its resonance fluorescence. Nat. Phys. **4**(1), 60–66 (2008). doi:10.1038/nphys812
8. Pinotsi, D, Imamoglu, A: Single photon absorption by a single quantum emitter. Phys. Rev. Lett. **100**, 093603 (2008). doi:10.1103/PhysRevLett.100.093603
9. Drechsler, A, Lieb, M, Debus, C, Meixner, A, Tarrach, G: Confocal microscopy with a high numerical aperture parabolic mirror. Opt. Express. **9**(12), 637–644 (2001). doi:10.1364/OE.9.000637
10. Stadler, J, Stanciu, C, Stupperich, C, Meixner, A. J: Tighter focusing with a parabolic mirror. Opt. Lett. **33**(7), 681–683 (2008)
11. Lindlein, N, Maiwald, R, Konermann, H, Sondermann, M, Peschel, U, Leuchs, G: A new 4π geometry optimized for focusing on an atom with a dipole-like radiation pattern. Laser Phys. **17**(7), 927–934 (2007). doi:10.1134/S1054660X07070055
12. Maiwald, R, Leibfried, D, Britton, J, Bergquist, J. C, Leuchs, G, Wineland, DJ: Stylus ion trap for enhanced access and sensing. Nat. Phys. **5**(8), 551–554 (2009). doi:10.1038/nphys1311
13. Fischer, M, Bader, M, Maiwald, R, Golla, A, Sondermann, M, Leuchs, G: Efficient saturation of an ion in free space. Appl. Phys. B. **117**(3), 797–801 (2014). doi:10.1007/s00340-014-5817-y
14. Linke, NM, Allcock, DTC, Szwer, DJ, Ballance, CJ, Harty, TP, Janacek, HA, Stacey, DN, Steane, AM, Lucas, DM: Background-free detection of trapped ions. Appl. Phys. B. **107**(4), 1175–1180 (2012). doi:10.1007/s00340-011-4870-z
15. Leuchs, G, Mantel, K, Berger, A, Konermann, H, Sondermann, M, Peschel, U, Lindlein, N, Schwider, J: Interferometric null test of a deep parabolic reflector generating a hertzian dipole field. Appl. Opt. **47**(30), 5570 (2008). doi:10.1364/AO.47.005570
16. Bassett, IM: Limit to concentration by focusing. Optica Acta Intl. J. Opt. **33**(3), 279–286 (1986). doi:10.1080/713821943
17. Quabis, S, Dorn, R, Leuchs, G: Generation of a radially polarized doughnut mode of high quality. Appl. Phys. B. **81**(5), 597–600 (2005). doi:10.1007/s00340-005-1887-1
18. Golla, A, Chalopin, B, Bader, M, Harder, I, Mantel, K, Maiwald, R, Lindlein, N, Sondermann, M, Leuchs, G: Generation of a wave packet tailored to efficient free space excitation of a single atom. Eur. Phys. J. D. **66**(7), 190 (2012). doi:10.1140/epjd/e2012-30293-y. arXiv:1207.3215
19. Sondermann, M, Lindlein, N, Leuchs, G: Maximizing the electric field strength in the foci of high numerical aperture optics. ArXiv e-prints (2008). 0811.2098
20. Alber, G, Bernád, JZ, Stobińska, M, Sánchez-Soto, LL, Leuchs, G: Qed with a parabolic mirror. Phys. Rev. A. **88**, 023825 (2013). doi:10.1103/PhysRevA.88.023825. https://arxiv.org/abs/0811.2098
21. Meyer, HM, Steiner, M, Ratschbacher, L, Zipkes, C, Köhl, M: Laser spectroscopy and cooling of Yb$^+$ ions on a deep-UV transition. Phys. Rev. A. **85**(1), 012502 (2012). doi:10.1103/PhysRevA.85.012502
22. Olmschenk, S, Younge, KC, Moehring, DL, Matsukevich, D. N, Maunz, P, Monroe, C: Manipulation and detection of a trapped Yb$^+$ hyperfine qubit. Phys. Rev. A. **76**(5), 052314 (2007). doi:10.1103/PhysRevA.76.052314

23. Richards, B, Wolf, E: Electromagnetic diffraction in optical systems. II, structure of the image field in an aplanatic system. Proc. R. Soc. Lond. A Math. Phys. Eng. Sci. **253**(1274), 358–379 (1959). doi:10.1098/rspa.1959.0200

24. Sheppard, CJR, Choudhury, A: Image formation in the scanning microscope. Optica Acta Int. J. Opt. **24**(10), 1051–1073 (1977). doi:10.1080/713819421

25. Hell, SW: Far-field optical nanoscopy. Science. **316**(5828), 1153–1158 (2007). doi:10.1126/science.1137395

26. Horton, NG, Wang, K, Kobat, D, Clark, CG, Wise, FW, Schaffer, CB, Xu, C: In vivo three-photon microscopy of subcortical structures within an intact mouse brain. Nat. Photonics. **7**(3), 205–209 (2013). doi:10.1038/nphoton.2012.336

27. Denk, W, Strickler, JH, Webb, WW: Two-photon laser scanning fluorescence microscopy. Science. **248**(4951), 73–76 (1990). doi:10.1126/science.2321027

28. Andersen, ME, Muggli, RZ: Microscopical techniques in the use of the molecular optics laser examiner raman microprobe. Anal. Chem. **53**(12), 1772–1777 (1981). doi:10.1021/ac00235a013. http://dx.doi.org/10.1021/ac00235a013

29. Heugel, S, Fischer, M, Elman, V, Maiwald, R, Sondermann, M, Leuchs, G: Resonant photo-ionization of Yb^+ to Yb^{2+}. J. Phys. B Atomic Mol. Opt. Phys. **49**(1), 015002 (2016). doi:10.1088/0953-4075/49/1/015002

30. Tey, MK, Maslennikov, G, Liew, TCH, Aljunid, SA, Huber, F, Chng, B, Chen, Z, Scarani, V, Kurtsiefer, C: Interfacing light and single atoms with a lens. New J. Phys. **11**(4), 043011 (2009)

31. Eschner, J: Sub-wavelength resolution of optical fields probed by single trapped ions: Interference, phase modulation, and which-way information. Eur. Phys. J. D. **22**(3), 341–345 (2003). doi:10.1140/epjd/e2002-00235-7

32. Stenholm, S: The semiclassical theory of laser cooling. Rev. Mod. Phys. **58**, 699–739 (1986). doi:10.1103/RevModPhys.58.699

33. Eschner, J, Morigi, G, Schmidt-Kaler, F, Blatt, R: Laser cooling of trapped ions. J. Opt. Soc. Am. B. **20**(5), 1003–1015 (2003). doi:10.1364/JOSAB.20.001003

34. Chang, R, Hoendervanger, AL, Bouton, Q, Fang, Y, Klafka, T, Audo, K, Aspect, A, Westbrook, CI, Clément, D: Three-dimensional laser cooling at the doppler limit. Phys. Rev. A. **90**, 063407 (2014). doi:10.1103/PhysRevA.90.063407

35. Maiwald, R, Golla, A, Fischer, M, Bader, M, Heugel, S, Chalopin, B, Sondermann, M, Leuchs, G: Collecting more than half the fluorescence photons from a single ion. Phys. Rev. A. **86**(4), 043431 (2012). doi:10.1103/PhysRevA.86.043431

18

Wide-viewing integral imaging system using polarizers and light barriers array

Ying Yuan, Xiongxiong Wu, Xiaorui Wang[*] and Yan Zhang

Abstract

Background: Integral imaging is considered one of the most promising three-dimensional display technologies, while the limited viewing angle is regarded as a primary disadvantage of integral imaging display to reach a commercial level. This paper proposes a viewing angle enhancement method for both liquid crystal display (LCD) and the projection-type three-dimensional integral imaging system.

Methods: The proposed wide-viewing integral imaging system is established by using the lenslet array coupling with the polarizers, light barriers, and enlarged elemental images array. The size of light barrier and polarizer is equal to the pitch of each lenslet, the light barrier prevents light rays from passing through, and the polarizer controls the passage of light ray in the specific polarization direction. In the projection-type integral imaging system, two orthogonal elemental images arrays (EIA) are projected onto the projection screen simultaneously by the corresponding projectors. In LCD integral imaging system, two orthogonal EIAs are displayed by use of an LCD screen which can switch the polarization direction of the EIA by time-multiplexed technology within the time constant of the eyes' response time.

Results: The viewing angle can be enlarged dramatically by the improvement of the size of each elemental image according to the integral imaging principles. The experimental result shows that the proposed method exhibits approximately four times the viewing angle of conventional integral imaging with the same lens array.

Conclusions: The increment of viewing angle can be determined by the number of light barriers between two adjacently orthogonal polarizers, the more the light barrier, the larger the viewing angle.

Keywords: Integral imaging, Viewing angle, Polarizer, 3D Display, Light barrier

Background

Integral imaging, proposed by Lippman in 1908, is a promising three-dimensional (3D) technique for its full-parallax, continuous-viewing 3D images and without any special glasses [1–3]. Integral imaging uses lens array to collect and reproduce the light field of the three-dimensional scene, affording full color, full parallax and quasi-continuous viewing points within the viewing angle, working with incoherent light and can be viewed with the naked eye. Despite its many advantages, integral imaging suffers from the inherent drawbacks such as pseudoscopic effect [4], limited viewing angle [5–10], low viewing resolution [11–15] and small image depth [16, 17]. Focusing on the improvement method of field of view, the researchers have proposed a variety of solutions.

The viewing angle of integral imaging system is limited by the size of each elemental image and the distance between the lenslet and the EIA. In an integral imaging system, the lenslet array acts as a spatial beam splitter. Each lenslet unit is equivalent to a macro pixel so the lenslet unit is always not large. Most of the commercially available lenslet arrays are made of single-layer optical glass or optical plastic. They have no aberration correction capability and the monolayer lenslet array has a small viewing angle of about 30 degrees. To overcome the limitation of viewing angle, various methods have been proposed [5–10, 14, 15]. However, most of these algorithms have some limitations and their performances heavily depend on some specific conditions. The viewer tracking has a complex control system and modifying the configuration of the lens array can also extend the viewing angle, such as the fresnel lens array and negative index planoconcave lens array [18, 19]. Conventional

* Correspondence: xrwang@mail.xidian.edu.cn
School of Physics Optoelectronic Engineering, Xidian University, Xi'an 710071, China

fisheye lenses can map flat image onto wide field of view, but suppressing field curvature and their aberrations needs multi-layer components, and thus imposes harsh design tradeoffs.

In order to improve the viewing angle of the integral imaging system, this paper presents a method of placing the polarizers and light barriers array in front of the lenslet array. By jointly controlling LCD/projection screen, the lenslet array, the polarizer, and light barriers array, different polarized light ray can pass through corresponding lenslets. The proposed method can increase the size of the elemental images (EI), eliminate the crosstalk phenomenon of the light through the adjacent lenslets, and thus improve the viewing angle of the 3D display system.

Method
Viewing angle of the conventional integral imaging
A typical integral imaging system consists of two parts: the pickup process and the display process. In the pickup process, the EIs to record 3D object from different perspectives are captured by a camera with a lenslet array. In the display process, the captured EIs are back projected to reconstruct the 3D images either by optical or computational method. The basic configuration of the pickup process is that the recording medium has the same size as the lenslet array, as shown in Fig. 1.

The viewing angle of integral imaging is one of the key indicators of integral imaging system, which has a direct impact on the 3D viewing experience. The conventional integral imaging display system by refractive type lenslet array has a large aberration, the display area is small, and there is a jumping phenomenon at the edge of the viewing area. Since the lenslet array has multiple channels and the size of each lenslet is small, it is difficult to use multi-lens combination for image quality optimization which is often used in traditional optics. In conventional

integral imaging display, the pitch of the elemental lens is equal to the size of the corresponding elemental image, which is displayed by a flat-panel monitor directly. According to paraxial optics theorem, the viewing angle is limited by the elemental lens pitch and the gap between the lens array and the display device. Thus, the viewing angle of the integral imaging shown in Fig. 2 can be expressed as:

$$\theta = 2 \arctan \frac{p}{2g} \tag{1}$$

where p is the pitch of the elemental lens, and g is its focal length. According to Eq. (1), an easy way to improve the viewing angle is to enlarge the pitch of the elemental lens or shorten the focal length. However, enlarging the pitch of adjacent lenslets degrades the viewing resolution. Thus, the pitch of the elemental lenslet should be no larger than several millimeters to avoid observing the grid structure effect.

Wide-viewing integral imaging system
In order to improve the viewing angle of the integral imaging system, this paper presents a method of placing the polarizers and light barriers array in front of the lenslet array. By jointly controlling liquid crystal display, the polarizers and the light barriers array, different polarized light ray can pass through the corresponding lenslets. Theoretically, the size of the elemental image corresponding to each lenslet will be several times larger than conventional method. The proposed method, as shown in Fig. 3, can increase the size of the EIs, eliminate the crosstalk phenomenon of the light through the adjacent lenslets, and improve the viewing angle of the 3D display system. Figure 3(a) is the projection-type 3D integral imaging system and Fig. 3(b) is the LCD integral

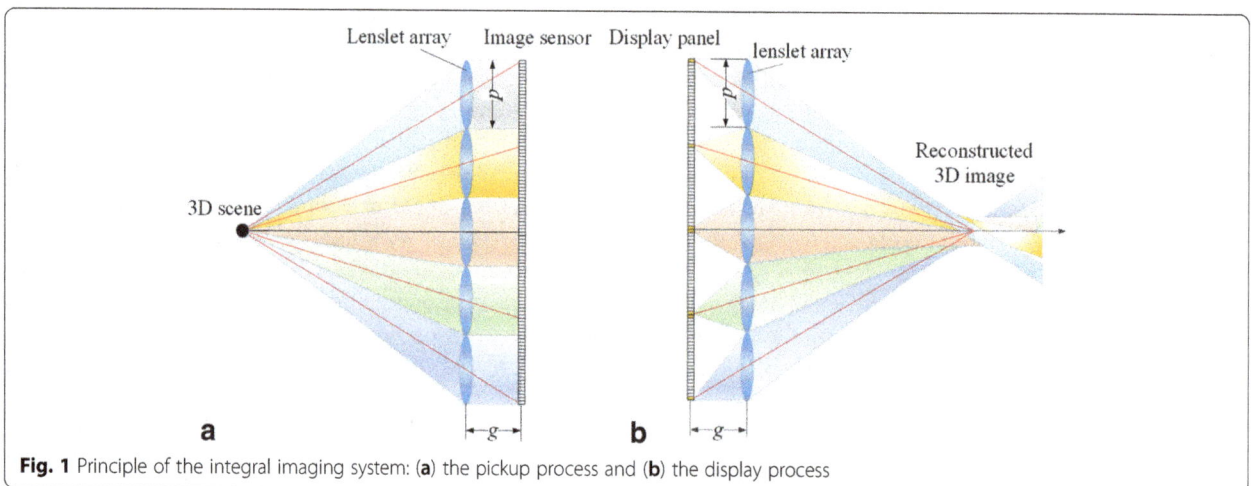

Fig. 1 Principle of the integral imaging system: (a) the pickup process and (b) the display process

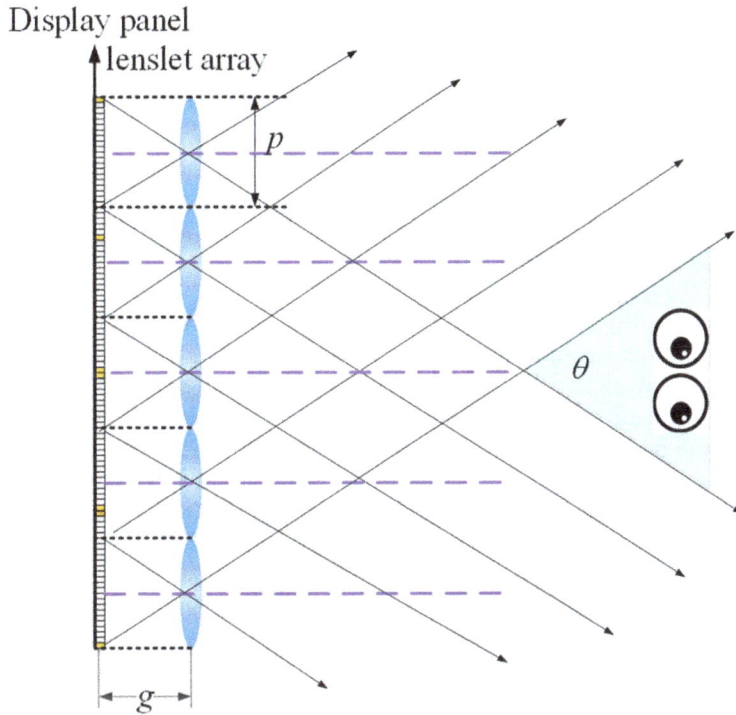

Fig. 2 Viewing angle of the conventional integral imaging

imaging system with the polarizers and light barriers array, respectively.

In Fig. 3(a), two orthogonal polarizers are placed in front of two projectors with the same parameters and then two EIAs with orthogonal polarization states are projected onto the projection screen by the corresponding projectors. The polarization direction of the EIA is the same as that of the polarizer in front of the projectors. The projection screen is a rear projection screen which is able to maintain the light polarization

characteristics. In addition, by adding a full light barrier between different polarizers before the lenslet array, we can increase the viewing angle further. The size of each light barrier is the same as that of each lenslet. The width of each EI is 4 times that of conventional integral imaging system, and its height remains unchanged. The polarizers and light barriers are placed in close proximity to the lenslet array and the polarization directions of adjacent polarizers are orthogonal. The light ray of the EIA is filtered through polarizers array and light barriers

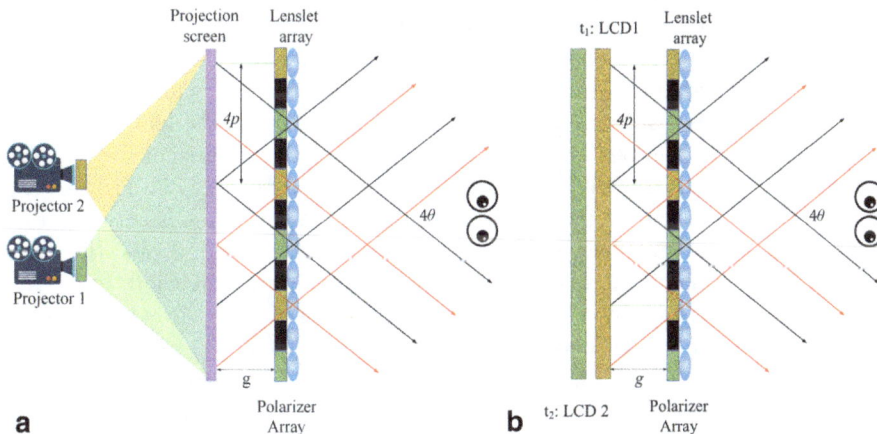

Fig. 3 Wide-viewing LCD/projection-type integral imaging system using the polarizers and light barriers array: (**a**) projection-type and (**b**) LCD integral imaging system

array and then reaches the lenslets array. The 3D reconstructed image can be obtained by the observer.

Figure 3(b) shows the variation of EIA with different polarization state on the polarization switchable LCD screen. At the time t1, the LCD screen display the EIA 1 which can pass through the yellow polarizers. At the time t2, the LCD screen display the EIA 2 which can pass through the green polarizers. The elemental images array for the corresponding mode is displayed in synchronization with the change of the polarization state. The different EIAs can be displayed repeatedly by time-multiplexed technology within the time constant of the eyes' response time. In practice, a polarization-switching device, such as the polarization shutter screen which is widely applied in the field of stereoscopy, electrically switchable polarization laser based on metasurface [20], and polarization switchable lens [21], can apply in the proposed wide-viewing integral imaging system for polarization-switching. If one light barrier is placed between the two orthogonal polarizers, the size of the elemental image should be magnified by a factor of 4 correspondingly.

The size of each light barrier and polarizer is equal to the pitch of lenslet, the light barrier prevents light ray from passing through, and the polarizer controls the passage of light ray in the specific polarization direction. By adjusting the number of full light barriers, you can increase the display viewing angle by any multiples. The more the light barrier, the lager the viewing angle is. There is the following relationship:

$$\theta = 2\arctan\frac{(N+1)p}{g} \tag{2}$$

where N represents the number of light barriers between two adjacent polarizers. Comparing the Eq. (1) and (2), it can be seen that the integral imaging display system based on the polarizers and light barriers array expands the display area of the EIs and increases the viewing angle. In both modes, the viewing angle which can be enlarged more than 2 times is the main difference between the proposed system and the previous methods.

One problem encountered with integral imaging is the pseudoscopic effect of the 3D reconstructed image when the captured elemental images used for display do not receive pre-processing. There is no direct relationship between the viewing angle and the pseudoscopic effect, and this paper cannot eliminate the pseudoscopic effect. Moreover, the imaging depth of integral imaging is different in real/virtual display mode and focused display mode, which is determined by the light wavelength, object distance, and the size of each lenslet [22]. Thus, the method proposed does not affect the imaging depth.

Hoshino et al. [23] define the viewing resolution of integral photography as the minimum value between the maximum viewing spatial frequency of elemental images and the Nyquist frequency determined by the distance between two adjacent lenslets q:

$$f_{max} = \min\left(f_{i\,max}, f_{nyq}\right) = \min\left(\alpha_{i\,max}\frac{z_i}{L-z_i}, \frac{L}{2q}\right) \tag{3}$$

where $f_{i\,max}$ is the maximum viewing spatial frequency of elemental images through the lenslet array, L is the distance from the lens array to the observer position, $\alpha_{i\,max}$ is the maximum projectable frequency of an elemental image through lenslet array, and z_i is the distance between the integral image and the observer; f_{nyq} is the Nyquist frequency. When the maximum viewing spatial frequency $f_{i\,max}$ is higher than f_{nyq}, the higher frequency component over the Nyquist frequency should be removed from the elemental image, and the viewing resolution of an integral image is determined by f_{nyq}. The viewing resolution will be decreased due to the increasement of the distance between two adjacent lenslets. Admittedly, more light barriers employed will cause the signal noise ratio to decrease and this is one of the main disadvantages of our proposed method.

Results and discussion

In this section, computational integral imaging reconstruction is used to verify the validity of the proposed method. The computational processing framework of integral imaging system using ray tracing are presented, which can generate EIs and reconstruct 3D images avoiding suffering from the diffraction and device limitations. Figure 4 illustrates the procedure of computational 3D integral imaging method, which can generate the enlarged EIAs according to different system parameters.

The 3D target object is a parrot positioned at depth of 8 cm. First, the planar EIA of a parrot is captured with our self-developed integral imaging pickup software. A lens array, 60×60, with 1.2 cm $\times$ 1.2 cm rectangular aperture is used in the pickup process. The focal length of the lenslet array is 1 cm and the pixel number of each elemental image is 75×75. Figure 5 shows the EIAs captured by the pickup system. In order to increase the viewing angle, we generate the EIAs with the same height and 4 times the width of conventional EIs, that is, the number of EIs is 15×60, the focal length of 1 cm, the number of pixels per elemental image is 300×75. Figure 5(a) is the EIA of the conventional integral imaging system, while Figs. 5(b) and (c) are the EIAs with different polarization states captured by the proposed integral imaging system. The viewing angle of traditional integral imaging system is 8.58° calculated by Eq. (1) and

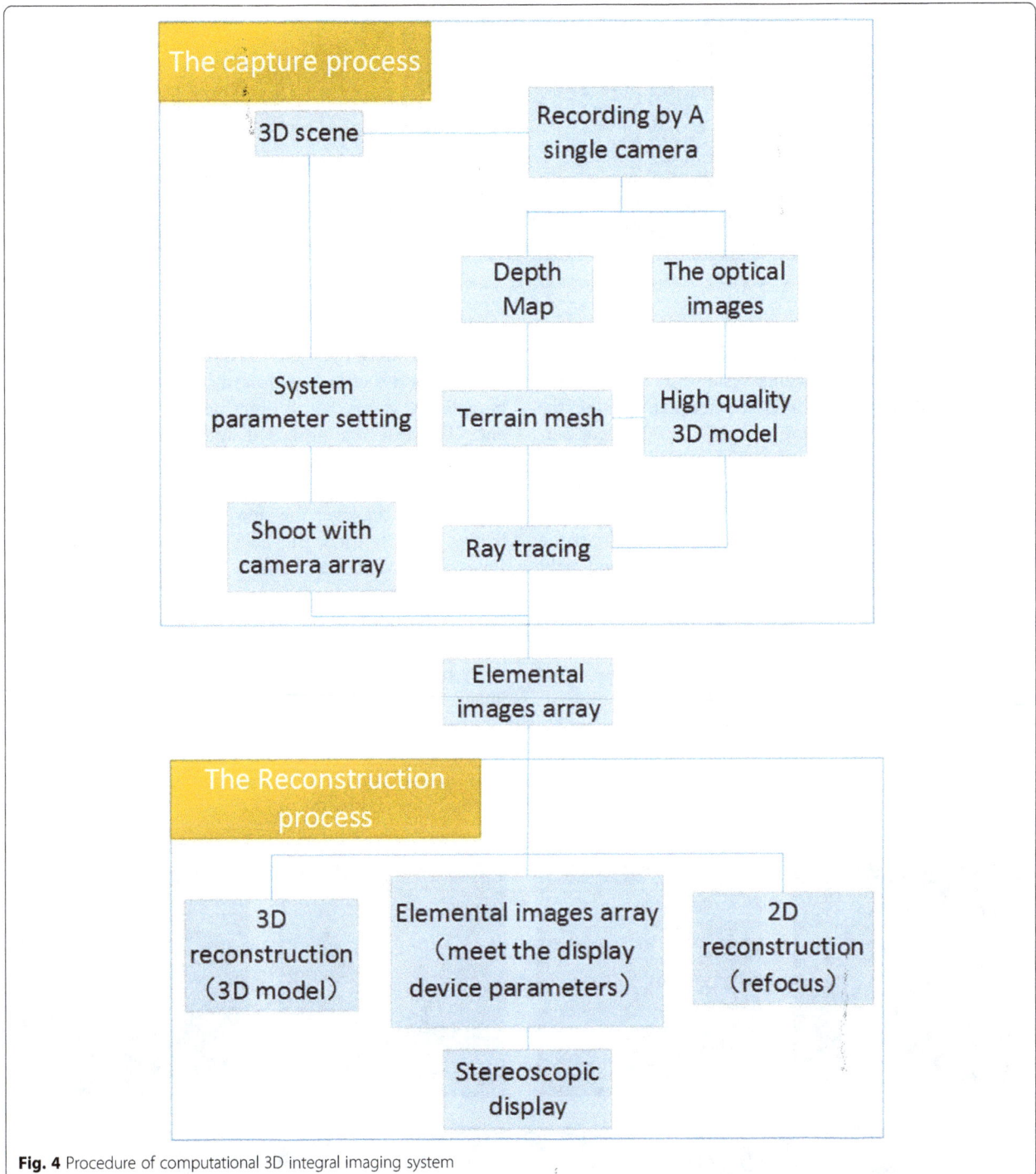

Fig. 4 Procedure of computational 3D integral imaging system

the viewing angle of the proposed integral imaging system with orthogonal polarizer array and light barriers is 33.40°.

Figure 6 shows the reconstructed display viewing angle using an integral imaging system with a polarizers and light barriers array. The observer observes the three-dimensional reconstructed image with respect to the center position (0°) of the central lens optical axis, the left side (angle is -), and the right side (angle +), respectively. It can be seen from the figure that the correct reconstructed image of the entire scene can be seen in the range of ±33°, while an error reconstructed image appears at ±34°. The wrong area is marked with a small yellow box. The perspectives can be observed

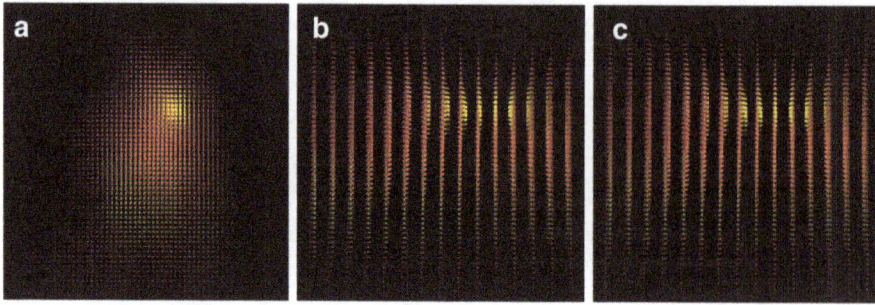

Fig. 5 (**a**) is the elemental images array of the conventional integral imaging system; (**b**) and (**c**) are the EIAs displayed in different polarization states

continuous within an angle of 8.58° by conventional computational integral imaging system. In contrast, the displayed 3D images on different viewpoints using the proposed method are shown in Fig. 6. The different perspectives of the images can be seen continuously up to 33°, which is in accordance with the theoretical result according to Eq. (2).

It can be seen in Fig. 6 that the viewing angle of the integral imaging system based on polarizers and light barriers array has been significantly improved. By comparing the different view images horizontally, different parallax information can be obtained. In the range of ±33°, you can get the correct reconstructed image of the whole scene. The proposed method is about approximately 4 times the viewing angle of the conventional integral imaging. The different wide-viewing integral three-dimensional imaging methods proposed by Refs. [9, 10] can improve the field angle to a certain extent,

but cannot more than 2 times the original viewing angle. In both modes of our manuscript, the viewing angle can be enlarged more than 2 times which is the main difference between the proposed system and the previous methods. The increment of viewing angle can be determined by the number of light barriers, the more the light barrier, the lager the viewing angle. When one light baffles array is used, the field angle will be enlarged 4 times as conventional method. The size of each elemental image can be enlarged as any positive integer multiples as the pitch of the lenslet. Thus, the viewing angle of the integral imaging system can be improved dramatically by the improvement of the size of each elemental image according to the integral imaging principles.

Conclusions

This paper proposes a viewing angle enhancement method for both the LCD and projection-type 3D

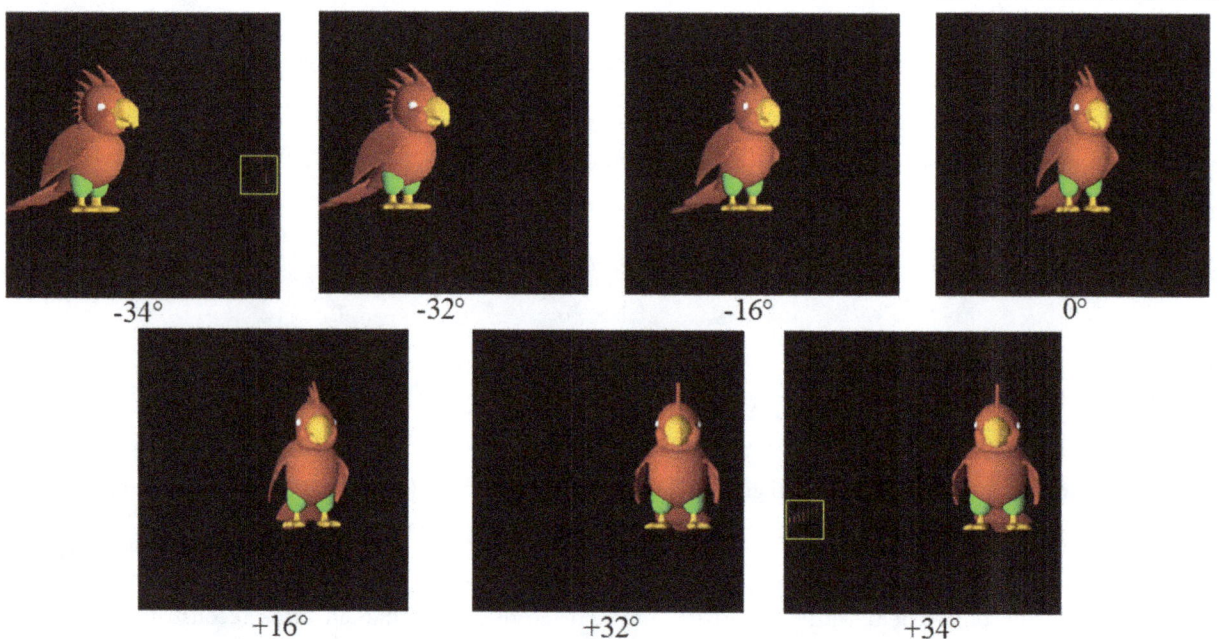

Fig. 6 Different viewpoints of the reconstruction image of the proposed 3D integral imaging system

integral imaging system. The proposed integral imaging system is established by using the lenslet array coupling with the polarizer array, light barrier, and the enlarged EIs. In projection-type integral imaging system, two orthogonal EIA are projected onto the projection screen simultaneously by the corresponding projectors. In LCD integral imaging system, two orthogonal EIA are displayed by the use of a LCD screen which can switch the polarization direction of the display images by time-multiplexed technology within the time constant of the eyes' response time. The increment of viewing angle can be determined by the number of light barriers, the more the light barriers, the lager the viewing angle. The size of each elemental image can be enlarged as any positive integer multiples. Thus, the viewing angle of the integral imaging system can be improved dramatically by the improvement of the size of each elemental image according to the integral imaging principles.

Abbreviations
3D: three-dimensional; EI: elemental images; EIA: elemental images array; LCD: liquid crystal display

Funding
A part of this study was supported by National Natural Science Foundation of China (NSFC) (61377007, 61575152).

Authors' contributions
YY and XRW engaged in the idea of the method. YY carried out the theoretical analysis and numerical simulation. The writing of the manuscript was done by YY and XXY. YZ participated in the discussion of experimental results. All authors read and approved the final manuscript.

Competing interests
The authors declare that they have no competing interests.

References
1. Hong, J., Kim, Y., Choi, H.J., Hahn, J., Park, J.P., Kim, H., Min, S.W., Chen, N., Lee, B.: Three-dimensional display technologies of recent interest: principles, status, and issues [invited]. Appl. Opt. **50**, H87–115 (2011)
2. Xiao, X., Javidi, B., Manual, M.C., Stern, A.: Advances in three-dimensional integral imaging: sensing, display, and applications [invited]. Appl. Opt. **52**, 546 (2013)
3. Park, J.H., Hong, K., Lee, B.: Recent progress in three-dimensional information processing based on integral imaging. Appl Opt. **48**, H77–H94 (2009)
4. Navarro, H., Martínez-Cuenca, R., Saavedra, G., Martínez-Corral, M., Javidi, B.: 3D Integral imaging display by smart pseudoscopic-to- orthoscopic conversion (SPOC). Opt. Express. **18**, 25573–25583 (2010)
5. Martínez-Cuenca, R., Navarro, H., Saavedra, G., Javidi, B., Martinez-Corral, M.: Enhanced viewing-angle integral imaging by multiple-axis telecentric relay system. Opt. Express. **15**, 16255–16260 (2007)
6. Park, G., Jung, J.H., Hong, K., Kim, Y., Kim, Y.H., Min, S.W., Lee, B.: Multi-viewer tracking integral imaging system and its viewing zone analysis. Opt. Express. **17**, 17895–17908 (2009)
7. Jang, J.S., Javidi, B.: Improvement of viewing angle in integral imaging by use of moving lenslet arrays with low fill factor. Appl. Opt. **42**, 1996–2002 (2003)
8. Jang, J.Y., Lee, H.S., Cha, S., Shin, S.H.: Viewing angle enhanced integral imaging display by using a high refractive index medium. Appl. Opt. **50**, B71–B76 (2011)
9. Jung, S., Park, J.H., Choi, H., Lee, B.: Viewing-angle-enhanced integral three-dimensional imaging along all directions without mechanical movement. Opt. Express. **11**, 1346–1356 (2003)
10. Jung, S., Park, J.H., Choi, H., Lee, B.: Wide-viewing integral three-dimensional imaging by use of orthogonal polarization switching. Appl. Opt. **42**, 2513–2520 (2003)
11. Jang, J.S., Javidi, B.: Three-dimensional integral imaging with electronically synthesized lenslet arrays. Opt. Lett. **27**, 1767–1769 (2002)
12. Kim, Y., Kim, J., Kang, J.M., Jung, J.H., Choi, H., Lee, B.: Point light source integral imaging with improved resolution and viewing angle by the use of electrically movable pinhole array. Opt. Express. **15**, 18253–18267 (2007)
13. Liao, H., Dohi, T., Iwahara, M.: Improved viewing resolution of integral videography by use of rotated prism sheets. Opt. Express. **15**, 4814–4822 (2007)
14. Navarro, H., Barreiro, J.C., Saavedra, G., Martínezcorral, M., Javidi, B.: High-resolution far-field integral-imaging camera by double snapshot. Opt. Express. **20**, 890–895 (2012)
15. Hyun, J.B., Hwang, D.C., Shin, D.H., Kim, E.S.: Curved computational integral imaging reconstruction technique for resolution-enhanced display of three-dimensional object images. Appl. Opt. **46**, 7697–7708 (2007)
16. Lee, B., Choi, H., Park, J.H., Kim, J., Cho, S.W., Kim, Y.: Depth-enhanced three-dimensional integral imaging by use of multilayered display devices. Appl. Opt. **45**, 4334–4343 (2006)
17. Shen, X., Wang, Y.J., Chen, H.S., Xiao, X., Lin, Y.H., Javidi, B.: Extended depth-of-focus 3D micro integral imaging display using a bifocal liquid crystal lens. Opt. Lett. **40**, 538 (2015)
18. Kim, H., Hahn, J., Lee, B.: The use of a negative index planoconcave lens array for wide-viewing angle integral imaging. Opt. Express. **16**, 21865–21880 (2008)
19. Min, S.W., Jung, S., Park, J.H., Lee, B.: Study for wide-viewing integral photography using an aspheric Fresnel-lens array. Opt. Eng. **41**, 2572–2576 (2002)
20. Xu, L., Chen, D., Curwen, C.A., Memarian, M., Reno, J.L., Itoh, T., Williams, B.S.: Metasurface quantum-cascade laser with electrically switchable polarization. Optica. **4**, 468–475 (2017)
21. Lee, Y.H., Peng, F., Wu, S.T.: Fast-response switchable lens for 3D and wearable displays, opt. Express. **24**, 1668 (2016)
22. Jang, J.S., Jin, F., Javidi, B.: Three-dimensional integral imaging with large depth of focus by use of real and virtual image fields. Opt. Lett. **28**, 1421–1423 (2003)
23. Hoshino, H., Okano, F., Isono, H., Yuyama, I.: Analysis of resolution limitation of integral photography. J. Opt. Soc. Am. A. **15**, 2059–2065 (1998)

19

Sparse kronecker pascal measurement matrices for compressive imaging

Yilin Jiang[*], Qi Tong, Haiyan Wang and Qingbo Ji

Abstract

Background: The construction of measurement matrix becomes a focus in compressed sensing (CS) theory. Although random matrices have been theoretically and practically shown to reconstruct signals, it is still necessary to study the more promising deterministic measurement matrix.

Methods: In this paper, a new method to construct a simple and efficient deterministic measurement matrix, sparse kronecker pascal (SKP) measurement matrix, is proposed, which is based on the kronecker product and the pascal matrix.

Results: Simulation results show that the reconstruction performance of the SKP measurement matrices is superior to that of the random Gaussian measurement matrices and random Bernoulli measurement matrices.

Conclusions: The SKP measurement matrix can be applied to reconstruct high-dimensional signals such as natural images. And the reconstruction performance of the SKP measurement matrix with a proper pascal matrix outperforms the random measurement matrices.

Keywords: Compressed sensing, Deterministic measurement matrix, Kronecker product, Pascal matrix

Background

Compressed sensing (CS) theory is a novel sampling scheme, which indicates that a sparse signal can be recovered from much fewer samples than conventional method [1, 2]. The sampling and the compression procedure are completed by the linear projection in CS. In matrix notation, it can be expressed as

$$y = \Phi x \tag{1}$$

where $x \in \mathbb{R}^N$ is the original signal, Φ is an $M \times N (M \ll N)$ measurement matrix, $y \in \mathbb{R}^M$ is the measurement vector. x is said to be K-sparse if $\|x\|_0 \leq K$. CS theory asserts that if the measurement matrix Φ satisfies some conditions, the signal x can be recovered from measurements y without distortion.

The emergence of CS provides a new inspiration for optical imaging. Actually most of the nature images are compressible in terms of some sparsity basis, such as Discrete cosine transform (DCT) and Discrete wavelet transform (DWT). The compressibility of the real-word images shows the potential for optical compressive imaging. In the past few years, CS technique has made great progress in many research fields, which include terahertz compressive imaging [3], spectral imaging [4], single pixel imaging [5] and infrared imaging [6]. Some optical imaging applications have been implemented in specific physical experiments.

Measurement matrix construction is a crucial problem in CS. The measurements obtained by measurement matrix are related to whether the signal can be accurately reconstructed. If there is enough information within the measurements, the signal can be recovered with high probability. Random measurement matrices are proved to have the merit of universality but suffer from several shortcomings. Firstly, random measurement matrices consume lots of storage resources. Secondly, there is no feasible algorithm to verify whether the random matrix satisfies the requirement as a measurement matrix [7, 8]. The research on deterministic sampling can be tracked back to the binary matrices via polynomials over finite field [9]. The deterministic measurement matrix has the superiority in physical implementation and the advantage of saving storage space. Therefore, many researches on

* Correspondence: jiangyilin@hrbeu.edu.cn
College of Information and Communication Engineering, Harbin Engineering University, Harbin 150001, China

the deterministic measurement matrix construction have been carried out. Lu introduced a construction of ternary matrices with small coherence [10]. Yao presented a novel simple and efficient measurement matrix named incoherence rotated chaotic matrix [11]. Huang proposed a symmetric Toeplitz measurement matrix [12]. Zhao introduced a deterministic complex measurement matrix to sample the signals in the single pixel imaging [13].

In this paper, we propose a new construction method of deterministic measurement matrix, termed sparse kronecker pascal (SKP) measurement matrix. The SKP measurement matrix combines the properties of the kronecker product and the pascal matrix. It is suitable for the reconstruction of natural images, which are usually high-dimensional signals. Simulations and analyses confirm that the SKP measurement matrices can reconstruct the natural images with a better performance.

Methods

The SKP measurement matrix construction

In mathematics, the kronecker product is an operation on two matrices of arbitrary size resulting in a block matrix [14].

Definition: If A is an $m \times n$ matrix, B is a $p \times q$ matrix, then the kronecker product $A \otimes B$ is the $mp \times nq$ block matrix. It can be expressed as

$$A \otimes B = \begin{bmatrix} a_{11}B \cdots a_{1n}B \\ \vdots \quad \ddots \quad \vdots \\ a_{m1}B \cdots a_{mn}B \end{bmatrix} \quad (2)$$

Pascal matrix is a symmetric positive definite matrix with integer entries taken from pascal's triangle [15]. The 4×4 truncations of these are shown below

$$P_4 = \begin{bmatrix} 1 & 1 & 1 & 1 \\ 1 & 2 & 3 & 4 \\ 1 & 3 & 6 & 10 \\ 1 & 4 & 10 & 20 \end{bmatrix} \quad (3)$$

it can be seen clearly that the entries near the diagonal of the pascal matrix increase with a geometric growth. It is effective to achieve sparse purpose by the kronecker product. Based on the pascal matrix and the kronecker product, we present the SKP matrix

$$H = k * I \otimes P = k \begin{bmatrix} P & 0 & \cdots & \cdots & 0 \\ 0 & \ddots & 0 & \cdots & \vdots \\ \vdots & 0 & P & 0 & \vdots \\ \vdots & \cdots & 0 & \ddots & 0 \\ 0 & \cdots & \cdots & 0 & P \end{bmatrix} \quad (4)$$

where H is the proposed SKP matrix, k signifies the scaling factor, I represents an identity matrix, P denotes the pascal matrix. Suppose that I is a $q \times q$ matrix, P is a $p \times p$ matrix, then H is a $pq \times pq$ matrix. We can get an $m \times n$ SKP measurement matrix Φ by selecting appropriately m rows from H for CS, here $n = p \times q$, m < n.

Now we describe how to select right rows from H to construct various dimensional measurement matrices. The selection method is to follow the principle of equal interval, which can improve the irrelevance between the selected row vectors. From the first row, we can construct the measurement matrix of multiple dimensions by choosing different interval lengths. If the interval length $d = 2, p = 4, q = 64$, the size of the SKP measurement matrix is 128×256. Similarly, when the interval length $d = 3$, the size becomes 86×256.

In CS, the measurement matrix must satisfy certain conditions. Candes and Tao propose a criterion named restricted isometry property (RIP) [16, 17]. A measurement matrix is said to satisfy the RIP of order K if there exists a constant $\delta_K \in (0, 1)$ such that

$$(1 - \delta_K)\|x\|_2^2 \leq \|\Phi x\|_2^2 \leq (1 + \delta_K)\|x\|_2^2 \quad (5)$$

for any K-sparse vector x. It is similar to that any K column vectors of the measurement matrix Φ are linearly independent. The RIP criterion guarantees that the sparse signal can be recovered exactly from the measurements.

The SKP measurement matrix is a particular matrix. The determinant of every P_n is 1 and the determinant of SKP matrix H is k_1 ($k_1 \neq 0$), which signify that any column vectors or row vectors from P_n and H are linearly independent. Therefore the SKP measurement matrix Φ is also a linear independent system between row vectors. The correlation among the resulting measurements is reduced, and the unique distribution of the SKP measurement matrix facilitates its implementation.

Results and discussion

In this part, we conduct numerical experiments to validate the performance of the SKP measurement matrix. The test images are of size 256×256 pixels. Orthogonal matching pursuit (OMP) algorithm is chosen as the recovery algorithm [18]. The sparsity basis Ψ is selected as the DCT matrix. Reconstruction processes are implemented in MATLAB R2016a. The size of the pascal matrix is 4×4, the scaling factor $k = 0.05$ and the identity matrix I is 64×64. Firstly the interval length is set to $d = 2$. We compare the reconstruction performance among the SKP measurement matrix, the random Gaussian measurement matrix and random Bernoulli measurement matrix. The quality of reconstructed images is measured by the peak signal-to-noise ratio (PSNR) in Eq. (7)

$$MSE = \frac{1}{N} \sum |x - x_{recons}|^2 \quad (6)$$

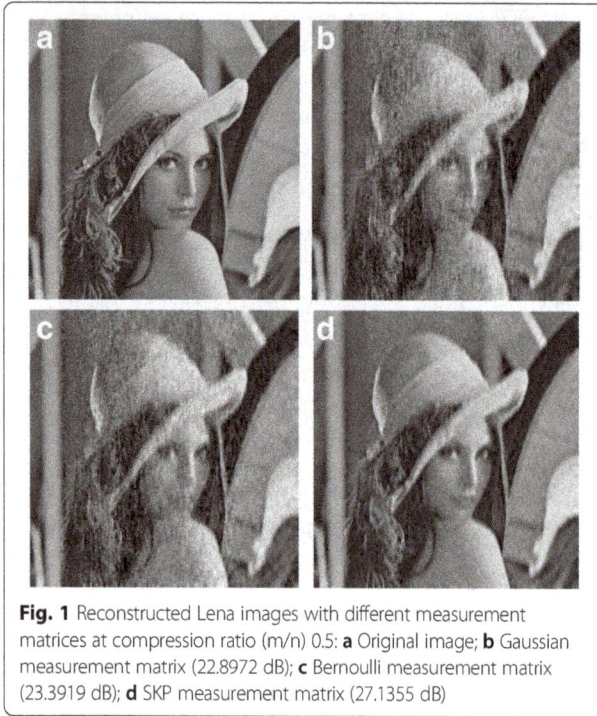

Fig. 1 Reconstructed Lena images with different measurement matrices at compression ratio (m/n) 0.5: **a** Original image; **b** Gaussian measurement matrix (22.8972 dB); **c** Bernoulli measurement matrix (23.3919 dB); **d** SKP measurement matrix (27.1355 dB)

$$PSNR = 20 \log \left(\frac{255}{\sqrt{MSE}} \right) \qquad (7)$$

Simulation results are shown in Figs. 1 and 2. It can be observed that the reconstructed images using the SKP

Fig. 2 Reconstructed Cameraman images with different measurement matrices at compression ratio (m/n) 0.5: **a** Original image; **b** Gaussian measurement matrix (20.1773 dB); **c** Bernoulli measurement matrix (20.0256 dB); **d** SKP measurement matrix (24.4332 dB)

Table 1 PSNR (in dB) values of reconstructed images under different experimental conditions

size	compression ratio 0.33			compression ratio 0.5		
	2	4	8	2	4	8
Lena						
Gaussian	20.2904	20.5879	20.2732	23.1280	23.1937	23.2459
Bernoulli	20.2536	20.6113	20.5530	22.8623	23.2264	23.1766
SKP	25.1923	23.8560	19.3820	28.3688	27.1355	21.9910
Cameraman						
Gaussian	17.7601	17.8477	17.8135	20.0125	20.3295	20.2115
Bernoulli	17.9861	17.9031	17.9366	20.0274	20.2517	20.6806
SKP	22.7272	21.9731	19.1054	25.5244	24.4332	19.6751

measurement matrix is the clearest among all the reconstructed images. The reconstructed images by the random Gaussian measurement matrices and random Bernoulli measurement matrices are blurry and lose some details compared to the SKP measurement matrix. In addition, the differences between the reconstructed images are also very obvious in terms of PSNR values. The PSNR values of reconstructed images by the SKP measurement matrix are almost 4 dB higher than that by the random measurement matrices. Figs. 1 and 2 demonstrate that the SKP measurement matrix outperforms the random Gaussian measurement matrices and random Bernoulli measurement matrices at the compression ratio of 0.5.

The further results present in Table 1. Table 1 shows more PSNR values of reconstructed images. And the measurement matrices include the random Gaussian measurement matrices, random Bernoulli measurement matrices and the SKP measurement matrices. In this part, the size of the pascal matrix is considered.

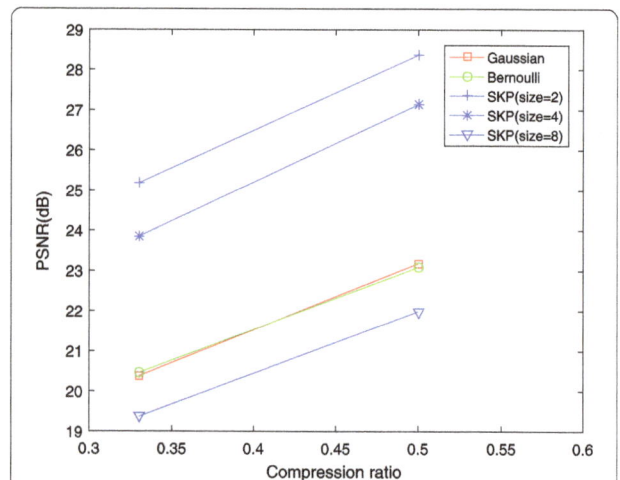

Fig. 3 The reconstruction accuracy of Lena image between each measurement matrix under different compression ratios

It can be seen from Table 1 when the size of the pascal matrix is 2 or 4, the reconstruction property of the SKP measurement matrices is better than that of the random Gaussian measurement matrices and random Bernoulli measurement matrices from PSNR values. When the size of the pascal matrix is 8, the reconstruction performance of the SKP measurement matrices has a serious decline or even less than the random Gaussian measurement matrices and random Bernoulli measurement matrices, which is caused by the further weakening of the orthogonality between row vectors of the SKP measurement matrices. Thus, the SKP measurement matrix construction needs to consider the influence of the pascal matrix dimension. The reconstruction accuracy of Lena image between each measurement matrix under different compression ratios is shown intuitively in Fig. 3.

Conclusions

In this paper, a new deterministic measurement matrix, SKP measurement matrix, is proposed for compressive imaging. The SKP measurement matrix has the advantages of simple structure, less storage space and convenient physical implementation, which offer great potential for compressive imaging applications. And we find that the size of the pascal matrix affects the reconstruction performance of the SKP measurement matrix. Simulation results demonstrate that the SKP measurement matrix with a proper pascal matrix can be used to effectively reconstruct the natural images and outperforms the random measurement matrices.

Abbreviations

CS: Compressed sensing; DCT: Discrete cosine transform; DWT: Discrete wavelet transform; OMP: Orthogonal matching pursuit; PSNR: Peak signal-to-noise ratio; RIP: Restricted isometry property; SKP: Sparse kronecker pascal

Funding

This work was supported by the National Natural Science Foundation of China (No.61571146), the Natural Science Foundation of Heilongjiang Province (No. F201407) and the Fundamental Research Funds for the Central Universities (HEUCF170802).

Authors' contributions

All the authors make contribution to this work. YJ and QT conceived the idea and wrote the manuscript; HW designed the experiments and analyzed the data; QJ revised the manuscript. All authors read and approved the final manuscript.

Competing interests

The authors declare that they have no competing interests.

References

1. Donoho, D.L.: Compressed sensing. IEEE Trans. Inf. Theory 52(4), 1289–1306 (2006)
2. Candes, E.J., Wakin, M.B.: An introduction to Compressive sampling. IEEE Signal Process. Mag. 25(2), 21–30 (2008)
3. Chan, W.L., Moravec, M.L., Baraniuk, R.G., et al.: Terahertz imaging with compressed sensing and phase retrieval. Opt. Lett. 33(9), 974–976 (2008)
4. Arguello, H., Arce, G.R.: Colored Coded Aperture Design by Concentration of Measure in Compressive Spectral Imaging. IEEE Trans. Image Process. 23(4), 1896–1908 (2014)
5. Duarte, M.F., Davenport, M.A., Takhar, D., et al.: Single-Pixel Imaging via Compressive Sampling. IEEE Signal Process. Mag. 25(2), 83–91 (2008)
6. Xiao, L.L., Liu, K, Han, D.P., et al.: A compressed sensing approach for enhancing infrared imaging resolution. Opt. Laser Technol. 44(8), 2354–2360 (2012)
7. DeVore, R.A.: Deterministic constructions of compressed sensing matrices. J. Complex 23(4), 918–925 (2007)
8. Li, S.X., Ge, G.N.: Deterministic construction of sparse sensing matrices via finite geometry. IEEE Trans. Signal Process. 62(11), 2850–2859 (2014)
9. Calderbank, R., Howard, S., Jafarpour, S.: Construction of a large class of deterministic sensing matrices that satisfy a statistical isometry property. IEEE J. Sel. Top Sign Process. 4(2), 358–374 (2010)
10. Lu, W.Z., Xia, S.T.: Construction of ternary matrices with small coherence for compressed sensing. Electron. Lett. 52(6), 447–448 (2016)
11. Yao, S., Wang, T., Shen, W., Pan, S., Chong, Y.: Research of incoherence rotated chaotic measurement matrix in compressed sensing. Multimedia Tools Appl. 1–19 (2015)
12. Huang, T., Fan, Y.Z., Zhu, M.: Symmetric Toeplitz-Structured Compressed Sensing Matrices. Sens. Imaging. 16(1), 1–9 (2015)
13. Zhao, M., Liu, J., Chen, S., Kang, C., Xu, W.: Single-pixel imaging with deterministic complex-valued sensing matrices. J. Eur. Opt. Soc. 10, 15041 (2015)
14. Loan, C.F.V.: The ubiquitous Kronecker product. J. Comput. Appl. Math. 123(1-2), 85–100 (2000)
15. Edelman, A., Strang, G.: Pascal Matrices. Am Math Mon 111(3), 189–197 (2004)
16. Candes, E.J., Tao, T.: Decoding by linear programming. IEEE Trans. Inf. Theory 51(12), 4203–4215 (2005)
17. Candes, E.J.: The restricted isometry property and its implications for compressed sensing. Comp. Rendus Math. Acad. Sci. Paris. 346(9–10), 589–592 (2008)
18. Tropp, J.A., Gilbert, A.C.: Signal Recovery From Random Measurements Via Orthogonal Matching Pursuit. IEEE Trans. Inf. Theory 53(12), 4655–4666 (2007)

20

PN-PAM scheme for short range optical transmission over SI-POF — an alternative to Discrete Multi-Tone (DMT) scheme

Linning Peng[1], Ming Liu[2*], Maryline Hélard[3] and Sylvain Haese[3]

Abstract

Background: How to deal with the time-dispersive channel is the main challenge faced by the short-range optical communication systems. In this work a novel pseudo-noise sequence (PN) assisted pulse-amplitude modulated (PN-PAM) transmission scheme for short-range optical communication systems is proposed in this work. With the help of the PN based channel estimation, minimum-phase pre-filter and reduced-state sequence estimation based equalizer, the proposed PAM transmission scheme can significantly reduce the training overhead for channel estimation in the classical PAM systems using decision feedback equalizer (DFE). In addition, the proposed PAM transmission scheme can effectively avoid the error propagation phenomenon in the classical DFE.

Results: Theoretical study shows that the proposed PAM scheme can achieve a 1.5 dB SNR gain with PAM-8 modulation over 50 m step-index polymer optical fiber (SI-POF) channel at a desired BER level of 1×10^{-3} and 1.2 Gbps transmission rate. Furthermore the hardware experiment using commercially available components proves the improvements in the proposed PAM transmission scheme.

Conclusion: The novel PAM transmission scheme is compared to the optimized discrete multi-tone (DMT) transmission with bit-loading. The experimental results show that for a transmission distance less than 50 m over SI-POF, DMT systems outperform the PAM systems. However, for a SI-POF transmission over the distance longer than 50 m, the proposed scheme can reach a better performance than the DMT systems thanks to the advantage of a lower peak-to-average-power ratio.

Keywords: PAM, Single-carrier modulation, DMT, Multi-carrier modulation, POF, PN sequence, Channel estimation, Minimum-phase pre-filter, Reduced-state sequence estimation

Background

Recently, high-speed transmission over plastic optical fiber (POF) has attracted many research interests [1, 2]. POF owns the advantages of easier connection and band insensitive, which could be an economic solution for home networking. However, the available transmission bandwidth in POF is limited due to the significant modal dispersion in large core diameter POF. The intensive modal dispersion could be modeled as multi-path channel delay which causes low-pass channel frequency response [1, 2]. In addition, for different types of short range optical

communication systems, such as single-mode fiber (SMF), multi-mode fiber (MMF), and optical wireless with visible light communications (VLC), there have similar channel characteristics as POF's, which can be modeled as a low-pass frequency response with very high signal-to-noise ratio (SNR) [3–6]. Therefore, the increasing traffic demands require better bandwidth utilization in the existing short range systems. In the state-of-the-art short range optical transmission systems, the used transmission bandwidth is far wider than the system's 3 dB bandwidth [3–6]. In order to explore the best transmission performance in these situations, several advanced modulation schemes have been investigated in recent works [7–9].

Multi-carrier modulation (MCM) schemes are widely used for the short range optical transmission systems. The

*Correspondence: mingliu@bjtu.edu.cn
[2]School of Computer and Information Technology, Beijing Jiaotao University, No.3 ShangYuanCun, 100044 Beijing, China
Full list of author information is available at the end of the article

discrete multi-tone (DMT) with bit-loading technique can effectively approach the channel capacity. However, MCM schemes have the drawbacks of high peak-to-average-power ratio (PAPR) and computational complexity [9, 10]. Compared to the MCM schemes, single carrier modulation (SCM) schemes enjoy the advantages of computational simplicity and low PAPR. The Not-Return-to-Zero (NRZ) coding with equalization offers satisfactory achievable link power budget margin [8, 9]. The pulse-amplitude modulation (PAM) is another SCM scheme that provides better spectrum efficiency than NRZ coding. As the short range optical transmission systems usually have high SNR, the PAM transmission scheme shows notable advantages over the classical NRZ transmission scheme [8, 11].

However, sophisticated equalization techniques are needed for most SCM schemes in order to restore the distorted SCM signal after the transmission over the bandwidth-limited channels [8]. An equalizer is commonly adopted at the receiver side to equalize the received signal. In addition, the decision-feedback equalizer (DFE) is proved to outperform the feed-forward equalizer (FFE) [12]. Moreover, in case of adaptive equalization, long training sequences are required for good convergence, which consequently reduces the overall system efficiency. Moreover, DFE has the drawback of the error propagation which reduces the system performance in the practical implementations. Although some works show that the Tomlinson-Harashima precoding technique can resolve the error propagation problems in DFE [13], it requires a prior information at transmitter side and increases the overall system complexity [13].

Concerning practical PAM transmissions over SI-POF, [14] and [15] summarized available transmission rates with different SCM schemes, such as 1 Gbps over 20 m SI-POF, 400 Mbps over 50 m SI-POF, and 170 Mbps over 115 m SI-POF in [15] and 10 Mbps over 425 m SI-POF, 100 Mbps over 275 m SI-POF and 1 Gbps over 75 m SI-POF in [14]. Additionally, recent works reported higher transmission rate over 50 m SI-POF with new designed front-end receivers [16, 17]. In this work, we focus on a system architecture research of increasing SI-POF transmission rate with existing off-the-shelf components. In order to improve the transmission efficiency of the PAM transmission and avoid the error propagation problem in DFE, we propose a novel pseudo-noise sequence assisted pulse-amplitude modulated (PN-PAM) transmission scheme for short range optical communication systems. This novel scheme is a SCM in nature and is incorporated with a minimum phase pre-filter and a simplified trellis-based equalizer at the receiver side. The coefficients of minimum phase pre-filter could be obtained from a pseudo noise (PN) sequence-based channel estimation with very short training symbol length.

A comparison between the capacities of the novel PAM scheme and the existing DMT scheme over SI-POF is carried out with the same experimental setups. It is shown that this proposed PAM transmission scheme is suitable for other short range optical communication systems due to its advantage of low PAPR.

The remainder of this paper is organized as follows: In "Methods" section, the method of PN-PAM transmission scheme is introduced. In "Theoretical analysis of the PN-PAM transmission" section, a theoretical analysis for PAM transmission over short range step index-POF (SI-POF) system which is modeled as Gaussian low-pass filter. The PAPR of PAM transmission with root-raised-cosine (RRC) filters is also evaluated. In "Numerical analysis of the PN-PAM transmission" section, the proposed PAM transmission scheme is compared with the classical one via simulations. In "Results and discussion" section, a real SI-POF transmission system with commercially available components is used to further prove the merits of the proposed scheme. The comparisons between the PAM and DMT transmissions over practical channels are also presented. Finally, conclusions are highlighted in "Conclusion" section.

Methods

In optical fiber communication systems, PAM transmission with adaptive DFE has been widely used. However, as the optical fiber channels are quite stable, it is possible to obtain the optimal DFE coefficients with a reduced transmission overhead. Furthermore, as DFE has the drawback of the error propagation, it is necessary to introduce a novel decision mechanism to replace the direct decision in DFE. In this section we propose a novel PAM transmission scheme for optical fiber communications. To achieve satisfactory performance, the reception of this novel scheme includes three main components: the PN sequence based channel estimator, a minimum-phase receiver filter, and a reduced-state sequence estimation(RSSE) based equalizer. More details of the proposed scheme is presented in following part of the section.

PN sequence based channel estimation

The SCM based transmissions enjoy the advantage of low PAPR over MCM ones and are therefore attractive for optical fiber transmissions. In the meantime they normally require high channel estimation accuracy in order to prevent the error propagation phenomenon in the reception. The PN sequence-based channel estimation was known for its merits of low complexity and high accuracy in wireless communication scenarios [18]. It was recently proved to be very efficient for in DMT transmission over optical fiber [19]. This motivates us to investigate the application of PN sequence based channel estimation for SCM-based optical fiber transmissions.

In classical PAM systems, the transmitted symbols are expressed as:

$$\bar{S}_{\text{PAM}} = \left[\bar{S}_{\text{T}}, \bar{S}_{\text{D}}\right]^T = [S_{\text{T}}(0), \cdots S_{\text{T}}(N_{\text{T}} - 1), S_{\text{D}}(0), \cdots$$
$$S_{\text{D}}(N_{\text{D}} - 1)]^T, \tag{1}$$

where $\bar{S}_{\text{T}}$ is the length-N_{T} vector of training symbols for the equalizer; $\bar{S}_{\text{D}}$ is the length-N_{D} vector of PAM data symbols.

When the PN sequence is inserted to the PAM symbols to assist the channel estimation, the transmitted PN-PAM symbols are written as:

$$\bar{S}_{\text{PN-PAM}} = \left[\bar{\rho}_{\text{PN}}, \bar{S}_{\text{D}}\right]^T = [\rho_{\text{PN}}(0), \cdots \rho_{\text{PN}}(N_{\text{PN}} - 1),$$
$$S_{\text{D}}(0), \cdots S_{\text{D}}(N_{\text{D}} - 1)]^T, \tag{2}$$

where $\bar{\rho}_{\text{PN}}$ is the vector of PN sequence for channel estimation. The length of $\bar{\rho}_{\text{PN}}$ is N_{PN}.

At the receiver side, the received PN sequence is used to perform channel estimation. The PN sequence based channel estimation for optical communications has been initially introduced in [19]. The m-sequences are selected as the PN sequences for channel estimations due to their ease of generation and their associated low complexity. The most significant benefit of using m-sequence for channel estimation is its special circular autocorrelation property. The circular autocorrelation of the m-sequence is known as:

$$\text{CR}_j = \frac{1}{N_{\text{PN}}} \sum_{i=0}^{N_{\text{PN}}-1} m_i m^*_{[i+j]N_{\text{PN}}} = \begin{cases} 1 & j = 0 \\ -\frac{1}{N_{\text{PN}}} & else \end{cases} \tag{3}$$

where m is the m-sequence, $(\cdot)^*$ is the complex conjugate, $[\cdot]_{N_{\text{PN}}}$ denotes modulo-N_{PN} operation. With the help of circular autocorrelation property shown in (3), the channel estimation can be simply obtained by performing time domain correlation of known and received PN sequences:

$$\widetilde{h} = \frac{1}{N_{\text{PN}}} \sum_{i=0}^{N_{\text{PN}}-1} \left(\sum_{l=0}^{N_{\text{H}}-1} h_l \rho_{i-l} + w_i \right) \cdot m^*_{[i+j]N_{\text{PN}}} \tag{4}$$

where w is the noise, N_{H} is the channel length. In POF channel model, the massive multi-path delay could be modeled as discrete filter taps. Therefore, the maximal channel multi-path delay in real POF channel could be denoted by channel length N_{H} with number of filter taps.

Finally, the accurate estimate of the channel impulse response (CIR) $\widetilde{h} = [\widetilde{h}_0, \widetilde{h}_1, \cdots \widetilde{h}_{N_{\text{H}}-1}]^T$ can be easily obtained at a very low complexity cost [20]. According to the analysis carried out in [19], the overall complexity of the PN sequence-based channel estimation is $\mathcal{O}(N_{\text{PN}} \cdot \log N_{\text{PN}})$, which is determined by the PN sequence length.

Minimum-phase pre-filtering

In communication systems, trellis-based equalizer can effectively eliminate the inter-symbol interference (ISI) after transmission over the channel. The maximum-likelihood sequence estimation (MLSE) is recognized as the optimal equalization algorithm in the sense of sequence detection. As the decision is based on a sequence of symbols, it can effectively avoid the error propagation problem of DFE. However, it is worth noting that for PAM with high orders modulations, the computational complexity of MLSE equalizer dramatically increases. The full MLSE equalizer becomes computationally prohibitive when the modulation order is high and/or when the channel length is long. To avoid the prohibitive complexity, a sub-optimal trellis-search based equalizer, namely RSSE, is commonly used for its simplicity in the hardware implementation.

Studies in [21, 22] show that, in order to obtain the sub-optimal performance after trellis-based equalization, discrete time minimum-phase overall impulse response needs to be carried out previously. We employ an FFE pre-filter to achieve the minimum-phase overall impulse response. As an accurate CIR is obtained directly from the PN sequence-based channel estimation, it is feasible to calculate the filter coefficients from the estimated CIR. The coefficients of the pre-filter can be calculated in closed-form from the estimated CIR $\widetilde{h}$.

In [22], the coefficients of the minimum-phase pre-filter are calculated by the linear prediction from the estimated CIR. The linear prediction is realized by the well-known Levinson-Durbin algorithms. Concretely, the pre-filter is determined as follows:

$$\widetilde{F}(z) = z^{-N_{\text{H}}} H^* \left(1/z^*\right) (1 - P(z)), \tag{5}$$

where $H^*(1/z^*)$ is the time-reversed conjugated CIR, $(1 - P(z))$ is the calculated linear prediction filter, $z^{-N_{\text{H}}}$ introduces a delay of N_{H} samples.

The analysis of this minimum-phase pre-filter calculation shows that the overall computational complexity of linear prediction method is significantly lower than that of the minimum mean-squared error (MMSE)-DFE method [22].

RSSE based equalization

In contrast to the MLSE equalizer where all possible combinations of symbol sequences are compared with received signal sequence, RSSE dramatically reduces the number of candidates to be compared by applying constellation partitioning and decision-feedback with early decisions [23]. With a proper choice of the number of survivor states, the RSSE based equalizer can approach the optimal performance offered by the MLSE equalizer.

More concretely, the entire symbol alphabet is divided into subsets, and the search trellis is built based on these subsets. The subsets need to be chosen such that the Euclidean distance of symbols within each subset is maximized. Once the survival path is determined, the hard decision within the subset is made directly so that only one survival candidate is reserved in each set, while others are simply discarded for the following search. The selection among subsets is not taken in the current step. Therefore, the number of overall trellis states involved in the search is reduced to $Z = \prod_{k=1}^{N_C} J_k$, where J_k is the number of subsets preserved for the symbol k steps before the currently detecting symbol, N_C is the constraint length which is chosen according to the number of significant channel paths, and can be less than the overall channel length N_H. It is worth noting that through the choice of J_k, the performance and complexity of RSSE equalizer can achieve arbitrary trade-off between the optimal MLSE equalizer and the simple DFE equalizer [22]. For example, when $J_k = M, 1 \leq k \leq N_C$, the RSSE equalizer becomes the MLSE equalizer for PAM-M modulation. Similarly, when $J_k = 1, 1 \leq k \leq N_C$, since there is only one subset preserved for decision, the RSSE equalizer is turned into a DFE equalizer.

Theoretical analysis of the PN-PAM transmission

Gaussian low-pass filter channel model

In most short range optical transmission systems, the channels show similar features such as a low-pass frequency response and very good channel condition (very high SNR) at low frequency part [3–5]. The Gaussian low-pass filter channel model has been proved to be suitable for the POF systems [24]. In this section, it will be used to model a 50-m SI-POF system for the theoretical investigation of the proposed PN-PAM transmission.

The channel frequency response (CFR) of the Gaussian low-pass filter channel model is written as:

$$H(f) = A \cdot \exp\left[-\left(\frac{f}{f_0}\right)^2\right], \quad f_0 = \frac{f_{3dB}}{\sqrt{\ln(2)}}, \quad (6)$$

where $H(f)$ is the CFR at f Hz; A is the optical fiber loss; f_{3dB} is the 3 dB bandwidth of the considered 50-m SI-POF system.

PAM transmission over Gaussian low-pass filter channel

Using the Gaussian low-pass filter channel model introduced in (6), we can theoretically analyze the PAM transmission over a 50-m SI-POF with different equalizers. It is well known that in additive white Gaussian noise (AWGN) channel, channel capacity is calculated as:

$$C = W \cdot \log_2\left(1 + \frac{P}{N_0 W}\right), \quad (7)$$

where W is the used bandwidth, P is the signal power, N_0 is the power spectral density of the noise. When signal is

transmitted over a multi-path channel, ISI will be introduced among consecutive symbols due to the time dispersion of signal. At the receiver side, noise can be boosted after the equalization depending on the equivalent SNR over the whole signal band. Therefore, the performance of the PAM system can be determined through the corresponding SNR after equalization [25]. Based on the post-equalization SNR, we can derive the achievable rate of the PN-PAM transmission in the Gaussian low-pass filter channel.

As the RSSE equalization used in the proposed scheme is a non-linear process, it is difficult to derive the capacity of the proposed PN-PAM scheme in a straightforward manner. Alternatively, we analytically investigate the PAM transmission system using DFE which is an extreme case of the RSSE equalizer. For PAM transmission system using DFE, we assume that no incorrect decision is fed back. Following the results given in [25], the post-equalization SNR of PN-PAM transmission system is written as:

$$SNR_{DFE} = \frac{1 - \xi_{DFE}}{\xi_{DFE}}, \quad (8)$$

where

$$\xi_{DFE} = \exp\left[\frac{T}{2\pi}\int_{-\pi/T}^{\pi/T}\ln\frac{N_0}{H\left(e^{j\omega T} + N_0\right)}d\omega\right]. \quad (9)$$

The integral in the expression of ξ_{DFE} in (9) can be mathematically approximated by discrete summation. Employing (6), ξ_{DFE} is estimated by:

$$\xi_{DFE} \approx \exp\left[\frac{1}{N_p}\sum_{i=1}^{N_p}\ln\frac{N_0 \cdot N_p}{A \cdot \exp\left[-\left(\frac{i\cdot\Delta f}{\sqrt{\ln(2)}\cdot f_{3dB}}\right)^2\right] + N_0 \cdot N_p}\right], \quad (10)$$

where Δf is the frequency spacing of the discrete calculation; N_p is the number of intervals in order to approximate the integral function. The bandwidth of the PAM system is therefore $W_{PAM} = \Delta f \cdot N_p$. The normalized noise energy within the Δf frequency spacing is N_0.

We assume that the Nyquist bandwidth ($1/2T$) is used. Introducing the post-equalization SNR in (8) and the channel capacity in (7), we can obtain the capacity of PAM transmission with DFE equalization:

$$C = 2 \cdot W_{PAM} \cdot \log_2\left(1 + SNR_{DFE}\right). \quad (11)$$

Introducing the linear approximation method proposed in [26], we can compute the achievable PAM modulation order b as:

$$b = \frac{SNR(dB) - A_2}{A_1}, \quad (12)$$

where A_1 and A_2 are coefficients defined in the linear approximation. The exact values of A_1 and A_2 for PAM modulation can be obtained by following similar manipulations as in [26] for different desired BER levels. The obtained A_1 and A_2 are listed in Table 1.

Applying (11) and (12), the available transmission rate can be calculated as:

$$R_s = 2 \cdot W_{\text{PAM}} \cdot \frac{10 \cdot \log_{10}\left(\frac{1-\xi_{\text{DFE}}}{\xi_{\text{DFE}}}\right) - A_2}{A_1}. \tag{13}$$

Take the 50-m SI-POF transmission system as an example, where the channel model parameters are the same as in [26]. We can get that $f_{3\text{dB}}$ is 90 MHz and the measured noise power spectral density is −113.7 dBm/Hz. We numerically calculate the PAM transmission rate with the frequency spacing Δf of 1 MHz and number of intervals N_p equal to 1000. With a target BER of 1×10^{-3} (a level that the residual error can be effectively corrected by channel coding), the relationship between PAM transmission rate and used bandwidth is depicted in Fig. 1.

As shown in the figure, the narrower bandwidth, the higher PAM modulation orders. However, concerning transmission rate, there is an optimized used bandwidth and PAM modulation order. For instance, in our POF system model, the optimal used bandwidth in Fig. 1b is around 220 MHz. The achievable PAM modulation order with this bandwidth is PAM-8 accordingly in Fig. 1a.

RRC filters for PAM transmissions

In practical communication systems, the square RRC filters are normally used at the transmitter and receiver in order to reduce the required bandwidth for the transmission. For the RRC filters, a roll-off factor β is a measure of the excess bandwidth of the filter, which represents the bandwidth occupied beyond the Nyquist bandwidth of $1/2T$. β can be chosen between 0 and 1. Therefore in a practical system, the real required bandwidth W_{PAM} is denoted as:

$$W_{\text{PAM}} = \frac{1}{2}R_s(1 + \beta), \tag{14}$$

where R_s is the PAM symbol rate.

In a pure PAM system without the RRC filter, the PAPR is directly related to the PAM modulation order. However, the use of RRC filter at the transmitter will increase the PAPR with a reduced excess bandwidth or increased filter length [27]. The relationship between the PAPR of the QAM modulation and the RRC filter roll-off factor was studied for PAM-2 and QAM-32 in [27]. Here we investigate the relationship between β and PAPR

for PAM-2, PAM-4, PAM-8, PAM-16, and PAM-32. The maximal PAPR of the PAM signal after the RRC filtering is evaluated by the following equation:

$$\text{PAPR}_{\text{max}} = 10 \cdot \log_{10}\left(\frac{x_{\text{max}}^2}{\mathbb{E}[x^2]}\right), \tag{15}$$

where x_{max} is the maximal amplitude of the PAM-M signal. $\mathbb{E}[x^2]$ is the average power of the generated signal. The measured results with a RRC filter length of 8 taps are presented in Fig. 2. From the figure, we can find that the PAPR of the PAM modulated signal after RRC filtering significantly increases when the roll-off factor is lower than 0.4. The PAPR of the PAM-2, PAM-4 and PAM-8 with a RRC filter roll-off factor of 0.2 is obtained around 4.3, 6.9 and 8.0 dB, respectively.

Numerical analysis of the PN-PAM transmission

In this section, we simulate the PN-PAM transmission over the 50-m SI-POF system. The modulation is chosen to be PAM-8 which has been shown to be the best modulation scheme for the given channel condition. The symbol rate is set to be 440 Mega symbols per second. Therefore the theoretically minimal required bandwidth is 220 MHz when roff-off factor in RRC filter is assumed at 0. In the practical simulation system, we use a square RRC transceiver filter to reduce the used bandwidth of the PAM transmissions. The roll-off factor is set to be 0.2. Therefore, the actual used bandwidth in simulations is 264 MHz. The estimated SNR with the noise PSD of −113.7 dBm/Hz is about 29.5 dB.

PN-PAM transmission with different equalization methods

Firstly, in Fig. 3, we present the simulation results of PAM-8 transmission over the 50-m SI-POF channel with different equalization methods. For the classical adaptive DFE equalizations, two typical algorithms referred to as recursive least square (RLS) and least mean square (LMS) are adopted. In genie-aided mode, we initially transmit 10,000 training symbols to get the DFE coefficients and subsequently using the known symbols as the decided symbols for the DFE feedback. Alternatively, in direct-decision mode, we use decided symbols for the DFE feedback. In both cases, the number of decision feedback taps is 6 and the number of FFE taps is 24. The simulation results show that DFE-RLS and DFE-LMS provide similar performance in the 50-m SI-POF system with PAM-8 transmission. However, in both DFE-RLS and DFE-LMS schemes, due to the error propagation in direct-decision

Table 1 Parameters for the relationship between SNR and available PAM modulation order with different desired BER

BER	10^{-1}	10^{-2}	10^{-3}	10^{-4}	10^{-5}	10^{-6}	10^{-7}	10^{-8}
A_1	4.926	5.848	6.020	6.090	6.120	6.150	6.156	6.158
A_2	-5.307	-1.087	1.214	2.741	3.913	4.814	5.620	6.670

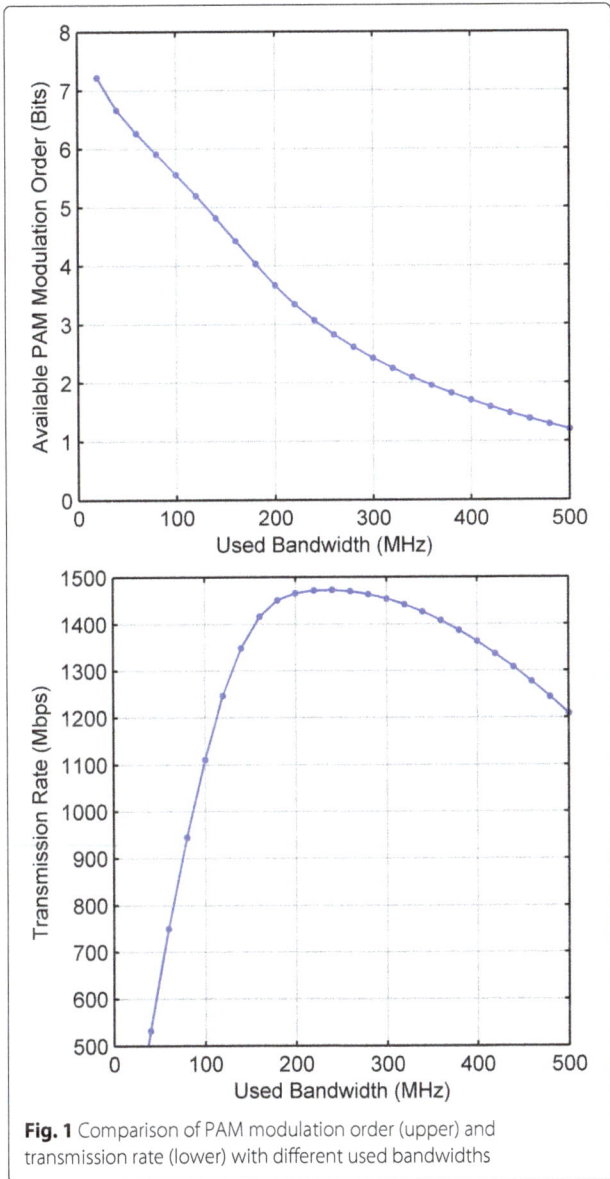

Fig. 1 Comparison of PAM modulation order (upper) and transmission rate (lower) with different used bandwidths

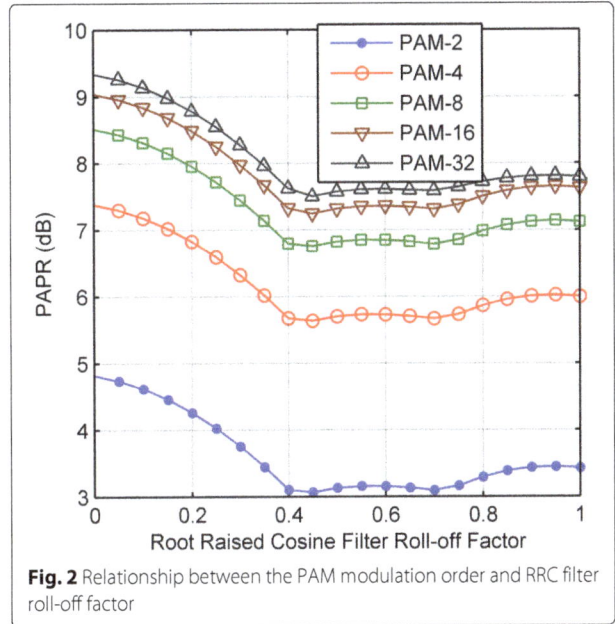

Fig. 2 Relationship between the PAM modulation order and RRC filter roll-off factor

mode, there is an approximate 1.5 dB degradation at the BER level of 1×10^{-3} compared to the genie-aided mode. Therefore it is crucial to avoid the error propagation problem in the practical systems.

Then we simulate the proposed PN-PAM transmission scheme with RSSE. The PN sequence length for channel estimation is set to be 255 symbols and a cyclic-prefix with the length of 33 symbols is added to the PN sequence in order to prevent the ISI from previous transmitted symbols. The length of the estimated CIR is 7 taps. The first 6 most significant taps are used to calculate the minimum-phase pre-filter. An illustration of the multi-path CIR in original Gaussian low-pass filter channel model and after pre-filtering is presented in Fig. 4. As shown in the figure, a main path is recovered at the first

tap after pre-filtering. Then the most significant multi-path component is located at the second tap, which owns an amplitude about 0.86. Therefore, we design and compare two RSSE structures in 50-m POF transmissions. The first RSSE structure uses the first 2 most significant delay taps for symbols decision. Hence the length of the RSSE trellis is selected to 2. The number of subsets for the first delay tap J_1 is set to 2 and the number of subsets for the second delay tap J_2 is set to 1. The second RSSE structure uses all of the 5 delay taps for symbols decision. For each delay tap, the number of subsets J_k is set to 1. In the second RSSE structure, as each delay tap only has one subset, the RSSE equalizer turns into a DFE equalizer with 5 taps. Obviously, for both of the RSSE structure, the overall complexity is very low.

After simulations of the transmission over 50-m SI-POF channel, performances of RSSE equalizers are depicted in Fig. 3. As been shown in the figure, the RSSE $(2,1)$ equalizer can achieve similar performance to classical DFE equalizers in genie-aided mode, which has 1.5 dB improvement with the practical DFE equalizers in direct-decision mode. Additionally, the RSSE $(1,1,1,1,1)$ equalizer has a very close performance to classical DFE equalizers. Therefore, it is clear that in the 50-m SI-POF channel, using RSSE $(2,1)$ equalizer can solve the error propagation problem in classical DFE equalizers.

PAM transmission with respect to different training sequence lengths

In addition, as mentioned in "Methods" section, the novel PAM transmission scheme leads to an improved

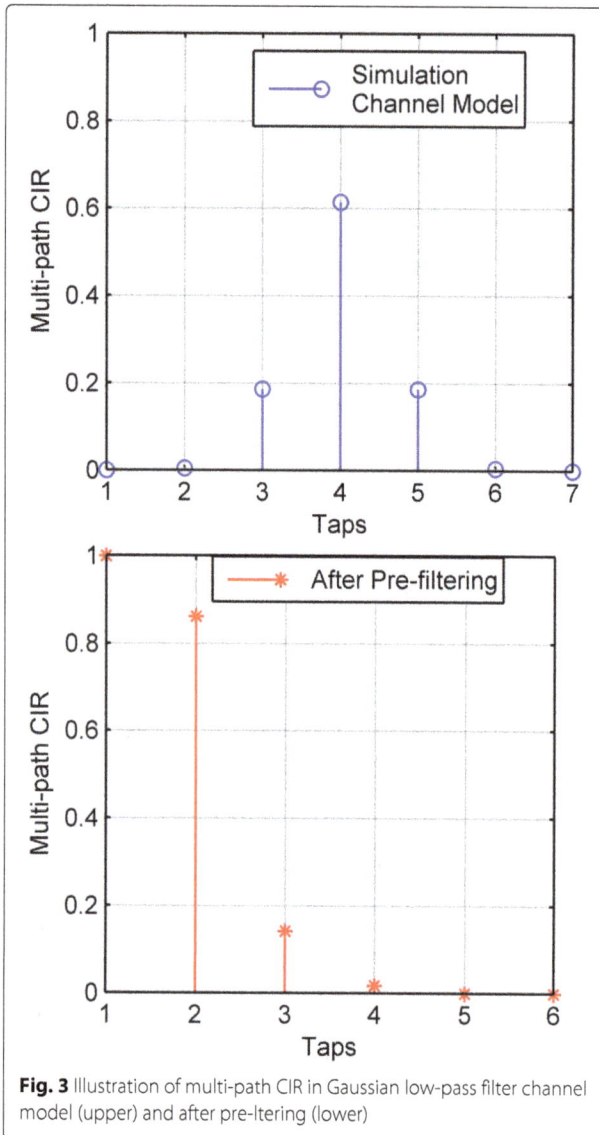

Fig. 3 Illustration of multi-path CIR in Gaussian low-pass filter channel model (upper) and after pre-ltering (lower)

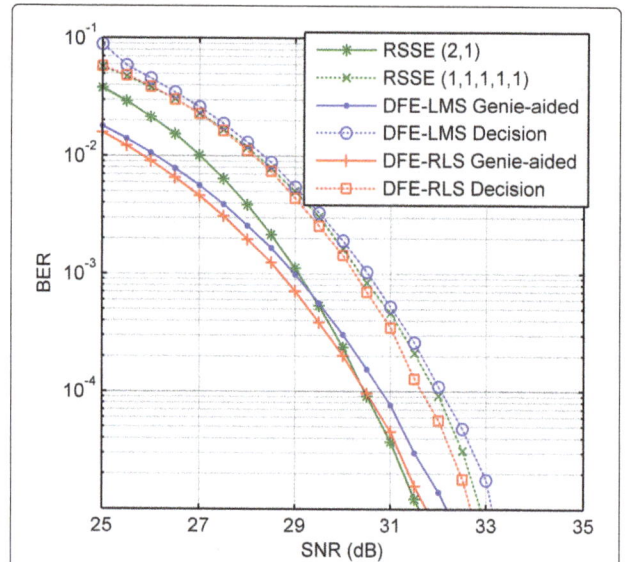

Fig. 4 Simulation results of the PAM transmission over 50 meters SI-POF with different equalization methods

message. Therefore, the genie-aided mode is only used to theoretical optimal performance evaluations.

For the proposed scheme, the training sequence is the PN sequence inserted before the data symbols. We simulate the PAM-8 transmission over the 50-m SI-POF channel with different training sequence lengths for both classical DFE and the proposed scheme. The SNR is fixed to 29.5 dB. The BER performance is depicted in Fig. 5.

With a PN sequence length longer than 255 symbols, the proposed system achieves the best BER performance of around 4.6×10^{-4}. However, the classical DFE based

transmission efficiency. The training sequence takes up a certain amount of transmission energy and hence reduces the overall system efficiency. For the classical DFE-RLS and DFE-LMS equalizers, the predefined training sequence is inserted before the data symbols. The equalizers work in training mode when the training symbols are received and then switch to the equalization mode when data symbols are received. In equalization mode, there are two feed-back methods. In general transmissions, the feed-back bits are obtained from decisions, which is called as direct-decision mode. However, in this mode, decision error will be propagated and cause penalties. Besides, it is also possible to use known pilot bits as the feed-back bits. In this mode, error propagation could be avoided due to the known pilot bits at receiver. This mode is called as genie-aided mode. In practical system, the known pilot bits cannot be used to transmit

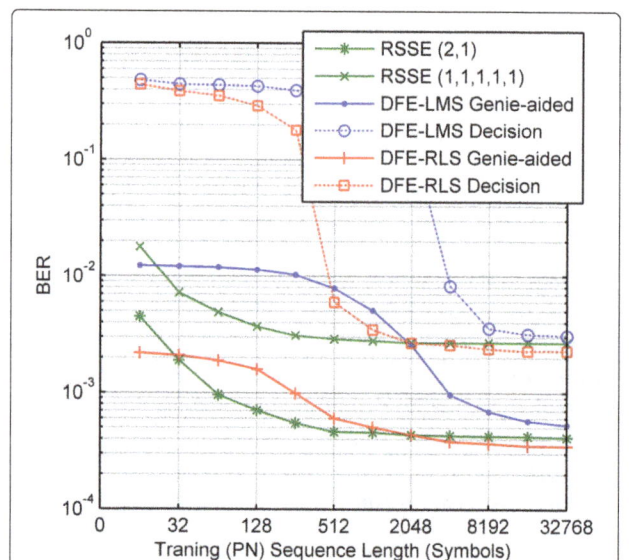

Fig. 5 Simulation results of the PN-PAM transmission over 50-m SI-POF with different training sequence lengths

equalizers require much longer training sequences. Simulation results show that even in the genie-aided mode, in order to get converged BER performance, the DFE-RLS requires a training sequence longer than 4096 symbols and the DFE-LMS requires a training sequence longer than 32,768 symbols. Moreover, when DFE works in direct-decision mode, even extremely long training sequence is adopted, they can only reach the best BER performance of around 2.0×10^{-3}.

Summary

From simulations shown above, it is obvious that the proposed PN-PAM transmission scheme can achieve higher system efficiency. The improvement comes from two aspects. On the one hand, the PN-based channel estimation requires very little training overhead, roughly 0.01 to 0.05 compared to the classical DFE schemes. On the other hand, the RSSE equalization leads to an improved BER performance with the help of the reduced error propagation compared to classical DFE schemes.

Results and discussion

In order to verify the proposed PAM transmission schemes, we setup a practical system for experimental comparisons. We also compared PAM transmissions to DMT transmissions with different POF lengths.

Experimental system setups

In contrast to most experimental POF transmission systems, we use a commercially available digital to analog convertor (DAC) and analog to digital convertor (ADC). Both DAC and ADC are provided from Texas Instruments. The DAC (DAC5681) has 1 Giga samples per second sampling rate and 16 bits resolution. The ADC (ADC12D1800RFRB) has 1.8 Giga samples per second sampling rate and 12 bits resolution. In order to drive the resonant-cavity light emitting diode (RCLED), we designed an amplifying circuit at the transmitter using

a commercially available amplifier (OPA695) also from Texas Instruments. The designed transmission bandwidth is 250 MHz.

At the transmitter, RCLED is provided from Firecomms (FC300R-120) with a 3 dB bandwidth of 100 MHz. The biasing current is 20 mA and the coupled power into the SI-POF is −0.2 dBm. PMMA Φ 1 mm SI-POFs (Eska MEGA) are prepared with different lengths (15, 30, 50, 75 and 100 m). At the receiver, a photodiode combined with a trans-impedance amplifier (FC-1000D-120) is employed to detect the received optical signal. The 3 dB bandwidth of the receiver is 625 MHz. The block diagram of the experimental system is presented in Fig. 6.

Both the novel PN-PAM transmission scheme and classical PAM transmission scheme is generated off-line in computer. For the novel PN-PAM transmission scheme, the PN length is set to be 255 symbols and for classical PAM transmission scheme, the training sequence length is chosen as 5000 symbols. The roll-off factor of the RRC filter is selected as 0.25. Therefore, PAM symbol rate is set to 400 Mega symbols per second. The total used bandwidth is $400 \div 2 \times 1.25 = 250$ MHz. As 2 times oversampling is adopted at the transmitter, the DAC sampling rate is 800 Mega samples per second. The ADC works at 1.8 Giga samples per second. At the receiver, for classical PAM transmission with DFE, the number of decision feedback taps is 6 and the number of FFE taps is 24. For the proposed PN-PAM transmission scheme, the first 13 prominent taps in the CIR estimate are used for the calculation of minimum-phase pre-filter. The performance is evaluated with 20 frames, each containing 14,500 symbols. Therefore, in our experimental systems, the overall training overhead is approximately 25.6% for PAM with DFE (5000 training symbols, 14,500 data symbols) and 1.7% for the proposed PAM transmission scheme (255 PN symbols, 14,500 data symbols).

In the meantime, the DMT transmission signals are also generated for the comparison. For the DMT system, we setup the system with the same 250 MHz used bandwidth

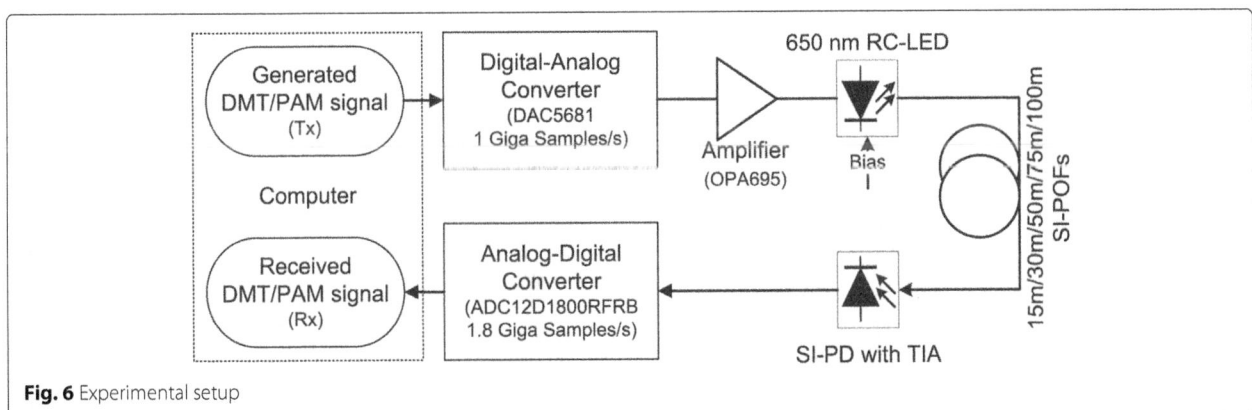

Fig. 6 Experimental setup

as the PAM system. The DAC sampling rate is set to 1 Giga samples per second indicating 4 times oversampling. The number of available subcarrier is 512. The first subcarrier is closed to avoid the direct current component. Bit-loading algorithm with linear approximation [26] is employed to allocate bits and power to each subcarrier. The DMT signal is digitally clipped with an optimal clipping ratio of 10 dB. The hybrid pseudo-noise and zero-padding DMT scheme [19] is adopted for its advantages compared to the classical DMT scheme. The PN sequence length is 255 symbols and the ZP length is 32 symbols. The ADC works at 1.8 Giga samples per second. The performance is evaluated with 20 frames, each consisting of 7 DMT symbols.

Experimental results

For the proposed PN-PAM transmission scheme and DMT transmission scheme, we can estimate channel response using the received PN sequence. The estimated CIR of 15, 30, 50, 75 and 100 m SI-POF for the PAM transmissions are presented in Fig. 7. As shown in the figure, the CIRs exhibit significant time dispersion which causes ISI among consecutive symbols. In addition, the system overall non-linearity causes multi-path delay spreading, which is shown as the additional delay path at 9th and 10th taps in Fig. 7. Based on the CIR, we can calculate the coefficients of the minimum-phase pre-filter using (5). The after pre-filtering are presented in Fig. 8. As we can see from the figure, the normalized impulse responses have significantly reduced the pre-interference before the main path. This is essential for the subsequent DFE or RSSE equalizations.

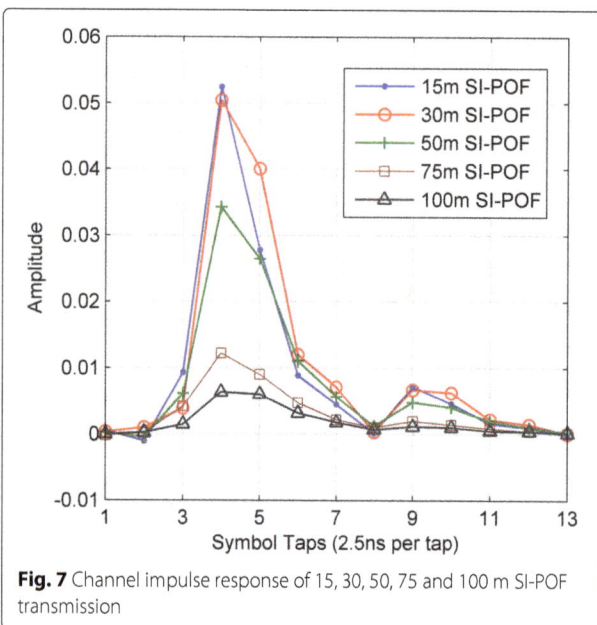

Fig. 8 Normalized impulse responses of 15, 30, 50, 75 and 100 m SI-POF transmission after pre-filtering

According to the channel conditions, we evaluate several PAM modulations for each transmission length. Both classical PAM with DFE scheme and the proposed PN-PAM transmission scheme are tested. The comparison of the BER performances is presented in Fig. 9. In addition, the BER performance of PAM-4 transmission over 15, 30 and 50 m SI-POF, the PAM-2 transmission over 15, 30, 50, 75 and 100 m SI-POF is measured with error-free (i.e., BER$< 1 \times 10^{-6}$). Therefore, these results are not depicted in Fig. 9. For PAM-8 transmissions, the proposed PN-PAM scheme can achieve the BER performance of around

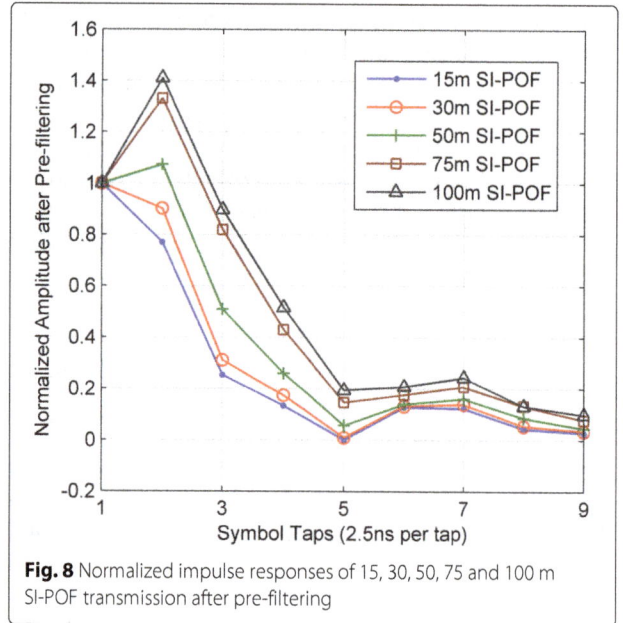

Fig. 7 Channel impulse response of 15, 30, 50, 75 and 100 m SI-POF transmission

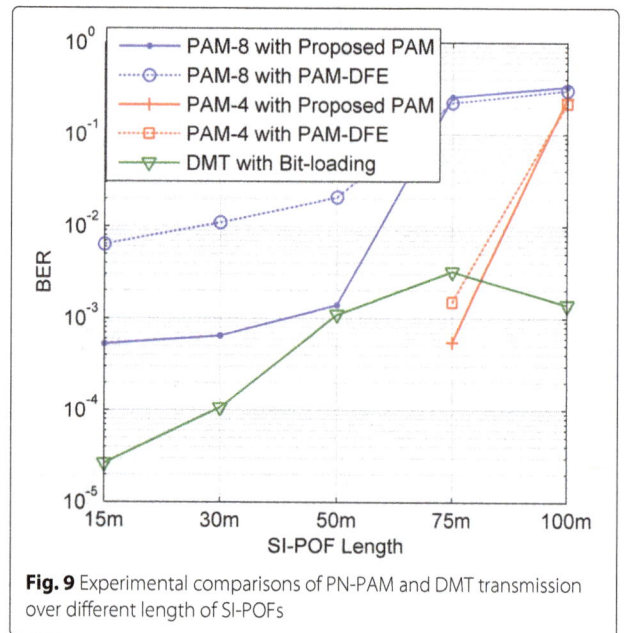

Fig. 9 Experimental comparisons of PN-PAM and DMT transmission over different length of SI-POFs

1×10^{-3} over 15, 30 and 50 m SI-POF. However, for classical PAM with DFE, the BER performance is only around 1×10^{-2} with the same conditions. This BER degradation is mainly due to the error propagation when DFE switches to the direct-decision mode after 5000 training symbols. In addition, it is worth noting that in the proposed PAM scheme, the length of the training sequence is only 255 symbols.

In summary, it can be concluded that using the proposed PN-PAM transmission scheme, PAM-8 is suitable for 15, 30 and 50 m SI-POF transmission with a desired BER around 1×10^{-3}. As the symbol rate of the PAM transmission is 400 Mega symbols per second, the total achievable bit rate is 1.2 Gbps. For 75 m SI-POF, PAM-4 is a reasonable choice and 800 Mbps can be achieved. Finally, PAM-2 can support a quasi-error-free transmission (BER$< 1 \times 10^{-6}$) over 100 m SI-POF, which leads to a transmission rate of 400 Mbps.

Comparison and discussion

In this section, we introduce the DMT transmissions for comparisons. In order to implement a fair comparison with the PAM transmissions, we employ bit-loading algorithms and allocate bits for total transmission rates of the DMT system at 1.2 Gbps over 15, 30, and 50 m SI-POF, 800 Mbps over 75 m SI-POF and 400 Mbps over 100 m SI-POF.

The BER performances of DMT are also depicted in Fig. 9 and a summary of experimental results in different transmission schemes with fixed transmission bit rate is presented in Table 2. As shown in the results, we can find that the proposed PN-PAM transmission scheme achieves similar performance to the DMT transmission in the 50-m SI-POF system.

In addition, for the transmission length shorter than 50 m, the DMT systems present better performance. This is mainly due to the fact that for transmission lengths shorter than 50 m, due to the high SNR at receiver, DMT can fully approach the channel capacity with the help of bit-loading. However, when high order PAM modulations are employed for short range communications, the transmitter non-linearity distortion will seriously affect overall performance.

Furthermore, for the transmission lengths longer than 50 m, PN-PAM transmission systems achieve better performance. From the aforementioned PAPR results in Fig. 2, we can find that PAM-2 and PAM-4 present lower PAPR than DMT with 10 dB clipping. Therefore, with the fixed DAC output dynamic range, the PN-PAM-based transmission systems can achieve higher transmitting power and obtain better performance.

Conclusion

In this paper, we introduced a novel PAM transmission scheme for short range optical transmission systems. The novel PN-PAM transmission scheme includes PN sequence-based channel estimation, minimum-phase receiver filter and RSSE based equalizer. A theoretical analysis for the PN-PAM transmission over Gaussian low-pass frequency response channel was also presented. From the theoretical analysis, the PAM-8 modulation is proved to be the best modulation choice for the 50-m SI-POF system.

The simulation results show that, compared to the classical PAM transmission with DFE equalization, the proposed PN-PAM transmission scheme can totally achieve 1.5 dB gain in a 50-m SI-POF system with an affordable complexity increase. Furthermore, simulation results reveal that the proposed PAM scheme requires less than 0.01 to 0.05 cost of system training symbol overhead compared to the classical DFE schemes, and can largely increase overall system efficiency.

Experimental systems with commercially available components are also investigated for both PN-PAM and DMT transmissions with the same used bandwidths. Experimental results show that, for the SI-POF lengths longer than 50 m, the proposed PN-PAM transmission scheme outperforms the classical PAM transmission with DFE equalization as well as the DMT transmissions that has been optimized in a previous work [19]. In the meantime, for the transmission lengths shorter than 50 m, the proposed PN-PAM transmission scheme performs better than the classical PAM with DFE equalization but worse than the DMT transmissions. This degradation can be understood by the fact that in order to achieve the same transmission rate as DMT, the PN-PAM based

Table 2 Summary of experimental results in different transmission schemes with fixed transmission bit rate

| POF Length | Bit Rate | BER | | |
		Proposed PAM	PAM with DFE	DMT
15 m	1.2 Gbps	5.3×10^{-4}	6.4×10^{-3}	2.7×10^{-5}
30 m	1.2 Gbps	6.3×10^{-4}	1.1×10^{-2}	1.1×10^{-4}
50 m	1.2 Gbps	1.4×10^{-3}	2.1×10^{-2}	1.1×10^{-3}
75 m	800 Mbps	5.5×10^{-4}	1.5×10^{-3}	3.2×10^{-3}
100 m	400 Mbps	$< 10^{-6}$	$< 10^{-6}$	1.4×10^{-3}

systems need to employ higher order modulations when the channel condition is good. However, the higher order modulations are more vulnerable to the non-linearity of the SI-POF system.

Therefore, it can be concluded that the proposed PN-PAM transmission scheme outperforms the classical PAM with DFE equalization in terms of both BER performance and overall system throughput. The complexity increase due to the use of RSSE equalizer is also acceptable. For short range optical transmissions with POF length longer than 50 m, the proposed PN-PAM transmission scheme is a better choice than the DMT transmission scheme.

Abbreviations
ADC: Analog to digital convertor; AWGN: Additive white Gaussian noise; CFR: Channel frequency response; CIR: Channel impulse response; DFE: Decision feedback equalizer; DAC: Digital to analog convertor; DMT: Discrete multi-tone; FFE: Feed-forward equalizer; ISI: Inter-symbol interference; LMS: Least mean square; MCM: Multi-carrier modulation; MLSE: Maximum-likelihood sequence estimation; MMF: Multi-mode fiber; MMSE: Minimum mean-squared error; NRZ: Not-return-to-zero; PAM: Pulse-amplitude modulated; PAPR: Peak-to-average-power ratio; PN: Pseudo-noise; PN-PAM: Pseudo-noise sequence assisted pulse-amplitude modulated; RCLED: Resonant-cavity light emitting diode; RLS: Recursive least square; RRC: Root-raised-cosine; RSSE: Reduced-state sequence estimation; SCM: Single carrier modulation; SI-POF: Step-index polymer optical fiber; SMF: Single-mode fiber; SNR: Signal-to-noise ratio; VLC: Visible light communications

Acknowledgements
Not applicable

Funding
Peng's work is supported by the National Natural Science Foundation of China (Grant No. 61601114) and the Natural Science Foundation of Jiangsu Province (BK20160692). Liu's work is supported by the National Natural Science Foundation of China (Grant No. 61501022) and the Fundamental Research Funds for the Central Universities (2017JBM028).

Authors' contributions
LP contributed to entire studies of this work, manuscript preparation and manuscript editing. ML contributed to system design, manuscript revision and manuscript editing. MH contributed to theoretical studies and manuscript revision. SH contributed to experimental studies and manuscript revision. All authors read and approved the final manuscript.

Authors' information
Linning Peng received his PhD degrees from IETR (Electronics and Telecommunications Institute of Rennes) laboratory at INSA (National Institute of Applied Sciences) of Rennes, France, in 2014. From 2014, he is a research associate at Southeast University. His research interests are in design and optimization for communication systems.
Ming Liu received the B.Eng. and M.Eng. degrees from the Xi'an Jiaotong University, China, in 2004 and 2007, respectively, and the Ph.D. degree from the National Institute of Applied Sciences (INSA), Rennes, France, in 2011, all in electrical engineering. He was with the Institute of Electronics and Telecommunications of Rennes (IETR) as a postdoctoral researcher from 2011 to 2015. He is now with Beijing Jiaotong University, China, as an Associate Professor. His main research interests include multicarrier transmissions, MIMO techniques, space-time coding and Turbo receiver.
Maryline Hélard received the M.Sc and PhD degrees from INSA (National Institute of Applied Sciences) of Rennes and the Habilitation degree from Rennes 1 University in 1981, 1884 and 2004 respectively. In 1985, she joined

France Telecom Research Laboratory as a research engineer and since 1991 she carried out physical layer studies in the field of digital television and wireless communications. In 2007, she joined INSA as a professor and she is now the co-director of the Communication Department at IETR (Electronics and Telecommunications Institute of Rennes). She is co-author of 22 patents and several papers (Journal and conferences). Her current research interests are in the areas of digital communications such as equalization, synchronization, iterative processing, OFDM, MC-CDMA, channel estimation, and MIMO techniques applied to wireless communications and more recently to wire communications (ADSL, optical).
Sylvain Haese received the engineer and PhD degrees in electrical engineering from INSA (National Institute of Applied Sciences) Rennes, France, in 1983 and 1997 respectively. From 1984 to 1993 he was an analog IC designer for automotive and RF circuits. In 1993, he joined INSA and carried out research activity at IETR laboratory where he conducted research for automotive powerline applications and for RF wideband channel sounder circuitry. He is currently involved in hardware analog implementation for RF and optical circuits with the Communication Department at IETR (Electronics and Telecommunications Institute of Rennes).

Competing interests
The authors declare that they have no competing interests.

Author details
[1]Institute of Information Science and Engineering, Southeast University, No.2 SiPaiLou, 210096 Nanjing, China. [2]School of Computer and Information Technology, Beijing Jiaotao University, No.3 ShangYuanCun, 100044 Beijing, China. [3]IETR (Institute of Electronic and Telecommunications in Rennes), INSA-Rennes (Institut National des Sciences Appliquées de Rennes), 20 Avenue des Buttes de Coësmes, 35708 Rennes, France.

References
1. Okonkwo, CM, Tangdiongga, E, Yang, H, Visani, D, Loquai, S, Kruglov, R, Charbonnier, B, Ouzzif, M, Greiss, I, Ziemann, O, Gaudino, R, Koonen, AMJ: Recent Results from the EU POF-PLUS Project: Multi-Gigabit Transmission over 1 mm Core Diameter Plastic Optical Fibers. IEEE/OSA J. Light. Tech. **29**(2), 186–193 (2011)
2. Popov, M: Recent Progress in Optical Access and Home Networks: Results from the ALPHA Project. In: Proc. ECOC. IEEE, Geneva, (2011)
3. Kai, Y, Nishihara, M, Tanaka, T, Takahara, T, Lei, L, Zhenning, T, Bo, L, Rasmussen, JC, Drenski, T: Experimental comparison of pulse amplitude modulation (PAM) and discrete multi-tone (DMT) for short-reach 400-Gbps data communication. In: Proc. ECOC. IEEE, London, (2013)
4. Lee, SCJ, Breyer, F, Randel, S, Cardenas, D, van den Boom, HPA, Koonen, AMJ: Discrete multitone modulation for high-speed data transmission over multimode fibers using 850-nm VCSEL. In: Proc. OSA/OFC/NFOEC. IEEE, USA, (2009)
5. Kottke, C, Hilt, J, Habel, K, Vucic, J, Langer, KD: 1.25 Gbit/s visible light WDM link based on DMT modulation of a single RGB LED luminary. In: Proc. ECOC. IEEE, Amsterdam, (2012)
6. Joncic, M, Kruglov, R, Haupt, M, Caspary, R, Vinogradov, J, Fischer, UHP: Four-Channel WDM Transmission Over 50 m SI-POF at 14.77 Gb/s Using DMT Modulation. IEEE Photonics. Tech. Letters. **26**(13), 1328–1331 (2014)
7. Randel, S, Breyer, F, Lee, SCJ, Walewski, JW: Advanced Modulation Schemes for Short-Range Optical Communications. IEEE Sel. Top. Quant. Electron. **16**(5), 1280–1289 (2010)
8. Loquai, S, Kruglov, R, Schmauss, B, Bunge, C-A, Winkler, F, Ziemann, O, Hartl, E, Kupfer, T: Comparison of Modulation Schemes for 10.7 Gb/s Transmission Over Large-Core 1 mm PMMA Polymer Optical Fiber. IEEE/OSA J. Light. Tech. **31**(13), 2170–2176 (2013)
9. Schmogrow, R, Winter, M, Meyer, M, Hillerkuss, D, Wolf, S, Baeuerle, B, Ludwig, A, Nebendahl, B, Ben-Ezra, S, Meyer, J, Dreschmann, M, Huebner, M, Becker, J, Koos, C, Freude, W, Leuthold, J: Real-time Nyquist pulse generation beyond 100 Gbit/s and its relation to OFDM. Optics Express. **20**(1), 317–337 (2012)
10. Armstrong, J: OFDM for Optical Communications. IEEE/OSA J. Light. Tech. **27**(3), 189–204 (2009)

11. Szczerba, K, Westbergh, P, Agrell, E, Karlsson, M, Andrekson, PA, Larsson, A: Comparison of Inter symbol Interference Power Penalties for OOK and 4-PAM in Short-Range Optical Links. IEEE/OSA J. Light. Tech. **31**(22), 3525–3534 (2013)

12. Zeolla, D, Antonino, A, Bosco, G, Gaudino, R: DFE Versus MLSE Electronic Equalization for Gigabit/s SI-POF Transmission Systems. IEEE Photon. Technol. Lett. **23**(8), 510–512 (2011)

13. Rath, R, Rosenkranz, W: Tomlinson-Harashima Precoding for Fiber-Optic Communication Systems. In: Proc. of ECOC. IEEE, London, (2013)

14. Straullu, S, Abrate, S: Overview of the performances of PMMA-SI-POF communication systems. In: Proc. SPIE 8645, Broadband Access Communication Technologies VII. SPIE, San Francisco, (2013)

15. Atef, M, Zimmermann, H: Optical Communication over Plastic Optical Fibers, Springer Series in Optical Sciences, Vol. 172. Springer-Verlag, New York (2013)

16. Gimeno, C, Guerrero, E, Sanchez-Azqueta, C, Royo, G, Aldea, C, Celma S: A new equalizer for 2 Gb/s short-reach SI-POF links. In: Proc. SPIE 9520, Integrated Photonics: Materials, Devices, and Applications III. SPIE, Barcelona, (2015)

17. Gimeno, C, Guerrero, E, Sanchez-Azqueta, C, Aguirre, J, Aldea, C, Celma, S: Multi-Rate Adaptive Equalizer for Transmission Over Up to 50-m SI-POF. IEEE Photon. Technol. Lett. **29**(7), 587–590 (2017)

18. Song, J, Yang, Z, Yang, L, Gong, K, Pan, C, Wang, J, Wu, Y: Technical Review on Chinese Digital Terrestrial Television Broadcasting Standard and Measurements on Some Working Modes. IEEE Trans. Broadcast. **53**(1), 1–7 (2007)

19. Peng, L, Hélard, M, Haese, S, Liu, M, Hélard, J-F: Hybrid PN-ZP-DMT Scheme for Spectrum-Efficient Optical Communications and Its Application to SI-POF. IEEE/OSA J. Light. Technol. **32**(18), 3149–3160 (2014)

20. Liu, M, Crussière, M, Hélard, J-F: Improved Channel Estimation Methods based on PN sequence for TDS-OFDM. In: Proc. International Conference on Telecommunications (ICT). Jounieh, (2012)

21. Gerstackerand, WH, Schober, R: Equalization Concepts for EDGE. IEEE Trans. Wirel. Commun. **1**(1), 190–199 (2002)

22. Gerstacker, WH, Obernosterer, F, Meyer, R, Huber, JB: On Prefilter Computation for Reduced-State Equalization. IEEE Trans. Wirel. Commun. **1**(4), 793–800 (2002)

23. Eyuboglu, MV, Qureshi, SUH: Reduced-State Sequence Estimation with Set Partitioning and Decision Feedback. IEEE Trans. Commun. **36**(1), 13–20 (1988)

24. Ziemann, O, Krauser, J, Zamzow, PE, Daum, W: POF Handbook: Optical Short Range Transmission Systems. 2nd Edition. Springer, Heidelberg (2008)

25. Proakis, J: Digital Communications. Fourth Edition. McGraw Hill, New York (2001)

26. Peng, L, Hélard, M, Haese, S: On Bit-loading for Discrete Multi-tone Transmission over Short Range POF Systems. IEEE/OSA J. Light. Technol. **31**(24), 4155–4165 (2013)

27. Chatelain, B, Gagnon, F: Peak-to-average power ratio and inter symbol interference reduction by Nyquist pulse optimization. In: Proc. of IEEE 60th Vehicular Technology Conference. vol. 2, pp. 954–958. SPIE, Los Angeles, (2004)

21

Exclusive and efficient excitation of plasmonic breathing modes of a metallic nanodisc with the radially polarized optical beams

Mengjun Li[1], Hui Fang[2]* 🆔, Xiaoming Li[2] and Xiaocong Yuan[2]

Abstract

Background: The plasmonic breathing modes of a metallic nanodisc are dark plasmonic modes and thus also have the advantage of much smaller radiation loss. These modes emerged previously in electron energy loss spectroscopy experiments and were then pursued with optical excitation methods.

Results: In this paper, through the finite element method type numerical simulations, we show evidence of the pure excitation of these modes under the illumination of two counter-propagating, radially-polarized optical beams. The obtained near-field spectrum of a single Au or Ag nanodisc shows the plasmonic resonant peaks at which the electric field distributions, as well as the plasmonic dispersion relationship both clearly exhibit the characteristics of the plasmonic breathing modes.

Conclusion: We expect that the method that we have proposed, due to operating conveniently in the manner of far field excitation, will open many possibilities in practical applications based on the interaction between a single metal nanodisc with the radially polarized optical beam or between a metal nanodisc array with the beam array.

Keywords: Surface plasmon, Plasmonic breathing mode, Radially polarized optical beam, Metallic nanodisc

Background

As predicted [1], the plasmonics of metallic nanostructures remain a robustly growing research theme, as reflected, for instance, in the fast progress of quantum plasmonics [2], plasmon-induced hot carrier science and technology [3], and plasmon-enhanced Raman spectroscopy [4]. Among the many forms of metallic nanostructures [5], nanodiscs stand out as a very useful model for studying plasmonic properties. A metal nanodisc not only allows the coexistence of plasmonic resonances at all dimension levels, i.e., the volume, surface, and edge plasmonics [6, 7] but also possesses various types of surface plasmonic modes with well-defined two dimensional symmetries [8–10], in particular the interesting case of the plasmonic breathing modes [6, 7, 10].

The plasmonic breathing (PB) modes of a metal nanodisc are very special in the sense that they are radially symmetric and correlate with the intriguing collective radial oscillation of free electrons. Their appearance was first revealed in an electron energy loss spectroscopy study [10]. Their excitation by optical methods was considered to be difficult, because they are categorized as the dark modes owing to the null net electric dipole moment. Therefore, a recent experimental study [11] that showed that these modes can be excited under the oblique illumination of a linear polarized optical beam with the help of the retardation effect is notable. However, under this excitation approach, the PB modes are only weakly excited compared with the dominant plasmonic dipolar mode. Moreover, the other high-order plasmonic modes such as quadrupole and hexapole modes are also excited, which could overlay the PB modes. These situations were similarly encountered in the previous studies of optically excitation of the plasmonic dark modes

* Correspondence: fhui79@szu.edu.cn
[2]Nanophotonics Research Centre & Key Laboratory of Optoelectronic Devices and Systems of Ministry of Education and Guangdong Province, College of Optoelectronic Engineering, Shenzhen University, Shenzhen 518060, China
Full list of author information is available at the end of the article

based either on using the symmetry-breaking nanostructures or on employing the retardation effect [12, 13].

Considering the unique symmetry and the low radiation loss of a dark mode [14–16], it will be desirable to exclusively and efficiently excite the PB modes with an optical method, which will certainly boost the application of metal nanodiscs in research fields such as surface enhanced Raman scattering, plasmonic lasing, localized refractive index sensing [17], local heating [18], and plasmonic trapping [19].

In this paper, we carry out a finite element method (FEM) numerical investigation to examine the interaction of an Au or an Ag nanodisc with two counter-propagating, radially polarized optical (RPO) beams. We find that only the PB modes will be excited if the axis of the incident beams coincides with that of the nanodisc. We systemically studied the near-field spectra, the electric field distributions correlated with the PB modes, and the dispersion relationship of these modes. To the best of our knowledge, such clear identification and detailed characterizations for exciting this type of mode with the RPO beam have not been reported. Previously, there were a few related demonstrations that either proved that the radially pointed plasmonic dark modes of the specifically patterned metal nanoparticle assembles can be excited by the RPO beam [20–22] or showed that the angular momentum is conserved when the high-order plasmonic modes of a metal nanodisc are excited by a focused optical vortex beam [23].

Methods

We performed the FEM simulation in the Comsol Multiphysics software under the geometry configuration displayed in Fig. 1(a). For simplicity, an isolated metal nanodisc in vacuum is considered. As we focused on the situation where the axis of the RPO beam coincides with that of the metal nanodisc (defined as the z-axis), the 2D rotationally symmetric model was applied. In our simulation, the RPO beam is created according to the paraxial expression given by

$$\overrightarrow{E} = E_0 \frac{\pi w_0 r}{w_1^2} G W \exp(-ikz + iwt) \overrightarrow{e_r,}$$

$$\text{with } w_1 = w_0 \sqrt{1 + \left(\frac{z}{z_0}\right)^2},$$

$$G = \exp\left(-\frac{r^2}{w_1^2}\right),$$

$$W = \exp\left[\frac{-ikzr^2}{2(z_0^2 + z^2)} - i2\operatorname{atan}\left(\frac{z}{z_0}\right)\right],$$

$$z_0 = \frac{\pi w_0^2}{\lambda},$$

(1)

where w_0 represents the beam waist located at $z = 0$, G is the Gaussian envelope form, and W is the overall phase term including the Guoy phase. This formula [24, 25] is derived by considering that the RPO beam is actually composed of a right circularly polarized optical vortex beam with the topological charge of +1 and a left circularly polarized optical vortex beam with the topological charge of –1. We set $E_0 = 1$ V/m, $w_0 = 10$ µm, and the incident wavelength λ to be variable. The typical shape of the RPO beam with the null electrical field at the axis can be visualized clearly in Fig. 1(a) (calculated at $\lambda = 532$ nm).

In our FEM simulation, we launched two RPO beams propagating face to face, as evidenced from the interference pattern shown on the side cross section of Fig. 1(a). This was done to obtain a volume zone containing zero z-component of electrical field (E_z) so that a metal nanodisc can be contained there and the condition for the interaction with a pure radial polarized electrical field can be fulfilled. We found that the above condition cannot be achieved by launching only one RPO beam: the electrical field always has some z-component even at the beam axis (which actually agrees with the theoretical prediction [26, 27]). Figure 1(b) and 1(c), respectively, plot the magnified views of the electrical field distributions E_r (r-component of the electrical field) and E_z at the neighbouring area of the rotational axis and $z = 0$ plane. Obviously, there is a zone of approximately 20 nm height satisfying $E_z = 0$ (in both figures, the colour background show the amplitude of the net electrical field). Therefore, we confined our study to the metal nanodisc with the thickness of 10 nm and with its mirror symmetrical plane on the $z = 0$ plane such that only E_r is effective in exciting the PB modes of the nanodisc. Note that the total incident optical beam now has the symmetry of the point group $D_{\infty h}$, the same as the nanodisc, whereas the single RPO beam can only be described by the point group $C_{\infty v}$.

Results and discussion

We first considered the Au-nanodiscs with various diameters. To follow the scaling law already demonstrated for the PB modes [7, 10], we varied the set diameters as $d = 1200$ nm, 600 nm, 400 nm, and 300 nm. To represent real conditions, we have used the Au dielectric functions reported by Johnson and Christy [28].

Figure 2 shows the results of the near-field spectra taken at the spatial point located on the Au-nanodisc axis and 2 nm above the nanodisc. Figure 2(a) plots the electric field amplitude $|E|$ versus λ the wavelengths of the incident RPO beam, Fig. 2(b) plots the phase ϕ versus λ, while the table in Fig. 2(c) lists all λ correlated to the surface plasmonic resonance (SPR) peaks appearing in Fig. 2(a) and zero-phase points in Fig. 2(b).

As shown in Fig. 2(a), for the 1200 nm Au-nanodisc, there are clear three SPR peaks, whereas two peaks are

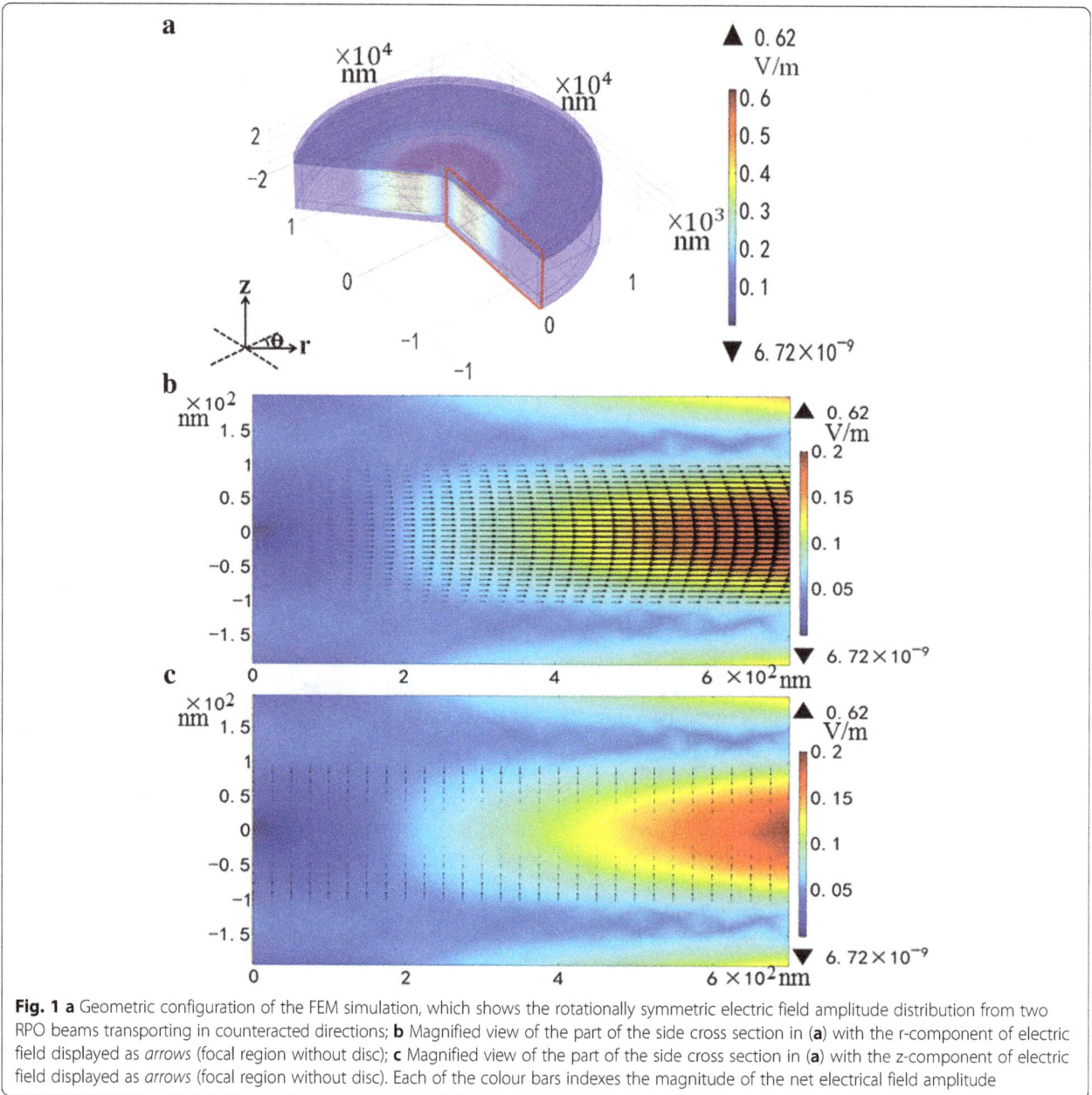

Fig. 1 a Geometric configuration of the FEM simulation, which shows the rotationally symmetric electric field amplitude distribution from two RPO beams transporting in counteracted directions; **b** Magnified view of the part of the side cross section in (**a**) with the r-component of electric field displayed as *arrows* (focal region without disc); **c** Magnified view of the part of the side cross section in (**a**) with the z-component of electric field displayed as *arrows* (focal region without disc). Each of the colour bars indexes the magnitude of the net electrical field amplitude

observed for the 600 nm Au-nanodisc and single peaks are observed for both 400 nm and 300 nm Au-nanodiscs. Going from the long λ to short λ (i.e., from low energy to high energy), it is obvious that the first SPR peak of the 600 nm Au-nanodisc coincides with the second SPR peak of the 1200 nm Au-nanodisc, the first SPR peak of the 400 nm Au-nanodisc coincides with the third SPR peak of the 1200 nm Au-nanodisc, and the first SPR peak of the 300 nm Au-nanodisc coincides with the second SPR peak of the 600 nm. These results clearly manifest the scaling law, which states that the SPR wavelengths λ_{SPR} of the PB modes scale with the nanodisc diameter according to [7, 10].

$$\lambda_{SPR} = d/n, \text{with n} = 1, 2, 3... \tag{2}$$

Figure 2(b) shows more interesting phenomena. The phase curves exhibit a series of π abrupt phase jumps near the SPR peaks shown in Fig. 2(a). This is actually in agreement with the criterion for characterizing a general resonance [29]. Now, the fourth and the fifth SPR peaks of the 1200 nm Au-nanodisc, the third peak of the 600 nm Au-nanodisc, and the second peak of the 400 nm Au-nanodisc, which are all barely visible in Fig. 2(a), can be easily identified. From Fig. 2(c), we can see that each of the SPR phase jumps λ (same as the zero-phase λ) is blueshifted relative to the corresponding SPR

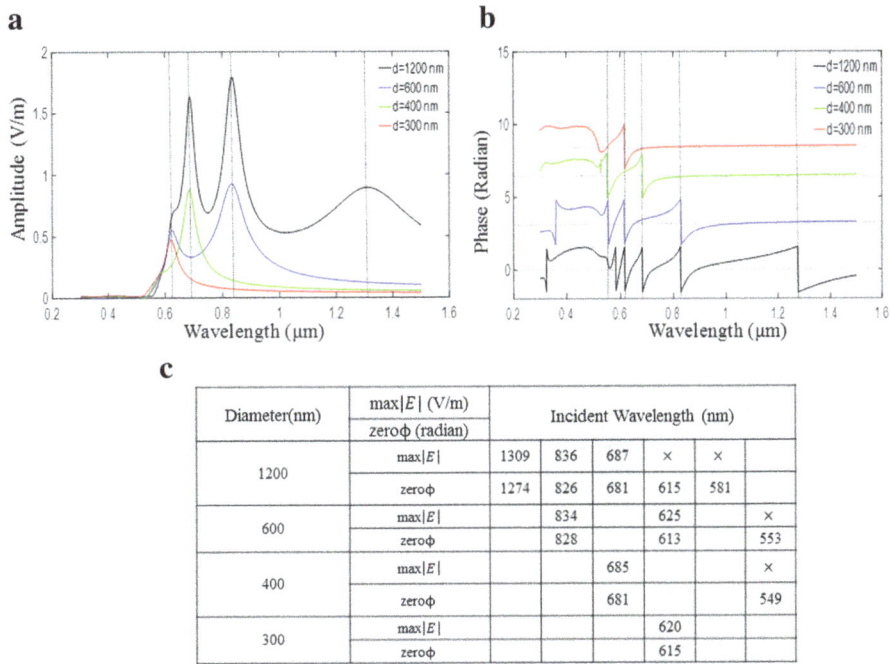

Fig. 2 Near-field spectra for various Au-nanodiscs with the same height of 10 nm but different diameters and obtained at the spatial point located on the nanodisc axis and 2 nm above the nanodisc. **a** Electric field amplitude versus incident wavelengths; **b** Electric field phase versus incident wavelengths; **c** Values of all incident wavelengths correlated with the SPR modes identified from (**a**) and (**b**). When plotting the curves in (**b**), vertical shift was applied for better visibility, and the zero-phase *lines* for each of the curves are plotted as the horizontal *dashed lines*

amplitude peak λ. From the data presented in this table, it is easy to deduce that the scaling law of Eq. (2) holds for all SPR modes defined by the SPR phase jumps.

To confirm that the SPR modes excited by the RPO beam are indeed the PB modes, we then computed their electric field distributions. Several examples are displayed in Fig. 3. Figure 3(a)-(d) show the $n = 4, 3, 2, 1$ SPR modes for the 1200 nm Au-nanodisc, while Fig. 3(e) shows the $n = 2$ SPR mode for the 600 nm Au-nanodisc and Fig. 3(f) shows the $n = 1$ SPR mode for the 300 nm Au-nanodisc. Here, we illustrate the various SPR modes excited for a same Au-nanodisc using Fig. 3(a)-(d), and compare the SPR modes for a different Au-nanodisc excited at a same incident wavelength using Fig. 3(a), (e), and (f).

In each of these figures, the coloured background draws the snapshot of E_z and the distributed arrows illustrate the snapshot of $\vec{E}$. Here, the electrical field takes the normalized value, with the normalization obtained by dividing with $|E_r|$ of the RPO beam at the spatial point 2 nm away from the edge of each Au-nanodisc. We note that by using such normalization, the resultant value of E_z and the arrow length of $\vec{E}$ both clearly reflect the enhancement factor and the excitation efficiency of these SPR modes excited by the RPO beam. At first glance, all electric field distributions display the apparent mirror symmetry with their vector nature. As the surface charge density is

determined by E_z at the surface (and the sign is determined by whether the direction of E_z is pointed outwards from the nanodisc), the surface charge distribution thus also shows mirror symmetry. Adding the radial symmetry, these symmetric properties fit well with the fingerprints of the PB modes.

Close examination of the electrical field distributions shown in Fig. 3(a)-(d) reveals their respective E_z patterns on the Au-nanodisc surface show 4, 3, 2, and 1 nodes, which well characterize the first four PB modes of the 1200 nm Au-nanodisc. These patterns also show the similarity to the zero-order Bessel functions. Inspection of Fig. 3(a), (e), and (f) shows that their respective E_z patterns show 4, 2, 1 nodes. As the modes in these three figures are excited at the same incident wavelength, they should have the same λ_{SPR} (considering the same dispersion relationship they should follow [7, 10]), and the scaling law of Eq. (2) is clearly verified once again. Finally, we can also compare Fig. 3(e) with Fig. 3(c) and (f) with Fig. 3(d), so that we can observe that the PB modes with the same index are merely rescaled by the changes in the nanodisc diameter.

It is then intriguing to further investigate whether these PB modes follow the dispersion relationship of the antisymmetric film surface plasmon (FSP) as has been claimed previously [7, 10]. The results are displayed in Fig. 4. Here, the discrete data points are plotted according to the values listed in the table of Fig. 2. The SPR

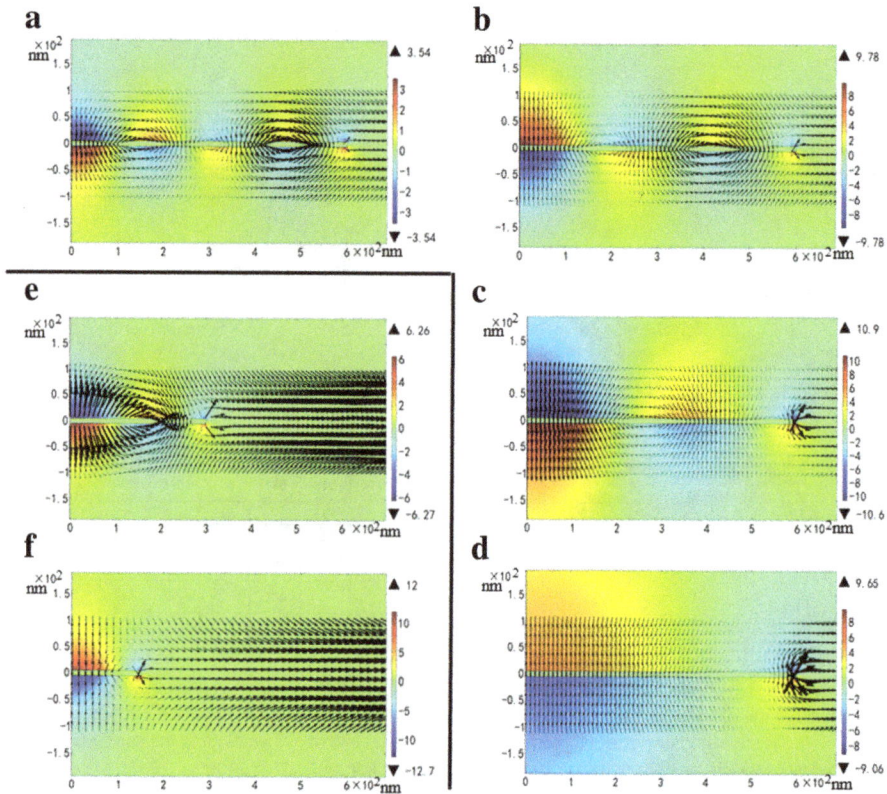

Fig. 3 Examples of computed electrical field distributions of the PB modes excited by the RPO beam (the electrical field is given in the normalized values by dividing with the electrical field amplitude of the incident beam at the point 2 nm away from the edge of each nanodisc). **a-d** PB modes of the 1200 nm Au-nanodisc excited at the incident wavelengths of 0.619 μm, 0.684 μm, 0.8304 μm and 1.291 μm, respectively. **e-f** PB modes excited at the incident wavelength of 0.619 μm for the 600 nm Au-nanodisc and the 300 nm Au-nanodisc, respectively. As can be determined in Fig. 2(c), all of these wavelengths are slightly larger than the zero phase wavelengths. From (**a**)-(**f**), the arrows are plotted with the relative length scale of 1:1:1:1.5:1:0.5 (the smaller scale means that the same length represents smaller electrical field amplitude)

energy is calculated as $hc/(\lambda e)$, and the SPR wavenumber takes the value of $2\pi/\lambda_{\text{SPR}}$ with λ_{SPR} determined by Eq. (2). The dispersion curve shown by the solid line is the result of the antisymmetric FSP calculated for the 10 nm thick Au film using [30, 31]:

$$\tanh(k_1) = -(k_1\varepsilon_2)/(k_2\varepsilon_1), \text{ with } k_i^2 = \beta^2 - k_0^2\varepsilon_i, \quad (3)$$

where β is the SPR wavenumber, k_0 is the incident wave wavenumber, ε_1 is the dielectric function of Au [28], and $\varepsilon_2 = 1$ is the dielectric constant of air. Here, the red dots represent the first five PB modes of the 1200 nm Au nanodisc, and the blue dot represents the third PB mode of the 600 nm Au nanodisc (the zero phase λ are used, as shown in Fig. 2(c) table). The data points for the other PB modes listed in Fig. 2(c) table will overlay with these red dots or the blue dot. Therefore, all obtained PB modes tightly follow the dispersion relationship of the Au antisymmetric FSP.

Finally, we also endeavoured to perform a similar systemic study on the 10 nm thick Ag nanodiscs. The

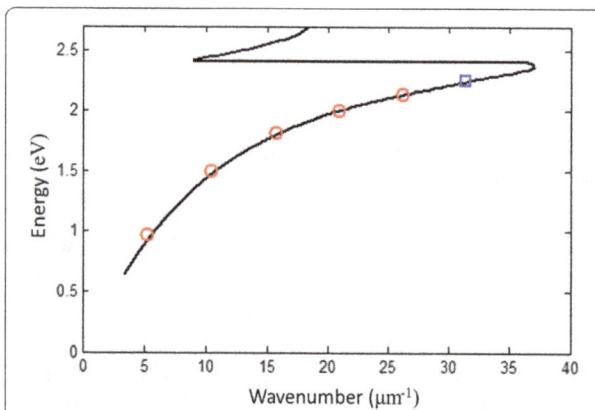

Fig. 4 Dispersion relationship of the PB modes compared with the dispersion curve of the antisymmetric FSP. *Red* and *blue dots* represent PB modes of the 10 nm thick Au-nanodiscs

results of the near field spectrum of the 1200 nm Ag nanodisc and the dispersion relationship of its PB modes are both plotted in Fig. 5. The Ag dielectric function from [28] is also applied. As seen, the obtained results are consistently similar to those obtained for the Au nanodisc.

In our current simulation, we considered the ideal simplified case that the RPO beam is not tightly focused. As the beam waist must be set to be much larger than the incident wavelength to satisfy the paraxial transport condition, the total portion of the light energy utilized to excite the PB modes of the metal nanodisc is limited. However, our intention here is to show the first step to prove that the pure radially pointed electrical field of the RPO beam can efficiently excite the PB modes where the excitation efficiency is defined relative to the beam intensity at the point close to the nanodisc edge instead of

to the maximum intensity across the whole beam. We will extend our study to the more practical case with the focused RPO beam in the next report where the effects of E_r and E_z on the PB mode excitation both need to be taken into account. Moreover, we will also upgrade to a 3-dimensional model, which allows us to investigate other realistic conditions such as the case when a supporting dielectric layer for the metal nanodisc is present, when the RPO beam is not perfectly aligned with the metal nanodisc and when the metal nanodisc is not perfectly shaped as a circle.

Conclusions

To summarize, we proposed the use of radially polarized optical vector beams to excite the plasmonic breathing modes of a metal nanodisc and confirmed this capability through the finite-element method simulation. The near

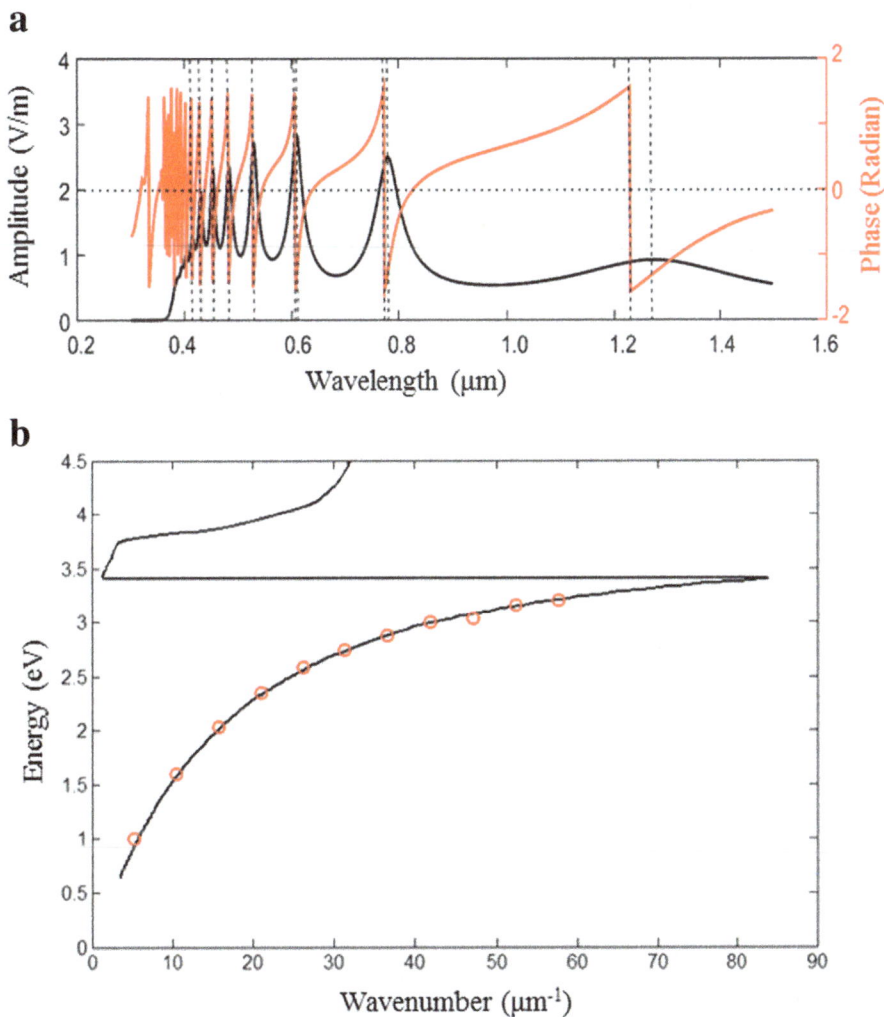

Fig. 5 Simulated results for the 10 nm thick Ag nanodisc with the diameter of 1200 nm. **a** Near-field spectrum taken at the spatial point located on the nanodisc axis and 2 nm above the nanodisc. **b** Dispersion relationship of the PB modes compared with the dispersion curve of the antisymmetric FSP

field spectra and the dispersion relationships presented in Figs. 2, 4, and 5 strongly supported the scaling law given by Eq. (2) for the metal nanodisc PB modes. Accordingly, one of the unique advantages of applying the metal nanodisc in plasmonic studies is that the plasmonic resonant wavelengths of its plasmonic breathing modes can be predictably tuned by adjusting the nanodisc size. Considering the fast expansion of the research directed at the application of optical vector beams to interact with nanostructures [32], we believe that the method demonstrated here will facilitate the study of dark mode surface plasmonics and will be applied to some practical designs such as the metal nanodisc based gap mode surface enhance Raman scattering and tip enhance Raman scattering, and the metal nanodisc array based plasmonic trapping and heating.

Abbreviations
FEM: finite element method; FSP: film surface plasmon; PB: plasmonic breathing; RPO: radially polarized optical; SPR: surface plasmonic resonance

Funding
This work is supported by the National Basic Research Program of China (Grant No. 2015CB352004), the National Natural Science Foundation of China under Grant No. 61427819, the Specialized Research Fund for the Doctoral Program of Higher Education of China (20130031110036), and the Tianjin Municipal Science and Technology Commission, China (14JCYBJC16600).

Authors' contributions
Hui Fang proposed the original idea, guided the research and wrote the manuscript. Mengjun Li implemented the simulations and drafted the manuscript. Xiaoming Li was involved in some of the simulations. Xiaocong Yuan advised the research and revised the manuscript. All of the authors read and approved the final manuscript.

Competing interests
The author(s) declare(s) that they have no competing interests.

Author details
[1]Institute of Modern Optics, College of Electronic Information and Optical Engineering, Nankai University, Tianjin 300071, China. [2]Nanophotonics Research Centre & Key Laboratory of Optoelectronic Devices and Systems of Ministry of Education and Guangdong Province, College of Optoelectronic Engineering, Shenzhen University, Shenzhen 518060, China.

References
1. Halas, NJ, "Plasmonics: An Emerging Field Fostered by Nano Letters," Nano Lett. 10(10), 3816–3822(2010)
2. Tame, MS, McEnery, KR, Ozdemir, SK, Lee, J, Maier, SA, Kim, MS: Quantum plasmonics. Nat. Phys. 9, 329–340 (2013)
3. Brongersma, ML, Halas, NJ, Nordlander, P: Plasmon-induced hot carrier science and technology. Nat. Nanotechnol. 10, 25–34 (2015)
4. Ding, SY, Yi, J, Li, JF, Ren, B, Wu, DY, Panneerselvam, R, Tian, ZQ: Nanostructure-based plasmon-enhanced Raman spectroscopy for surface analysis of materials. Nature Reviews Materials. 1, 1–16 (2016)
5. Zhang, JX, Zhang, LD: Nanostructures for surface plasmons. Adv. Opt. Photon. 4, 157–321 (2012)
6. Hobbs, RG, Manfrinato, VR, Yang, YJ, Goodman, SA, Zhang, LH, Stach, EA, Berggren, KK: High-energy surface and volume plasmons in nanopatterned sub-10 nm aluminum nanostructures. Nano Lett. 16, 4149–4157 (2016)
7. Schmidt, FP, Ditlbacher, H, Hohenester, U, Hohenau, A, Hofer, F, Krenn, JR: Universal dispersion of surface plasmons in flat nanostructures. Nat. Commun. 5, 3604 (2014)
8. Schmidt, FP, Ditlbacher, H, Hofer, F, Krenn, JR, Hohenester, U: Morphing a plasmonic nanodisk into a nanotriangle. Nano Lett. 14, 4810–4815 (2014)
9. Imura, K, Ueno, K, Misawa, H, Okamoto, H, McArthur, D, Hourahine, B, Papoff, F: Plasmon modes in single gold nanodiscs. Opt. Express. 22, 12189–12199 (2014)
10. Schmidt, FP, Ditlbacher, H, Hohenester, U, Hohenau, A, Hofer, F, Krenn, JR: Dark plasmonic breathing modes in silver nanodisks. Nano Lett. 12, 5780–5783 (2012)
11. Krug, MK, Reisecker, M, Hohenau, A, Ditlbacher, H, trugler, A, Hohenester, U, and Krenn, JR: "Probing plasmonic breathing modes optically," Applied Physics Letters 105, 171103 (2014)
12. Hao, F, Sonnefraud, Y, Dorpe, PV, Maier, SA, Halas, NJ, Nordlander, P: Symmetry breaking in plasmonic nanocavities: subradiant LSPR sensing and a tunable Fano resonance. Nano Lett. 8, 3983–3988 (2008)
13. Hao, F, Larsson, EM, Ali, TA, Sutherland, DS, Nordlander, P: Shedding light on dark plasmons in gold nanorings. Chem. Phys. Lett. 458, 262–266 (2008)
14. Liu, MZ, Lee, TW, Gray, SK, Sionnest, PG, Pelton, M: Excitation of dark plasmons in metal nanoparticles by a localized emitter. Phys. Rev. Lett. 102, 107401 (2009)
15. Chen, HY, He, CL, Wang, CY, Lin, MH, Mitsui, D, Eguchi, M, Teranishi, T, Gwo, S: Far-field optical imaging of a linear array of coupled gold nanocubes: direct visualization of dark plasmon propagating modes. ACS Nano. 10, 8223–8229 (2011)
16. D. Solis Jr, Willingham, BSL L, Nauert, LS, Slaughter, J, Olson, P, Swanglap, A, Paul, WS, Chang, and Link, S: "Electromagnetic energy transport in nanoparticle chains via dark plasmon modes," Nano Letter 12, 1349–1353 (2012)
17. Hafele, V, Trugler, A, Hohenester, U, Hohenau, A, Leitner, A, Krenn, JR: Local refractive index sensitivity of gold nanodisks. Opt. Express. 23, 10293–10300 (2015)
18. Donner, JS, Baffou, G, McCloskey, D, Quidant, R: Plasmon-assisted optofluidics. ACS Nano. 7, 5457–5462 (2011)
19. Righini, M, Volpe, G, Girard, C, Petrov, D, Quidant, R: Surface plasmon optical tweezers: tunable optical manipulation in the femtonewton range. Phys. Rev. Lett. 100, 186804 (2008)
20. Paramon, JS, Bosch, S: Dark modes and Fano resonances in plasmonic clusters excited by cylindrical vector beams. ACS Nano. 9, 8415–8423 (2012)
21. Gomez, DE, Teo, ZQ, Altissimo, M, Davis, TJ, Earl, S, Roberts, A: The dark side of plasmonics. Nano Lett. 13, 3722–3728 (2013)
22. Yanai, A, Grajower, M, Lerman, GM, Hentschel, M, Giessen, H, Levy, U: Plasmonic oligomers under radially and azimuthally polarized light excitation. ACS Nano. 8, 4969–4974 (2014)
23. Sakai, K, Nomura, K, Yamamoto, T, Sasaki, K: Excitation of multipole plasmons by optical vortex beams. Sci Rep. 5, 8431 (2015)
24. Andrews, DL, Babiker, M: Eds, the Angular Momentum of Light (Cambridge University Press, 2013)
25. Zhan, QW: Cylindrical vector beams: from mathematical concepts to applications. Adv. Opt. Photon. 1, 1–57 (2009)
26. Herrero, RM, Mejias, PM: Propagation of light fields with radial or azimuthal polarization distribution at a transverse plane. Opt. Express. 16, 9021–9033 (2008)
27. Kotlyar, VV, Kovalev, AA: Nonparaxial propagation of a Gaussian optical vortex with initial radial polarization. J. Opt. Soc. Am. A. 27, 372–380 (2010)
28. Johnson, PB, Christy, RW: Optical constant of nobel metals. Phys. Rev. B. 6, 4370–4379 (1972)
29. Joe, YS, Satanin, AM, Kim, CS: Classical analogy of Fano resonances. Phys. Scr. 74, 259–266 (2006)
30. Maier, SA: Plasmonics: Fundamentals and Applications (Springer, 2007)
31. Dionne, JA, Sweatlock, LA, Atwater, HA: Planar metal plasmon waveguides: frequency-dependent dispersion, propagation, localization, and loss beyond the free electron model. Phys. Rev. B. 72, 075404 (2005)
32. Wozniak, P, Banzer, P, Leuchs, G: Selective switching of individual multipole resonances in single dielectric nanoparticles. Laser Photonics Rev. 9, 231–240 (2015)

22

Plasmonic behavior of III-V semiconductors in far-infrared and terahertz range

Jan Chochol[1,2]* ⓘ, Kamil Postava[3], Michael Čada[2], Mathias Vanwolleghem[4], Martin Mičica[1,4], Lukáš Halagačka[1,3], Jean-François Lampin[4] and Jaromír Pištora[1]

Abstract

Background: In this article, III-V semiconductors are proposed as materials for far-infrared and terahertz plasmonic applications. We suggest criteria to estimate appropriate spectral range for each material including tuning by fine doping and magnetic field.

Methods: Several single-crystal wafer samples (n,p-doped GaAs, n-doped InP, and n,p-doped and undoped InSb) are characterized using reflectivity measurement and their optical properties are described using the Drude-Lorentz model, including magneto-optical anisotropy.

Results: The optical parameters of III-V semiconductors are presented. Moreover, strong magnetic modulation of permittivity was demonstrated on the undoped InSb crystal wafer in the terahertz spectral range. Description of this effect is presented and the obtained parameters are compared with a Hall effect measurement.

Conclusion: Analyzing the phonon/free carrier contribution to the permittivity of the samples shows their possible use as plasmonic materials; the surface plasmon properties of semiconductors in the THz range resemble those of noble metals in the visible and near infrared range and their properties are tunable by either doping or magnetic field.

Keywords: Surface plasmons, Semiconductor materials, Magneto-optical materials, THz-TDS, FTIR

Background

Utilizing the terahertz range for better and faster communications [1–3], sensing [4], medicine [5] and security [6] has created a need for devices, capable of operating in the desired frequency range of 0.1-30 THz. One of the principles that can serve as the basis of guiding, coupling and modulating THz waves is the surface plasmon - an interface wave propagating at the boundary of negative (conductive) and positive (dielectric) permittivity material. Traditional plasmonic materials usable in visible/near infrared range, noble metals, are unsuitable for uses in the THz regime due to low confinement to the metal;

the wave is weakly bound to the interface, a phenomenon sometimes called the Zenneck plasmon [7]. Semiconductors with their carrier levels have their metallic properties shifted to lower frequencies - microwave, terahertz and far infrared. They are therefore suitable as building blocks for THz devices. Furthermore, they allow for much needed control of their electromagnetic properties. In the manufacturing the carrier levels can be adjusted by doping and after the manufacturing the properties can be controlled by light [8], temperature [9], electric gating [10] and by external magnetic field [11].

This paper compares plasmonic behavior of several samples of doped III-V semiconductors (InP-n, GaAs-n,p, InSb-n,p) and undoped InSb through spectroscopic characterization using a terahertz time-domain spectrometer (THz-TDS) in the range of 2–100 cm^{-1} and a Fourier transform infrared spectrometer (FTIR) in the range of 50–7500 cm^{-1}. Appropriate figures of merit are estimated to establish suitable ranges for room temperature

*Correspondence: jan.chochol@vsb.cz; jan.chochol@dal.ca; jan.chochol@gmail.com
[1]Nanotechnology Centre, VSB – Technical University of Ostrava, 17. listopadu 15/2172, 708 33 Ostrava, Poruba, Czech Republic
[2]Department of Electrical and Computer Engineering, Dalhousie University, 6299 South St, Halifax NS B3H 4R2, Canada
Full list of author information is available at the end of the article

plasmonics applications. Furthermore, the undoped InSb sample is characterized in the presence of static magnetic field to explore the magnetic modulation.

The issue of semiconductor plasmonic properties in the far-infrared and terahertz range have been undertaken by several groups. The work of Palik and Furdyna [12] provides the necessary theory for optical and magneto-optical behavior of semiconductors. The experiments of Shubert et al. [13, 14] and Hofmann [15] show the potential of spectroscopic techniques in investigating that behavior, while further works [16–22] demonstrate the power of terahertz time-domain spectroscopy in determining the conductive and optical functions of semiconductors. Moreover, several papers [23–25] deal with the theory of the existence of surface plasmons in magnetically tunable materials.

Section "Optical functions of doped semiconductors" outlines the tools necessary for describing and modeling optical properties using the Drude-Lorentz model. Sections "Samples" and "Measurement" describe samples and the techniques used, and Section "Results and discussion" presents the measured spectra with the permittivity and parameters obtained from a reflectivity fit. Section "Semiconductors as plasmonic materials" discusses the suitability of these materials for plasmonic applications and Section "Magnetic modulation" provides data for magnetic modulation of plasmonic properties on undoped InSb.

Methods
Optical functions of doped semiconductors
The optical and conductive properties of III-V semiconductors in the far infrared and terahertz range are governed by three mechanisms, the free carrier absorption, lattice vibration and background permittivity, originating from high-frequency interband absorptions. These mechanisms are summarized in the Drude-Lorentz function, as

$$\varepsilon_r = \varepsilon_\infty - \underbrace{\frac{\omega_p^2}{\omega^2 + i\gamma_p\omega}}_{\varepsilon_D} + \underbrace{\frac{A_L\omega_L^2}{\omega_L^2 - \omega^2 - i\gamma_L\omega}}_{\varepsilon_L}, \quad (1)$$

which consists of three terms. The first one is the constant ε_∞, the background permittivity. The second one is the Drude term ε_D, originating from free carriers, where

$$\omega_p = \left(\frac{Ne^2}{\varepsilon_0 m^*}\right)^{\frac{1}{2}} \quad (2)$$

is the plasma frequency, N is the carrier concentration, e is the electron charge, ε_0 is the permittivity of free space and m^* is the effective mass of the charge carriers. γ_p is the damping constant, the inverse of the scattering time $\tau_p = 1/\gamma_p$.

The last term is the Lorentz term ε_L, describing a lattice vibration at the frequency ω_L, with the damping constant $\gamma_L = 1/\tau_L$ and the amplitude A_L. The amplitude can be understood as the difference between the permittivity limit below and above the oscillation at ω_L.

The parameters ε_∞, ω_p, τ_p, ω_L, τ_L, and A_L are the fitting parameters for the Drude-Lorentz model. The measured quantity is the reflectivity R, modeled using Berreman 4×4 matrix method [26].

The measurement of reflectivity allows us to obtain only the plasma frequency and the scattering time, in addition to the constant term and parameters of the Lorentz oscillator, as defined by Eq. (1). Using the plasma frequency and scattering time one can calculate the DC conductivity as

$$\sigma_0 = \frac{Ne^2\tau}{m^*} = \varepsilon_0\,\omega_p^2\,\tau\,. \quad (3)$$

Samples
We have measured six representative samples of single-crystal III-V semiconductors. All were polished on one side.

The GaAs samples were 2", 0.35 mm thick wafers made by AXT, Inc. One n-doped with Si dopants, with the reported electron concentration of $(0.8 - 4) \times 10^{18}$ cm^{-3} and the mobility of $(1 - 2.5) \times 10^3$ cm^2/Vs. One p-doped (Zn), with the reported hole concentration of $(0.5 - 5) \times 10^{19}$ cm^{-3} and the mobility of $50 - 120$ cm^2/Vs.

The InP sample was 2", 0.35 mm thick wafer also from AXT. It is n-doped with Sulfur; manufacturer reports values of $N = (0.8 - 8) \cdot 10^{18}$cm^{-3} and $\mu = (1 - 2.5) \cdot 10^3$cm^2/Vs.

Measured samples of InSb are n-doped (Te), p-doped (Ge) and undoped. All InSb samples were manufactured by MTI Corp as wafers of 2" diameter and a small square 10x10 mm of undoped InSb. The small sample was used for the Hall measurements by 4-contact van der Pauw method. The thickness of the wafers was 0.5 mm for the n-doped and 0.45 mm for the undoped and p-doped. The small sample has thickness of 0.45 mm. The n-doped samples have the manufacturer's reported carrier concentration of $(0.19 - 0.50) \cdot 10^{18}$ cm^{-3} and the mobility of $(3.58 - 5.60) \cdot 10^4$ cm^2/Vs, both at 77 K. The p-doped samples have the following reported parameters: $N = 0.5 - 5 \cdot 10^{17}$ cm^{-3} and $\mu = 4 - 8.4 \cdot 10^3$ cm^2/Vs, again at 77 K.

Measurement
We used two spectrometers to characterize the samples. The first one is the terahertz time-domain spectrometer TPS Spectra 3000 from TeraView Co., measuring in the THz range of 2-100 cm^{-1}. The second one is the Fourier transform infrared spectrometer Bruker Vertex

70v, measuring in the far-infrared range of 50–680 cm^{-1} and in the mid-IR range of 370–7500 cm^{-1}. All measurements were done in reflection configuration, with a fixed angle of incidence of 11 degrees, which doesn't allow measurements at smaller angles due to space limitations. On the other hand, such angle of incidence can be considered as near normal one and simplifies description of reflective phenomena. The data in overlapping ranges were averaged. We have used a thick gold layer as the reference. To avoid the influence of water vapor absorption, both reflection spectra were measured in vacuum.

Results and discussion

Spectroscopic characterization

Figure 1 shows the reflectivity spectra and the permittivity of the samples with the resulting parameters listed in Table 1 and the obtained permittivity in Fig. 2.

The sharp minima in reflectivity between 150 and 300 cm^{-1} corresponds to a crossing of the real part of the permittivity with the permittivity of vacuum due to the lattice vibrations. The fitted value of the Lorentz oscillator frequency is at the maximum of the imaginary part of the permittivity, called the transversal phonon [27]. The Lattice vibrations match those reported by other authors (InP [28], GaAs [11], InSb [12]).

The plasma edge, a region where the real part of the permittivity crosses zero and becomes negative for lower frequencies due to the free carries and the reflectivity rises is tied to the concentration and effective mass. For metals described by the Drude term, this would be where $\Re\{\varepsilon\} = 0$. Semiconductors however have a strong background permittivity, which from Eq. (1), places the

crossover frequency (reduced plasma freq.) between positive and negative at $\omega = \omega_p/\sqrt{\varepsilon_\infty}$ and the reflectivity minimum at $\omega = \omega_p/\sqrt{\varepsilon_\infty - 1}$; assuming no damping and negligible effect of the phonon. Real cases show effect of the phonon and damping, i.e. the n-doped GaAs the reflectivity minimum would be at 545.2 cm^{-1} but the real one is at 573.6 cm^{-1}. The effect of damping is strongly present in the p-doped samples, where the short scattering time of the holes makes the reflectivity spectra much shallower.

Semiconductors as plasmonic materials

There are two basic formulae describing the behavior of surface plasmon polaritons (SPP) on an interface between a dielectric and a metal/semiconductor. Those are the expressions for the wave vector components along the interface (y direction) and perpendicular to it (z direction). With the wave vector defined as $\mathbf{k} = \mathbf{x}k_x + \mathbf{y}k_y + \mathbf{z}k_z$ and assuming $k_x = 0$, the components of interest are

$$k_y = \frac{\omega}{c}\sqrt{\frac{\varepsilon_1\varepsilon_2}{\varepsilon_1 + \varepsilon_2}}, \tag{4a}$$

$$k_{zj} = \frac{\omega}{c}\sqrt{\frac{\varepsilon_j^2}{\varepsilon_1 + \varepsilon_2}}, \tag{4b}$$

where $j = 1, 2$ is the index of the respective media. For simplicity, let us assume the top medium is air, $\varepsilon_1 = 1$. To ensure a propagating surface plasmon polariton, two conditions must be fulfilled. The k_y (component along the interface) must be real and k_{zj} must be imaginary to ensure the localization of the plasmon (assuming just real permittivities). This occurs when the real part of ε_2 is negative

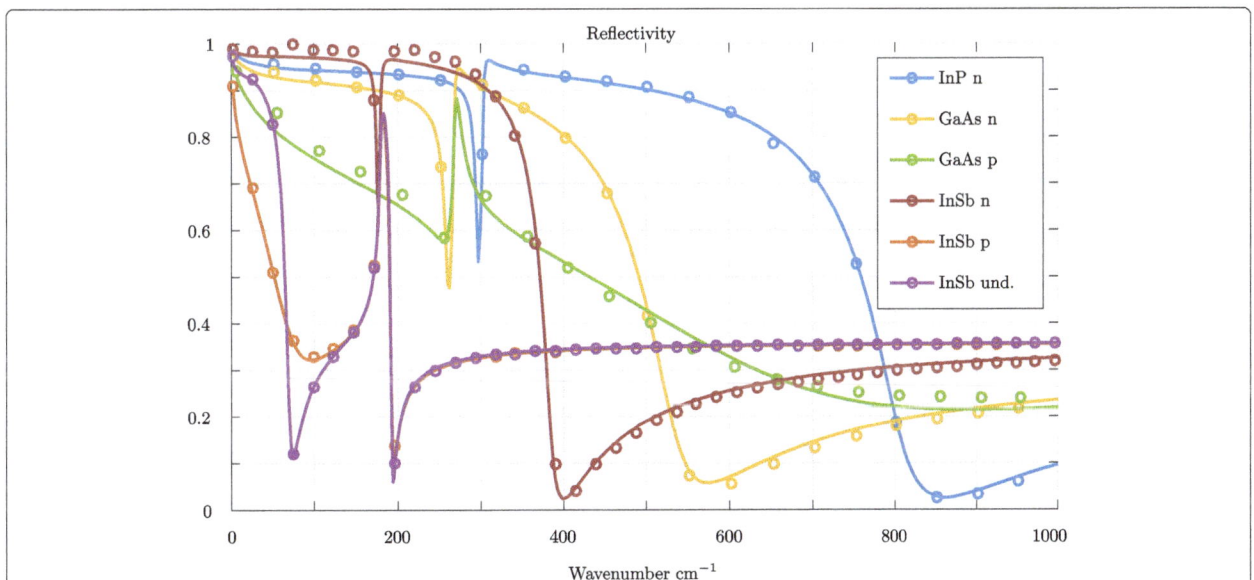

Fig. 1 Measured and fitted reflectivity spectra of the samples in the terahertz and far-infrared range, overlapping ranges were averaged. Data (*circles*) reduced for clarity

Table 1 Fitted parameters of GaAs, InP and InSb

	GaAs n-doped	GaAs p-doped	InP n-doped	InSb n-doped	InSb p-doped	InSb undoped
ω_p (10^{14} rad/s)	3.33 ± 0.01	4.44 ± 0.02	4.70 ± 0.01	2.82 ± 0.01	0.63 ± 0.01	0.57 ± 0.01
ω_p (cm^{-1})	1769.2 ± 1.9	2356.4 ± 9.1	2494.1 ± 1.8	1495.2 ± 1.83	332.6 ± 1.49	302.4 ± 0.33
τ_p (10^{-1} ps)	0.71 ± 0.01	0.09 ± 0.01	0.71 ± 0.01	2.65 ± 0.04	0.75 ± 0.01	5.16 ± 0.06
ω_L (10^{13} rad/s)	5.06 ± 0.01	5.06 ± 0.01	5.73 ± 0.01	3.38 ± 0.01	3.38 ± 0.01	3.38 ± 0.01
ω_L (cm^{-1})	268.4 ± 0.1	268.5 ± 0.2	303.9 ± 0.1	179.4 ± 0.13	179.4 ± 0.06	179.5 ± 0.06
τ_L (ps)	2.79 ± 0.27	1.95 ± 0.29	3.01 ± 0.24	1.81 ± 0.13	1.90 ± 0.04	1.99 ± 0.04
A_L	2.13 ± 0.03	2.15 ± 0.09	2.89 ± 0.04	2.02 ± 0.07	2.00 ± 0.01	2.02 ± 0.01
ε_∞	11.58 ± 0.01	11.34 ± 0.02	10.01 ± 0.01	15.68 ± 0.03	15.74 ± 0.01	15.86 ± 0.01
σ_0 (kS/m)	69.46 ± 0.54	16.54 ± 0.22	139.06 ± 0.77	186.35 ± 2.63	2.51 ± 0.04	14.83 ± 0.17

and is greater in absolute value than that of ε_1, which must be positive. That means $\Re\{\varepsilon_2\} < -\varepsilon_1$. A real case scenario has both components complex, meaning that the surface plasmon is decaying with propagation. The propagation length, denoted here as L_{SPP} when the electric field of the SPP drops to $1/e$ is simply $L_{SPP} = 1/\Im\{k_y\}$ and the penetration into the material (again, when the field drops to $1/e$) is $L_{1,2} = 1/\Im\{k_{z1,2}\}$.

An ideal material would allow a long propagation length of the SPP along the interface, yet sufficient confinement into the metallic (conductive) material; in other words short extension into the dielectric. When the difference between the permittivities ε_1 and ε_2 is large, the SPP can propagate many wavelengths, but is poorly guided by the interface (a small penetration depth into the conductor) and most of its energy is carried in the dielectric. The opposite is also valid - a heavily confined wave will have a lot of energy traveling in the absorbing material, and thus the propagation length is short.

Noble metals such as gold or silver are used for plasmonic applications in the visible and near-infrared range. By comparing the properties of the SPP on gold in the visible range and on semiconductors in the THz range, one can estimate how suitable the semiconductors are for plasmonic applications in the THz range. The comparison of the propagation length (along the interface) and the penetration (into the conducting material) normalized to the free space wavelength of light is shown in Fig. 3.

As Fig. 3 shows, the properties of semiconductors in the THz are almost identical to that of gold and silver in the visible range. For longer wavelengths, the trends on noble metals continue linearly to smaller confinement and longer propagation. The semiconductors do have several advantages. The adjustable doping concentration can significantly change the behavior of semiconductor, as can be seen from comparing the three samples of InSb. Even the p-doped sample is shown to be able of sustaining a surface

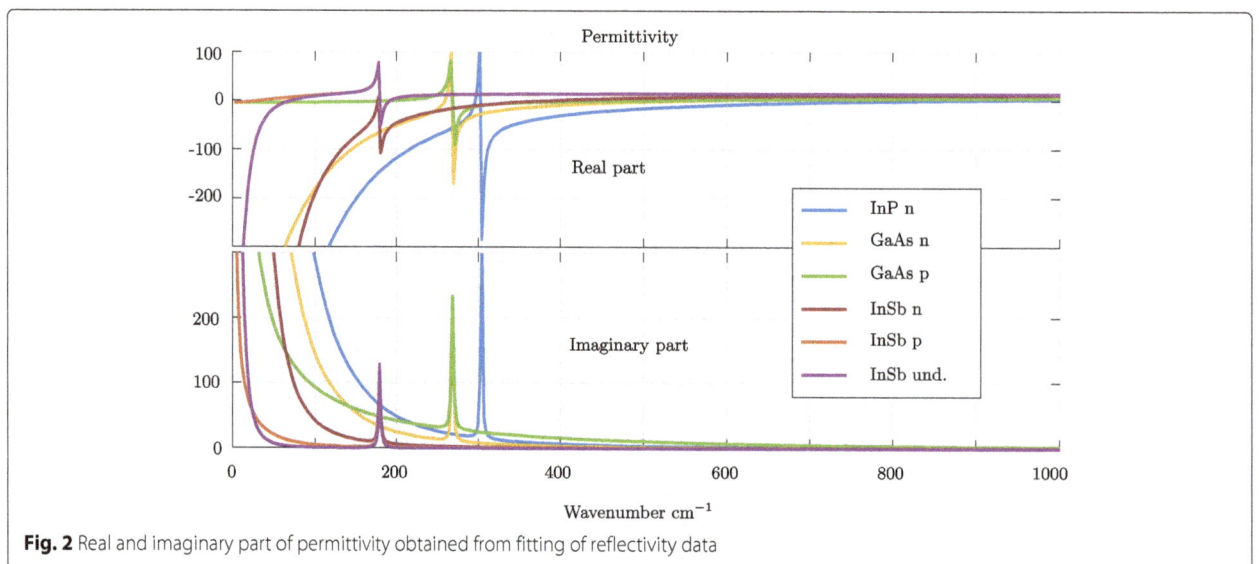

Fig. 2 Real and imaginary part of permittivity obtained from fitting of reflectivity data

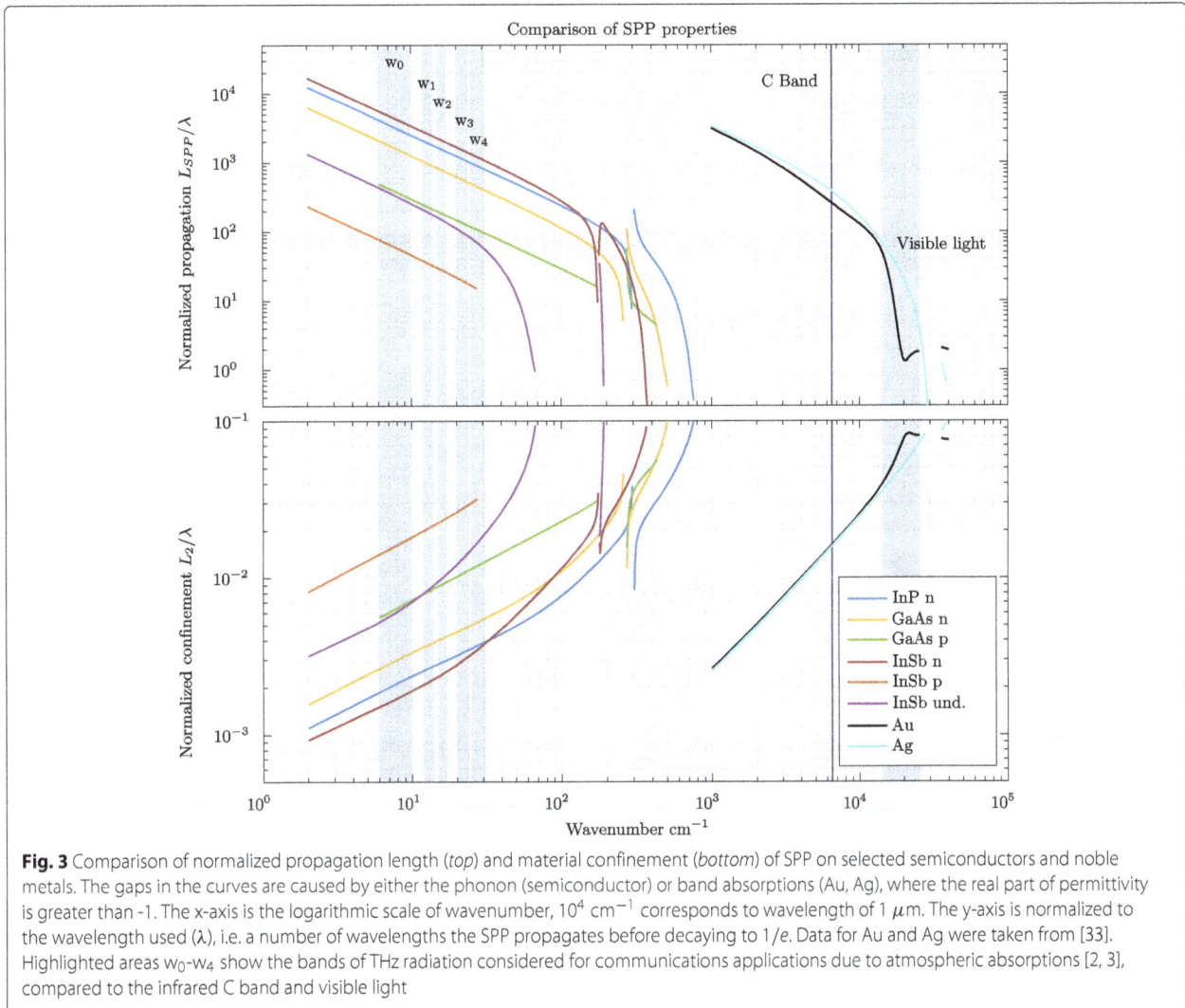

Fig. 3 Comparison of normalized propagation length (*top*) and material confinement (*bottom*) of SPP on selected semiconductors and noble metals. The gaps in the curves are caused by either the phonon (semiconductor) or band absorptions (Au, Ag), where the real part of permittivity is greater than -1. The x-axis is the logarithmic scale of wavenumber, 10^4 cm^{-1} corresponds to wavelength of 1 μm. The y-axis is normalized to the wavelength used (λ), i.e. a number of wavelengths the SPP propagates before decaying to $1/e$. Data for Au and Ag were taken from [33]. Highlighted areas w_0-w_4 show the bands of THz radiation considered for communications applications due to atmospheric absorptions [2, 3], compared to the infrared C band and visible light

plasmon for low energies. Therefore, doping can be used to fine-tune the plasmonic properties of semiconductors. Other techniques, such as optical pumping, electric gating, or as demonstrated in the next section, magneto-optics allow for further tuning, switching or modulation of surface plasmons on semiconductors. The gaps in the curves, caused by the phonon, and the rapid change of behavior around them, lead to a surface phonon-polariton (i.e. on the undoped InSb) or a combination of both, where the electromagnetic energy is stored not just in the collective oscillation of the free carriers, but also in the vibrations of the lattice.

Magnetic modulation

This section uses a simple THz reflectivity measurement to demonstrate the strength of this magnetic modulation on undoped InSb. Similar experiment has been done by Ino [21] on InAs. This type of measurement has been called the "Optical Hall effect" [11].

In the presence of the magnetic field the permittivity tensor becomes anisotropic. In our case, the magnetic field is in the z-direction (perpendicular to the interface in the x and y directions, $y - z$ is the plane of incidence). A derivation is presented in the [27]. The form of the tensor with the magnetic field applied in the z direction (polar configuration) is

$$\hat{\varepsilon_r} = \begin{bmatrix} \varepsilon_{xx} & \varepsilon_{xy} & 0 \\ \varepsilon_{yx} & \varepsilon_{yy} & 0 \\ 0 & 0 & \varepsilon_{zz} \end{bmatrix}. \tag{5}$$

The zz component stays the same as ε_r in (1) and xx, yy, xy, yx components of the Drude term (shown with the constant term) change to

$$\varepsilon_{D,xx} = \varepsilon_{D,yy} = \varepsilon_\infty - \frac{\omega_p^2(\omega^2 + i\gamma_p\omega)}{(\omega^2 + i\gamma_p\omega)^2 - \omega_c^2\omega^2}, \tag{6a}$$

$$\varepsilon_{D,xy} = -\varepsilon_{D,yx} = -i\frac{\omega_p^2\omega_c\omega}{(\omega^2 + i\gamma_p\omega)^2 - \omega_c^2\omega^2}, \tag{6b}$$

which contain an additional fitting parameter, proportional to the magnetic induction, the cyclotron frequency, defined as

$$\omega_c = \frac{eB}{m^*}. \tag{7}$$

The Lorentz term can also be affected by the magnetic field, however the oscillations correspond to lattice vibrations, with much heavier particles than free electrons with an effective mass of $\sim 0.02\, m_0$. No effect has been observed in GaAs at 8 T [11]. It is thus appropriate to neglect the effect of the magnetic field for the Lorentz term.

For measurements with the magnetic field, a small permanent magnet was placed on the backside of the sample. The magnet creates a magnetic field of 0.43T, which was measured by a Gaussmeter. Variable field was obtained using plastic spacers. One wire grid on polyethylene polarizer was used as both polarizer and analyzer (TE polarization). The phase information comes from three parts, $\varphi = \varphi_{sample} - \varphi_{reference} - \varphi_{shift}$. φ_{sample} is the phase angle of the complex reflection coefficient of the sample and φ_{shift} stems from the misalignment d of the sample and reference, as $\varphi_{shift} = 4\pi d \cos\alpha_i / \lambda$. The φ_{shift} is a fitting parameter in the data treatment (d is on the order of $1-100\ \mu m$) and is subtracted from the data for plotting.

The obtained parameters were verified using Hall effect [29] measurement - a standard Van der Pauw (VdP) measurement [30]. A good ohmic contact was obtained by placing the probes on the sample, so there was no need for soldering.

The bottom graph in Fig. 4 is the reflectivity and phase measured with the applied magnetic field 0.43 T and 0.29 T (nominal value 0.23 T due to the effect of spacer), and it shows a clear change in the TE reflectivity caused by the magnetically induced anisotropy. The fitting was done simultaneously with the results without magnetic field, so that the only difference is the cyclotron frequency and phase shift. The corresponding cyclotron frequency for 0.43 T is 4.4×10^{12} rad/s (23.4 cm^{-1}). Figure 5 shows the model of modulated permittivities with parameters obtained from this measurement. The change in the permittivity ε_{xx} is very responsive to the magnetic field and it is possible to change sign for lower frequencies even using small field. The ε_{xy} components also rapidly change with the strength of applied magnetic field and interestingly exhibit maximum for certain magnetic field.

In the metric of magneto-optics [31], the polar Kerr rotation is 26.1 degrees at its maximum is at 81 cm^{-1} for the field 0.43 T. Knowing the cyclotron frequency and magnetic field lets us calculate the effective mass as $m^* = 0.0169 m_0$, which is higher than the nominal value due to higher concentration of electrons due to thermal excitation. With the knowledge of effective mass, we can calculate the mobility $\mu = \frac{e\tau}{m^*}$, concentration $N = \frac{\omega_p^2 \varepsilon_0 m^*}{e^2} = \frac{\omega_p^2}{\omega_c} \frac{\varepsilon_0 B}{e}$ and Hall coefficient $R_H = -\frac{1}{Ne} = -\frac{\mu}{\sigma_0}$. The parameters are listed in Table 2. The differences in values obtained from electrical and spectroscopic measurement are due to different sensitivity of the measuring techniques to different mechanisms and their systematic errors. Generally, the electric VdP measurement is used with lithographically etched pattern, but if there is a good ohmic contact, it is possible to measure without it by placing the contact probes on to the sample. This measurement, used in our case, is prone to error due to possible misalignment of the contact probes. Moreover, the spectral characterization is sensitive only to the carriers with the highest plasma frequency, whereas VdP includes the effect of both. These effects combined explain the differences in obtained values. For measurement with higher doping levels see [32].

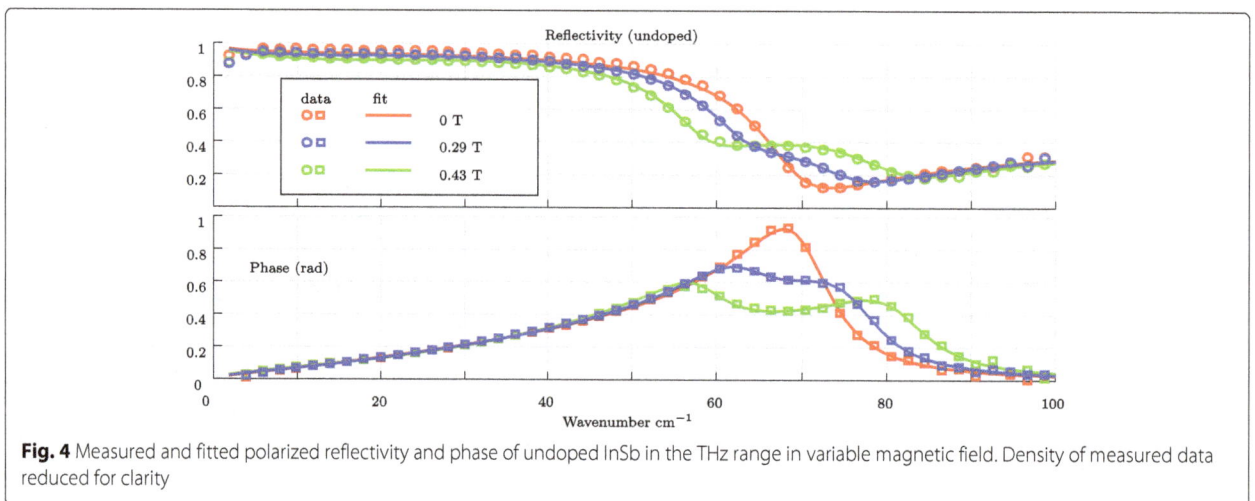

Fig. 4 Measured and fitted polarized reflectivity and phase of undoped InSb in the THz range in variable magnetic field. Density of measured data reduced for clarity

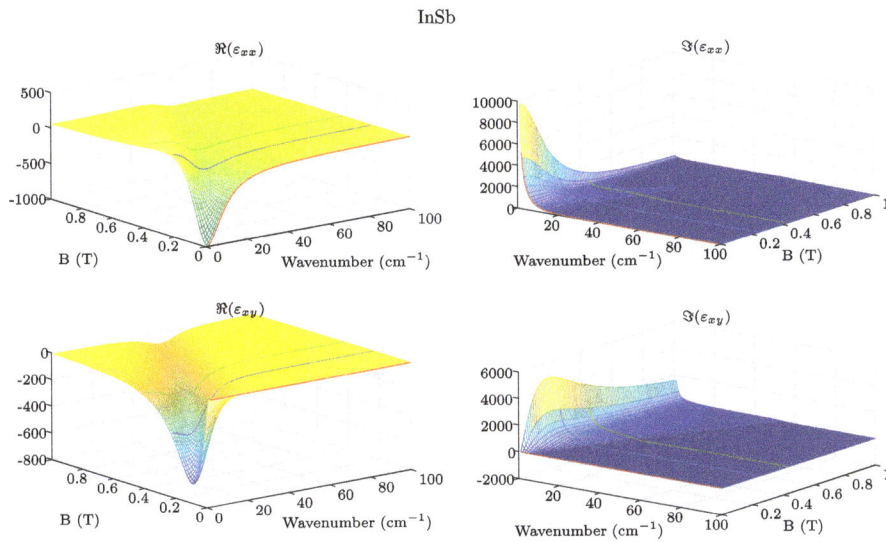

Fig. 5 Modeled permittivity of undoped InSb in variable magnetic field in the terahertz range. *Colored curves* represent measurement (same colors as in Fig. 4)

Conclusion

We have shown that the Drude-Lorentz model describes the optical properties of III-V semiconductors well. The THz-TDS and FTIR are suitable techniques for exploring properties of semiconductors in their respective ranges. The doping or intrinsic concentrations of free carriers in the measured ranges lead to a metal-like behavior. A surface plasmon polariton guided by these materials exhibits reasonable confinement and propagation length, similar to SPPs on noble metal-dielectric interface in the visible range. Thus semiconductors are suitable for the surface plasmon applications in the terahertz and far IR regime. An interesting property emerges with an applied magnetic field - a large anisotropy is induced, causing a huge magneto-optical effect. That can be used to significantly modulate the optical and guided wave properties using small magnetic field at room temperature. Free carrier magneto-optical effect is extremely weak in metals, due to their large plasma frequency and high effective mass and thus low cyclotron frequency. Coupled with low confinement of surface waves for THz frequencies in metals make semiconductors much more suitable for terahertz plasmonics.

Abbreviations
FTIR: Fourier transform infrared spectroscopy; IR: Infrared; SPP: Surface plasmon polariton; TE: Transversal electric; THz-TDS: Terahertz time-domain spectroscopy

Acknowledgements
Our thanks also go to Dominique Vignaud of IEMN, Lille 1 for Hall effect measurement.

Funding
This work was supported in part by projects GA15-08971S, "IT4Innovations excellence in science - LQ1602", "Regional Materials Science and Technology Centre - Feasibility program No. LO1203", SGS project SV 7306631/2101, CREATE ASPIRE Program supported by NSERC and research grant JCJC TENOR ANR-14-CE26-0006.

Authors' contributions
JC conducted the experiments and engaged in writing, KP engaged in the experiments and writing, MC provided samples and expert advice, MV provided expert advice and writing, MM engaged in the experiments, LH engaged in modeling, JL engaged in the experiment design, JP contributed to writing and coordinated the work. All the authors have read and approved the final manuscript.

Competing interests
The authors declare that they have no competing interests.

Author details
[1]Nanotechnology Centre, VSB – Technical University of Ostrava, 17. listopadu 15/2172, 708 33 Ostrava, Poruba, Czech Republic. [2]Department of Electrical and Computer Engineering, Dalhousie University, 6299 South St, Halifax NS B3H 4R2, Canada. [3]Department of Physics, VSB – Technical University of

Table 2 Comparison of parameters of undoped InSb, measured by Van der Pauw method and by spectral reflectivity measurement. The effective mass estimated from the cyclotron frequency as $0.0169 \pm 0.0001\ m_0$

Measurement	Hall	Spectral
N (10^{16}cm^{-3})	2.03	1.78 ± 0.01
μ (10^4 cm^2/Vs)	6.66	5.76 ± 0.03
σ_0 (kS/m)	21.7	16.4 ± 0.07
R_H (10^{-4} m^3/C)	-3.07	-3.51 ± 0.02

Ostrava, 17. listopadu 15/2172, 708 33 Ostrava, Poruba, Czech Republic.
[4]Institut d'Electronique, de Microélectronique et de Nanotechnologie, UMR CNRS 8520, Avenue Poincaré, F-59652 Villeneuve d'Ascq cedex, France.

References

1. Nagatsuma, T, Ducournau, G, Renaud, CC: Advances in terahertz communications accelerated by photonics. Nat. Photonics. **10**, 371–379 (2016)

2. Akyildiz, IF, Jornet, JM, Han, C: Terahertz band: Next frontier for wireless communications. Phy. Com. **12**, 16–32 (2014)

3. Seeds, AJ, Shams, H, Fice, MJ, Renaud, CC: TeraHertz Photonics for Wireless Communications. J. Lightwave Technol. **33**, 579–587 (2015)

4. O'Hara, JF, Withayachumnankul, W, Al-Naib, I: A Review on Thin-film Sensing with Terahertz Waves. J. Infrared Millim. Te. **33**, 245–291 (2012)

5. Yang, X, Zhao, X, Yang, K, Liu, Y, Liu, Y, Fu, W, Luo, Y: Biomedical Applications of Terahertz Spectroscopy and Imaging. Trends Biotechnol. **34**, 810–824 (2016)

6. Liu, H-B, Zhong, H, Karpowicz, N, Chen, Y, Zhang, X-C: Terahertz Spectroscopy and Imaging for Defense and Security Applications. P. IEEE. **95**, 1514–1527 (2007)

7. Jeon, T-I, Grischkowsky, D: THz Zenneck surface wave (THz surface plasmon) propagation on a metal sheet. Appl. Phys. Lett. **88**, 061113 (2006)

8. Cooke, DG, Jepsen, PU: Optical modulation of terahertz pulses in a parallel plate waveguide. Opt. Express. **16**, 15123–15129 (2008)

9. Gómez Rivas, J, Kuttge, M, Kurz, H, Haring Bolivar, P, Sánchez-Gil, JA: Low-frequency active surface plasmon optics on semiconductors. Appl. Phys. Lett. **88**, 082106 (2006)

10. Rahm, M, Li, J-S, Padilla, WJ: THz Wave Modulators: A Brief Review on Different Modulation Techniques. J. Infrared Millim. Te. **34**, 1–27 (2013)

11. Kühne, P, Herzinger, CM, Schubert, M, Woollam, JA, Hofmann, T: Invited Article: An integrated mid-infrared, far-infrared, and terahertz optical Hall effect instrument, Vol. 85 (2014)

12. Palik, ED, Furdyna, JK: Infrared and microwave magnetoplasma effects in semiconductors. Rep. Prog. Phys. **33**, 1193 (1970)

13. Schubert, M, Hofmann, T, Herzinger, CM: Generalized far-infrared magneto-optic ellipsometry for semiconductor layer structures: determination of free-carrier effective-mass, mobility, and concentration parameters in n-type GaAs. J. Opt. Soc. Am. A. **20**, 347–356 (2003)

14. Schubert, M, Hofmann, T, Šik, J: Long-wavelength interface modes in semiconductor layer structures. Phys. Rev. **B 71** (2005)

15. Hofmann, T, Herzinger, CM, Krahmer, C, Streubel, K, Schubert, M: The optical Hall effect. Phys. Status Solidi A. **205**, 779–783 (2008)

16. Mittleman, DM, Cunningham, J, Nuss, MC, Geva, M: Noncontact semiconductor wafer characterization with the terahertz Hall effect. Appl. Phys. Lett. **71**, 16 (1997)

17. Kadlec, F, Kadlec, C, Kužel, P: Contrast in terahertz conductivity of phase-change materials. Solid State Commun. **152**, 852–855 (2012)

18. Kužel, P, Němec, H: Terahertz conductivity in nanoscaled systems: effective medium theory aspects. J. Phys. D Appl. Phys. **47**, 374005 (2014)

19. Jeon, T-I, Grischkowsky, D: Characterization of optically dense, doped semiconductors by reflection THz time domain spectroscopy. Appl. Phys. Lett. **72**, 3032 (1998)

20. Grischkowsky, D, Keiding, S, Van Exter, M, Fattinger, C: Far-infrared time-domain spectroscopy with terahertz beams of dielectrics and semiconductors. J. Opt. Soc. Am. B. **7**, 2006–2015 (1990)

21. Ino, Y, Shimano, R, Svirko, Y, Kuwata-Gonokami, M: Terahertz time domain magneto-optical ellipsometry in reflection geometry. Phys. Rev. **B 70**(15) (2004). https://doi.org/10.1103/PhysRevB.70.155101

22. Stanislavchuk, TN, Kang, TD, Rogers, PD, Standard, EC, Basistyy, R, Kotelyanskii, AM, Nita, G, Zhou, T, Carr, GL, Kotelyanskii, M, et al.: Synchrotron radiation-based far-infrared spectroscopic ellipsometer with full Mueller-matrix capability. Rev. Sci. Instrum. **84**, 023901 (2013)

23. Palik, ED, Kaplan, R, Gammon, RW, Kaplan, H, Wallis, RF, Quinn, JJ: Coupled surface magnetoplasmon-optic-phonon polariton modes on InSb. Phys. Rev. B. **13**, 2497 (1976)

24. Brion, JJ, Wallis, RF, Hartstein, A, Burstein, E: Theory of Surface Magnetoplasmons in Semiconductors. Phys. Rev. Lett. **28**, 1455–1458 (1972)

25. Kushwaha, MS: Plasmons and magnetoplasmons in semiconductor heterostructures. Surf. Sci. Rep. **41**, 1–416 (2001)

26. Berreman, DW: Optics in stratified and anisotropic media: 4×4 -matrix formulation. J. Opt. Soc. Am. **62**, 502–510 (1972)

27. Yu, P, Cardona, M: Fundamentals of Semiconductor: Physics and Materials Properties. Springer, Berlin Heidelberg (2013)

28. Jamshidi, H, Parker, TJ: The far infrared optical properties of InP at 6 and 300 K. Int. J. Infrared Milli. **4**, 1037–1044 (1983)

29. Spitzer, WG, Fan, HY: Determination of optical constants and carrier effective mass of semiconductors. Phys. Rev. **106**, 882 (1957)

30. van der Pauw, L: A method of measuring specific resistivity and Hall effect of discs of arbitrary shape. Philips Res. Rep. **13**, 1–9 (1958)

31. Visnovsky, S: Optics in Magnetic Multilayers and Nanostructures (Optical Science and Engineering). CRC Press, Boca Raton (2006)

32. Chochol, J, Postava, K, Čada, M, Vanwolleghem, M, Halagačka, L, Lampin, J-F, Pištora, J: Magneto-optical properties of InSb for terahertz applications. AIP Adv. **6**, 115021 (2016)

33. Rakić, AD, Djurišić, AB, Elazar, JM, Majewski, ML: Optical properties of metallic films for vertical-cavity optoelectronic devices. Appl. Opt. **37**, 5271–5283 (1998)

23

Mechanical behavior study of laminate composite by three-color digital holography

M Karray[1,2*], C Poilane[3], M Gargouri[1] and P Picart[2]

Abstract

A method for real time 3D measurements based on three-color digital holographic interferometry is presented and applied to the investigation of fracture mechanisms in laminate composite submitted to a three point flexural loading. A convolution algorithm allows the three monochrome images to be superposed to provide simultaneous full-field 3D measurements. Experimental results are presented and exploited to obtain the evolution of the crack tip propagation during the test.

Keywords: Laminate composite, Three-color digital holography, Three point flexural loading, Displacement measurement

Background

Presently, composite materials are used very efficiently in structures aeronautics, automobile transport, and the shipbuilding or in the building clearly increased during these last ten year. Therefore, it is necessary to have a better understanding of the influence of damage on the mechanical behaviour of composite materials. Different methods are used to measure surface displacements, crack propagation or simply detect the presence of defects. Y. Y. Hung and W. Steinchen used a Shearographic non-destructive testing relies on measuring the reponse of a defect to stresses, in particular, of laminated composite structures [1–3]. Others authors developed methods based on speckle interferometry (SLBI), which are efficiently used in several interferometriy techniques as an information carrier of the macroscopic wave front distortion induced by the surface displacement field of the object under investigation [4–6]. Therefore some of these methods have limitations such as scale and segmentation and they deliver data that are abundant but partial. Furthermore, such an investigation needs a multidimensional deformation measurement.

Digital holography has become properly available since its confirmation was established in the 90's [7, 8]. Theory and reconstruction algorithms for digital holography, according to the different possible schemes for the recording, have been described by several authors [8]. Particularly, holographic techniques give a fruitful contribution to the analysis of mechanical structures under strain, by providing whole field information on displacement [9, 10].

Digital color holography provides therefore a very interesting opportunity for the simultaneous multidimensional deformation measurement [11–14].

In this way, we present in this paper a method of real-time 3D deformation measurements based on three-color digital holography and simultaneous recording with a stack of photodiodes image sensor. The method results in a 3D measurement of the deformations of laminate composite submitted to a three point flexural loading.

Experimental details

Materials

The materials used in this work are glass/epoxy unidirectional laminates. Our test specimen is a rectangular plate following proportioning: fiber 50%, epoxy resin 37.5% and 12.5% of hardener. The "Technique of Elaboration" consists of applying successively into a mould surface, a layer of resin (epoxy) a layer of reinforcement (glass fiber) and to impregnate the reinforcement by hand with the aid of a roller [15]. The stacking sequence consisted of 12 layers.

* Correspondence: mayssakarray@gmail.com
[1]Unité de l'état solide, Faculté des Sciences de Sfax, Route Soukra, 3018 Sfax, Tunisie
[2]LAUM UMR CNRS 6613, Université du Maine, Av. O. Messiaen, 72085 Le Mans, France
Full list of author information is available at the end of the article

Polymerization is achieved at room temperature during approximately 12 h period under pressure. After that, this plate is cut using a diamond wheel saw. The specimens obtained are 95 mm in length, and 30 mm in width with 4 mm of thickness. The type of samples is a 0°off-axis unidirectional. To evaluate the influence of the size of the body incorporated on the mechanical behavior of material, one metal patch in the form of disc is integrated within each test-beam, so we obtained tow test specimens: one simple laminate beam (SLB) and other incorporated laminate beam (ILB) with patch.

Optical set-up

The experimental set-up adapted to the investigation of the three point flexural loading, is described in Fig. 1. It uses three continuous diode-pumped solid-state lasers (red line R at $\lambda_R = 671$ nm, green line G at $\lambda_G = 532$ nm, and blue line B at $\lambda_B = 457$ nm). The color sensor consists of a color camera made up of three stacked layers of photodiodes with a single 8 bit per channel digital output, including (M, N) = (1060, 1420) pixels with pitches $p_x = p_y = 5$ µm. Each collimated laser beam illuminates the object under interest with θ_R, θ_G, and θ_B angles, respectively for the red, green, and blue lines leading to the three sensitivities. The R and G beams are included in the {x, z} plane, whereas the B beam is included in the {y, z} plane. The three beams are separated in reference and object beams by cubes PBS1, PBS2 and PBS3. The R and B reference beams are combined into an unique beam thanks to the use of a dichroic plate, which it reflects the B beam and transmits the R beam, while G is adjusted independently of the two other beams.

The smooth plane reference wave G is produced through the SF1, which includes an achromatic lens although the RB reference waves are, produced the SF2. Thus, the two reference beams RB impact the sensing area with the same incidence angle to produce spatial frequencies ($u_R = -28.54$; $v_R = -51.34$) mm^{-1} of the red hologram and ($u_B = -41.91$; $v_B = -75.39$) mm^{-1} of the Blue hologram. The green reference beam is regulated with an incidence angle leading to spatial frequencies ($u_G = 45.82$; $v_G = -30.73$) mm^{-1} of green hologram.

The reconstructing horizon is chosen equal K × L = 1024 × 1024 data points.

The reconstruction algorithm follows the method based on convolution method with an adjustable magnification [16]. The method is based on the image locations and magnification relations of holography when the illuminating beam is a spherical wave front. The curvature radius of the spherical wave front is R_c with the transversal magnification between the reconstructed object and the real one is given by:

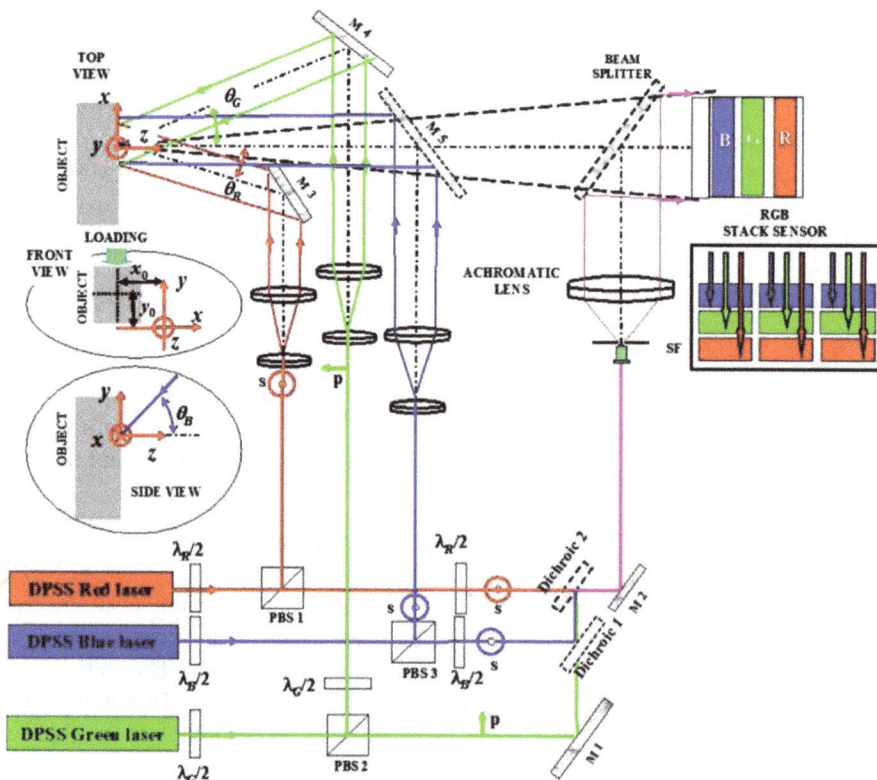

Fig. 1 Experimental setup

$$\gamma = -\frac{d_r}{d_0} \tag{1}$$

Thus the reconstruction distance d_r depends on R_c and the curvature of the spherical wave, as shown in this relation:

$$d_r = -\left(\frac{1}{d_0} + \frac{1}{R_c} + \frac{1}{d_s}\right)^{-1} \tag{2}$$

The transversal magnification and the number of points of the algorithm are linked by this expression.

$$\{K, L\} = |\gamma| \left\{ \frac{\Delta A_x}{p_x}, \frac{\Delta A_x}{p_x} \right\} \tag{3}$$

According to the potentialities of this strategy, the transverse magnification that must be put into the algorithm is $\gamma = 0.16$, the curvature radius of the numerical spherical reconstructing wave is $R_c = \gamma d_0/(\gamma-1) = -157.3481$ mm; and the effective reconstruction distance is $d_r = -\gamma d_0 = -132.27$ mm, for each wavelength.

The object is illuminated from three directions, which produces three sensitivity vectors and, subsequently, a three-sensitivity measurement. The relation between the displacement vector $\mathbf{U} = u_x\mathbf{i} + u_y\mathbf{j} + u_z\mathbf{k}$ and the illuminating geometry is $\Delta\phi_\lambda = 2\pi\mathbf{U}.\left(\mathbf{K}_e^\lambda - \mathbf{K_o}\right)/\lambda$ for each wavelength, where $\mathbf{K}_e^\lambda = -\cos\theta_{yz}^\lambda \sin\theta_{xz}^\lambda\mathbf{i} - \sin\theta_{yz}^\lambda\mathbf{j} - \cos\theta_{yz}^\lambda\cos\theta_{xz}^\lambda\mathbf{k}$ is the illumination vector and $\mathbf{K_o} \cong \mathbf{k}$ is the observation vector.

Using three wavelengths with three different lighting directions, we obtain a matrix relationship between the monochrome phase changes measured from the holograms and each component of the 3D displacement field.

$$\begin{pmatrix} \lambda_1\Delta\phi_{\lambda_1} \\ \lambda_2\Delta\phi_{\lambda_2} \\ \lambda_3\Delta\phi_{\lambda_3} \end{pmatrix} = 2\pi\mathbf{A}\begin{pmatrix} u_x \\ u_y \\ u_z \end{pmatrix} \tag{4}$$

The inversion of the matrix leads to the determination of the 3D displacement vector.

$$\begin{pmatrix} u_x \\ u_y \\ u_z \end{pmatrix} = \frac{1}{2\pi}A^{-1}\begin{pmatrix} \lambda_1\Delta\phi_{\lambda_1} \\ \lambda_2\Delta\phi_{\lambda_2} \\ \lambda_3\Delta\phi_{\lambda_3} \end{pmatrix} \tag{5}$$

To simplify the notation, we note: $\theta_R = \theta_{xz}^R$, $\theta_G = \theta_{xz}^G$ et $\theta_B = \theta_{yz}^B$. In the setup, we have:

$$\begin{aligned} K_e^R &= \sin\theta_R i - \cos\theta_R k \\ K_e^G &= -\sin\theta_G i - \cos\theta_G k \\ K_e^B &= -\sin\theta_B i - \cos\theta_B k \end{aligned} \tag{6}$$

According to Eq. (5) the calculation of the three components of the displacements field is given by the following relation:

$$\begin{pmatrix} u_x \\ u_y \\ u_z \end{pmatrix} = \frac{1}{2\pi\alpha}$$

$$\begin{pmatrix} 1 + \cos(\theta_G) & -1 - \cos(\theta_G) & 0 \\ \sin(\theta_G)(1 + \cos(\theta_B))/\sin(\theta_B) & \sin(\theta_R)(1 + \cos(\theta_B))/\sin(\theta_B) & -\alpha/\sin(\theta_R) \\ -\sin(\theta_G) & -\sin(\theta_R) & 0 \end{pmatrix}$$
$$\times \begin{bmatrix} \lambda_R\Delta\phi_R \\ \lambda_G\Delta\phi_G \\ \lambda_B\Delta\phi_B \end{bmatrix} \tag{7}$$

where $\alpha = \sin(\theta_R)(1 + \cos(\theta_G)) + \sin(\theta_G)(1 + \cos(\theta_R))$ and $\Delta\phi_R$, $\Delta\phi_G$ and $\Delta\phi_B$ are the optical phase changes between two deformation states, which are obtained from the reconstructed holograms, respectively, for the R, G, and B beams. In the setup, the angles are adjusted to $\theta_R = 30.17$, $\theta_G = 11.80$, and $\theta_B = 46.27$. Figure 2a indicates the illuminating geometry and Fig. 2b is a photograph illustrates the "white" illumination of the sample by the three laser wavelengths.

Results and discussion

The space specimens is glass/epoxy unidirectional laminates beam. The beam is 95 mm in length and 30 mm in width with a thickness of 4 mm, is placed at distance $d_0 = 830$ mm from the color sensor. In other space

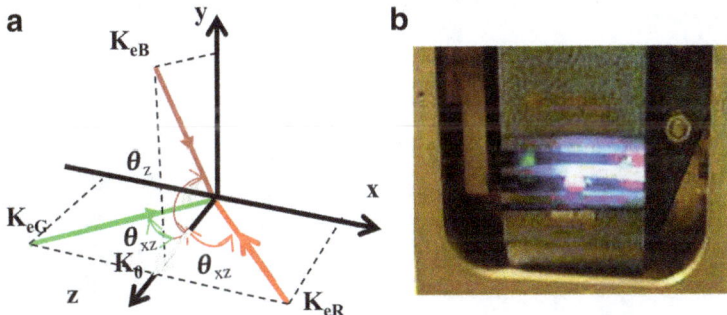

Fig. 2 a Illumination geometry, **b** photography of the mechanical head with a sample on test, white spot results of the mixing of the three R-G-B laser lines

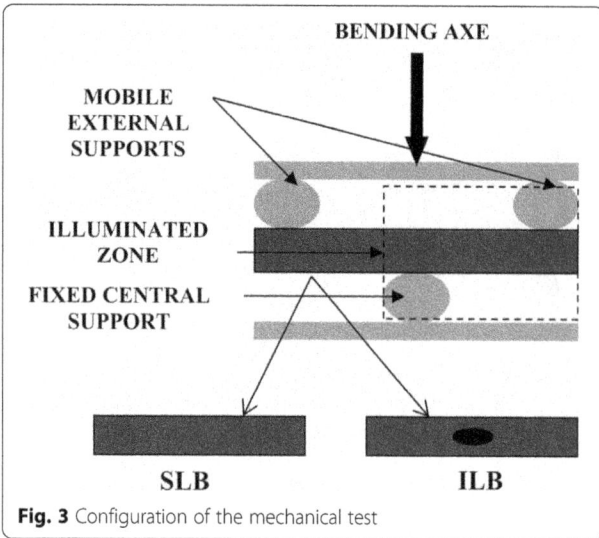

Fig. 3 Configuration of the mechanical test

specimens we incorporate metal patch inside the laminate beam to see the influence of these patch on the behaviour of the laminates.

The Fig. 3 shows how the beam is submitted to a three point flexural loading.

During the test, the span supports are mobile and the central one is fixed. At each step of the test, the deformation is limited to a low fringe number in order to keep compatibility with the spatial resolution on the image of the sample. With each increment of displacement, the hologram of the current state is recorded and this hologram is used as a reference for the following state. The test runs until the fracture of the sample occurs.

During the test, we recorded 250 holograms for each specimen. The phase differences are calculated between two states of successive stresses. Figure 4 shows two sets of red, green and blue phase difference maps. Figure 4a corresponding to the phase difference between the hologram no. 1 and the reference initial hologram, numbered 0. Figure 4b corresponding to the phase difference between numbered 99 and numbered 98.

We can note first for the series (a) the absence of the fringes for SLB since the constraint is weak at the beginning of the test. The fringes in ILB are related to a rigid body displacement.It is a first pitfall to avoid, the presence of fringes on the supports is useful to detect this problem. For series (b), more fringes were observed for both specimens. This means that the increment of deformation is higher for series (b) than for the series (a). For both series, the presence of noise between the fringes is observed for ILB, the experimental conditions are less good than for the SLB beam. A filtering of the fringes is therefore necessary.

To observe the influence of the patch, the whole of the phase maps were divided into 5 series and then these maps were unwrapped. Figure 5 shows the result obtained with ILB and SLB for series No. 3. The displacement maps are reset to compensate for the rigid body movements of the sample, which are inevitable at this measurement scale.

The background color corresponding to a zero displacement. The displacement maps computed for SLB are qualitatively validated by a MEF simulation of an embedded beam with a displacement imposed at 40 μm.

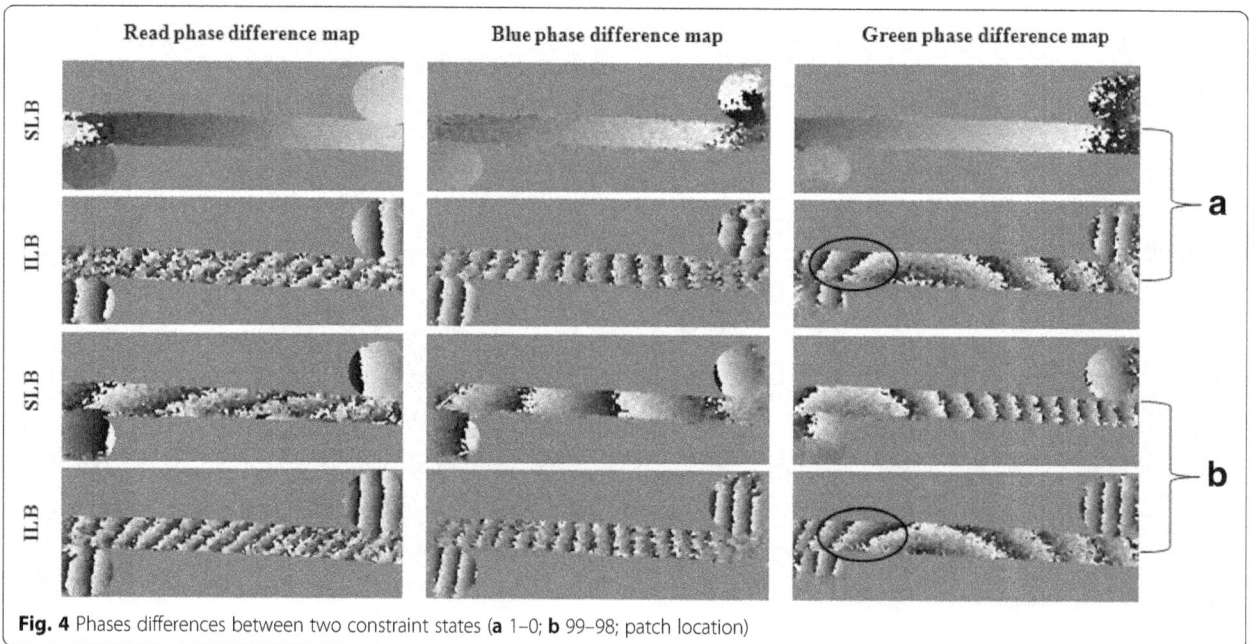

Fig. 4 Phases differences between two constraint states (**a** 1–0; **b** 99–98; patch location)

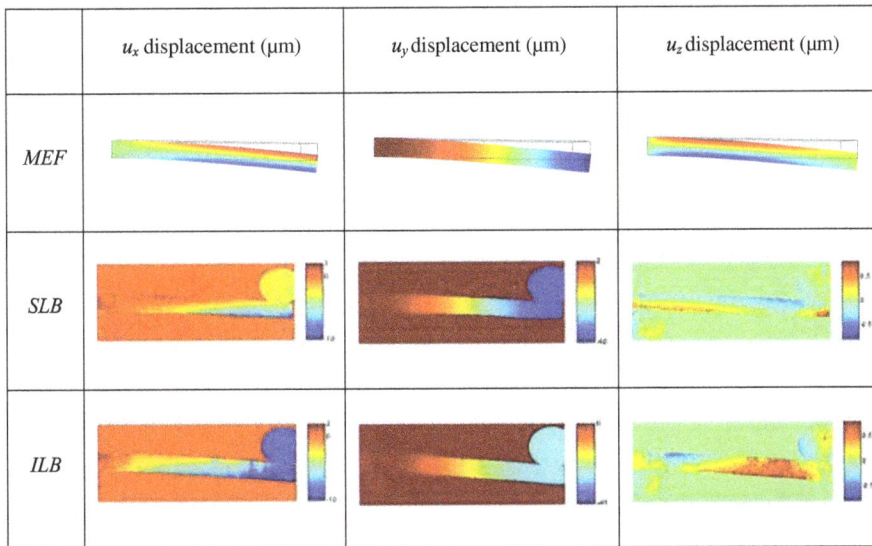

	u_x displacement (μm)	u_y displacement (μm)	u_z displacement (μm)
MEF			
SLB			
ILB			

Fig. 5 The 3D displacements field (series n ° 3)

The fine analysis of the components u_x and u_y of the displacement field reveals a difference in rigidity: the patch stiffens locally the sample. The component u_z also reveals a Poisson effect on the quarter of the specimen which is due to the presence of the piezoelectric patch.

This result demonstrates in the one hand, that the sample, for the loading analyzed here, has not yet undergone compensation of the interlaminar delamination type.

In the other hand, the use of three primary colors (red, green and blue for example) digital holography led to the absolute measurement of the displacement field or index variation [17], which was not the case in monochrome holography [18–20].

However, these applications are based on the use of a monochromatic source that greatly considerably complicates the experimental set-up (three different reference waves) and leads to a loss of spatial resolution [21].

With the advent of color cameras such as the CMOS sensor used, it is now possible to simultaneously record colors on the same sensor. The reference waves then have the same incidence and the segmentation of the colors is carried out by the sensor. Thus the experimental set-up becomes very simple and the spatial resolution is maximum because the orders share the same frequency space.

Finally, the reconstruction of the color holograms can be carried out independently for each wavelength by convolution, provided that the size of the monochromatic holograms remains identical for a perfect superposition to the ready pixel [22].

Conclusion

In conclusion, this Letter presents a three-color digital holographic method for real-time 3D deformation measurements. The setup is considerably simplified, because a unique reference beam can be used to simultaneously record the three-color digital holograms. Thanks to the reconstruction of each individual monochrome hologram using a convolution algorithm, the three images are superposed. The application of the method to real-time 3D measurement shows satisfactory results compared with a pure sequential monochrome measurement, which validates the concept. This paper demonstrates that three dimensional measurements make it possible to identify on the visible edge of the sample the displacements fields. In addition from this analysis we have shown the effect of the embedded pastilles which it stiffens locally the sample. Also we remark a Poisson effect on the quarter of the specimen but, has not yet undergone compensation of the interlaminar delamination type.

Consequently, three-color digital holographic method is a very efficient technique used in non-destructive testing and characterizing of composites, since many kinds of defects may be simultaneously detected over large areas, without any contact with the tested material.

Abbreviations
B: Blue line; d$_r$: Reconstruction distance; G: Green line; ILB: Incorporated laminate beam; M: Mirror; MEF: Model element fini; PBS: Polarizer beam splitter; R: Red line; R$_c$: Curvature radius; SLB: Simple laminate beam; θ$_B$: Angle blue; θ$_G$: Angle green; θ$_R$: Angle red

Acknowledgments
The authors are grateful to IUT members in Mans for his help in preparing the material studied.

Funding
The financial support of LAUM UMR CNRS of Mans (France) and scientific research of Tunisia is acknowledged.

Authors' contributions

C Poilane, P Picart and M Gargouri conceived the project. C Poilane, P Picart and M Karray designed and performed the experiments. C Poilane, P Picart and M Karray analyzed the data. M Karray wrote the manuscript. All the authors discussed the results and commented on the manuscript at all stages. All authors read and approved the final manuscript.

Competing interests

The authors declare no competing financial interests.

Author details

[1]Unité de l'état solide, Faculté des Sciences de Sfax, Route Soukra, 3018 Sfax, Tunisie. [2]LAUM UMR CNRS 6613, Université du Maine, Av. O. Messiaen, 72085 Le Mans, France. [3]CIMAP UMR6252, Université de Caen, 6 Boulevard du Maréchal Juin, 14050 Caen cedex 4, France.

References

1. Hung, Y.Y.: Comp: Part B. **30**, 765–773 (1999)
2. Hung, Y.Y., Ho, H.P.: Mat. Sci. Eng. **49**, 61–87 (2005)
3. Steinchen, W., Kupfer, G., Maâckel, P., Voâssing, F.: Measurement. **26**, 79–90 (1999)
4. Borza, D.N.: Comp. Part B. **29B**, 497–504 (1998)
5. Denisyuk, Y.N.: Opt. I Spec. **15**, 522–532 (1963)
6. Ambu, R., Aymerich, F., Ginesu, F., Priolo, P.: Comp. Sci. Tech. **66**, 199–205 (2006)
7. Schnars, U., Jüptner, W.: App. Opt. **179**, 33 (1994)
8. Kreis, T., Adams, M., Jüptner, W.: Proc. SPIE. **224**, 3098 (1997)
9. Pedrini, G., Tiziani, H.: Opt. Laser. Tech. **249**, 29 (1997)
10. Picart, B., Diouf, E., Lolive, J.-M.: Berthelor. Opt. Eng. **1169**, 43 (2004)
11. Yamaguchi, I., Matsumura, T., Kato, J.: Opt. Let. **1108**, 27 (2002)
12. Tankam, P., Picart, P., Mounier, D., Desse, J.M., Li, J.C.: App.Opt. **320**, 49 (2010)
13. Tankam, P., Picart, P.: Opt. Las. Eng. **1335**, 49 (2011)
14. Desse, J.M., Picart, P., Tankam, P.: Opt. Las. Eng. **18**, 50 (2012)
15. Stringer, L.G.: Compo. **20**, 441–452 (1989)
16. Li, J.C., Tankam, P., Peng, Z., Picart, P.: Opt. Lett. **34**, 572 (2009)
17. Desse, J.M., Picart, P., Tankam, P.: Opt. Exp. **16**, 5471–5480 (2008)
18. Gass, J., Dakoff, A., Kim, M.K.: Opt. Lett. **28**, 1141 (2003)
19. Wagner, C., Osten, W., Seebacher, S.: Opt. Eng. **39**, 79 (2000)
20. Mann, C.J., Bingham, P.R., Paquit, V.C., Tobin, K.W.: Opt. Exp. **16**, 9753 (2008)
21. Saucedo, T., Santoyo, F.M., De la Torre-Ibarra, M.: Opt. Exp. **14**, 1468 (2006)
22. Zhang, F., Yamaguchi, I., Yaroslavsky, L.P.: Opt. Lett. **29**, 1668–1670 (2004)

24

Development of multi-pitch tool path in computer-controlled optical surfacing processes

Jing Hou[1,2], Defeng Liao[1,2*] and Hongxiang Wang[1]

Abstract

Background: Tool path in computer-controlled optical surfacing (CCOS) processes has a great effect on middle spatial frequency error in terms of residual ripples. Raster tool path of uniform path pitch is one of the mostly adopted paths, in which smaller path pitch is always desired for restraining residual ripple errors. However, too dense paths cause excessive material removal in lower removal regions deteriorating the form convergence.

Methods: With this in view, we propose a novel tool path planning method named multi-pitch path. With the path, the material removal map is divided into several regions with varied path pitches according to the desired removal depth in each region. The path pitch is designed larger at low removal regions while smaller at high removal regions, and the feeding velocity of the tool is maintained at high level when scanning the whole surface.

Results and conclusions: Experiments were conducted to demonstrate this novel tool path planning method, and the results indicate that it can successfully restrain the residual ripples, and meanwhile guarantee favorable convergent rate of form error.

Keywords: Multi-pitch tool path, Middle spatial frequency error, Residual ripple, Removal regions

Background

Large optics has been widely used in interferometers, telescopes, high-power lasers and other optical systems. In these systems, the optics are required of stringent specifications of low, middle and high spatial frequency errors [1, 2]. Various CCOS processes have been developed which can provide good solutions for the fabrication of these optics because of their high convergence rates of low frequency error (i.e. surface form) [3–5]. Nowadays, more and more attentions have been paid to the middle spatial frequency (MSF) error, which is crucial for image performance and beam quality [6]. MSF error is primarily introduced during the CCOS processes, and it is hard to restrain. It is reported that MSF error is mainly affected by the initial surface error distribution (spatial and frequency domain), the removal function characters (profile, removal efficiency and stability) and the adopted paths [7, 8].

During CCOS, the tool is numerically controlled to traverse a path with a varied feeding velocity to obtain the desired removal map. The tool path plays an important role in the deterministic removal process, which has to cover the whole optic surface. There are several tool paths utilized in CCOS processes, such as the regular raster and spiral paths, and several kinds of random path [9–11]. The random path is claimed to be useful for reducing the MSF error, [12] but is hard to achieve a high precision surface form because of the difficulty in tool speed management [7]. The spiral and raster paths are more prone to generating MSF error in terms of residual ripples due to their inherent regular pattern. Spiral paths are suitable for circular optics as the tool is driven to traverse a radius while the optic mounted on a turntable rotates simultaneously [13].

Raster path is usually adopted for the fabrication of square-shaped optics. During polishing with raster path, the tool feeds along a straight line and then translates to another parallel line. This process is repeated to cover the whole surface. The pitch between adjacent path lines is commonly set identical (i.e. uniform pitch) on the whole surface, and the feeding velocity along each path

* Correspondence: defeng_liao@163.com
[1]School of Mechatronics Engineering, Harbin Institute of Technology, Harbin 150001, China
[2]Research Center of Laser Fusion, China Academy of Engineering Physics, Mianyang 621900, China

is instantaneously controlled based on the local removal [14]. It is obvious that for a uniform removal map of a certain removal amount, the smaller the path pitch, the larger the feeding velocity. However, if the desired feeding velocity is larger than the largest one allowed by the machine, the actual feeding velocity has to be changed to the largest one, which will introduce extra dwell time leading to material over-removal. As the tool feeds fast at lower region while slowly at higher region with the uniform pitch path, the smallest removal region (lowest region) will commonly bring with such over-removal. Since, the decreased pitch is propitious to restraint of MSF errors [15]. On the other hand, the decreased pitch will greatly increase over-removal especially in the lowest region deteriorating the form correction precision. Hence, it is needed to develop an optimized tool path planning method which can solve this problem.

A novel tool path planning method named multi-pitch path, is developed in this paper. With this method, the material removal map is divided into several regions with varied path pitches according to the desired removal depth in each region. The path pitch is designed larger at low removal regions, so as to bring much less over-removal; while smaller at high removal regions, so as to decrease MSF error in terms of the residual ripples. The path has obvious advantage in restraint of residual ripples, and meanwhile can guarantee convergent rate of form errors. In the following section II, the correlation between the ripple and MSF error is analyzed to verify the rationality of characterizing the MSF error by ripples. In section III, the factors impacting ripple errors, including the removal amount and path pitch are discussed. In section IV, the multi-pitch path and polishing procedure with the path are detailed and the experimental validation is conducted.

Method
Verification of characterizing MSF errors with the residual ripple
Spatial frequency of surface errors is divided into several separate bands in the field of high power lasers [2]: surface figure (>33 mm), MSF error (0.12 ~ 33 mm) and

surface roughness (0.01 ~ 0.12 mm). There are two types of specification for MSF error; one is RMS value after band pass filtering, and the other is a not-to-exceed line for the power spectral density (PSD) as a function of spatial frequency [16]. In the following, we select RMS after band pass filtering over 0.12 ~ 33 mm range for evaluation of the MSF error.

MSF errors induced by CCOS processes are commonly in form of residual ripples. Thus, in order to quantitatively specify the correlation between residual ripple error and MSF error, we formulated a series of sinusoidal surface forms with variable spatial frequency and magnitude (see Fig. 1). The surface forms are sinusoidal distributed in x direction, while are uniformly distributed in y direction. Surface forms of this shape are fairly similar to the local regions of surface forms practically corrected by CCOS processes, which are nearly sinusoidally-distributed in the scanning direction while nearly uniformly-distributed in the feeding direction. Herein, the MSF error, in terms of the RMS value in the mid-spatial frequency band (0.0303 ~ 8.33 mm^{-1}), is calculated for all the surface forms as shown in Fig. 2. It is revealed that the spatial frequency has little effect on the RMS value, while there has a good linear relationship between the spatial magnitude and the RMS value. Thus, we should focus on the residual ripple magnitude rather than the frequency while restraining MSF errors.

Influencing factors of residual ripple errors
As revealed above that residual ripple error can be characterized by the ripple amplitude, i.e. the peak-to-valley value of the ripple (PVe). We introduce a normalized PV value of residual ripple (PVn), which is derived from PVe divided by the average removal depth (r). PVn represents the residual error PVe while achieving unit removal (see Eq.1).

$$PV_n = PV_e / r \tag{1}$$

Primary factors impacting the residual ripple errors include the scanning pitch (i.e., path pitch), removal depth, and tool influence function (TIF) features. Without loss

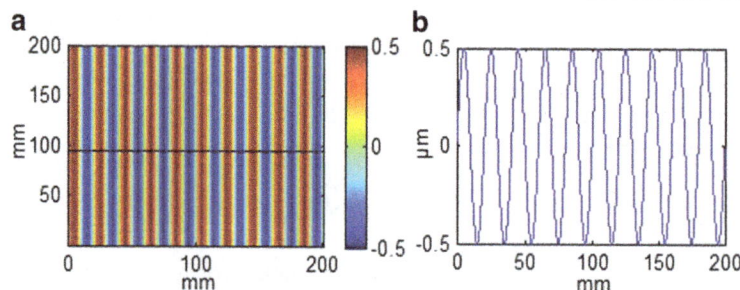

Fig. 1 One example of sinusoidal distributed residual ripples, (**a**) contour map and (**b**) one-dimensional distribution

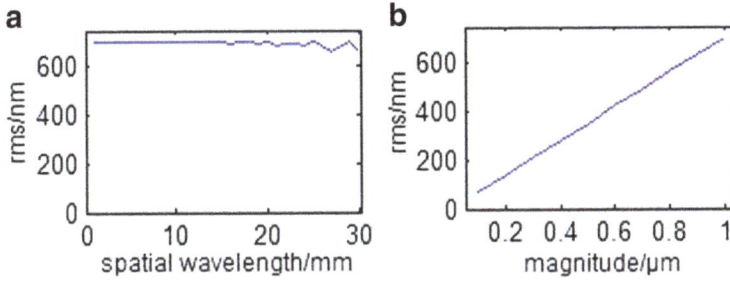

Fig. 2 The relationships between the MSF error and the ripple features. **a** Rms and ripple frequency and (**b**) Rms and ripple magnitude (Rms after band pass: 0.12-33 mm)

of generality, we modelled variable scanning pitch resulting in a uniform removal map as well as variable removal map under the same scanning pitch, to reveal the effects of the canning pitch and removal depth on the residual ripple and MSF error. Herein, we consider a Magnetorheological Finishing (MRF) TIF traversing a uniform pitch raster path, under the condition that the feeding direction is set in perpendicular to the fluid flow direction as shown in Figs. 3 and 4. As the TIF traverses a single line path with a constant feeding velocity of v, the removal is uniformly distributed along the feeding direction, while the removal distribution R_j in the perpendicular direction can be obtained by Eq.2 (see Fig. 5), in which TIF matrix (R, unit in um/s) has s row, k column elements as shown in Eq.3; and the pixel size is p (unit in mm).

$$R_j = p/v \cdot \sum_{i=1}^{k} r_{i,j}, \quad j = 1, ..., l. \qquad (2)$$

$$R = \begin{bmatrix} r_{11} & r_{12} \cdots & r_{1k} \\ r_{11} & r_{12} \cdots & r_{1k} \\ \cdots & \cdots & r_{ij} \cdots \\ r_{s1} & r_{s2} \cdots & r_{sk} \end{bmatrix} \qquad (3)$$

Figure 6 shows the local removal amount distribution in the scanning direction while correcting uniformly-distributed form errors. The blue sections represent the removal amount in independent single path and the red one is the convolved removal amount. It is obvious that the convolved removal amount is periodically distributed, and the spatial wavelength is identical to the scanning pitch. It is confirmed that surface form correction by small-sized TIF inevitably induces residual ripple error.

Figure 7a shows that PVe becomes a linear growth along with the increment of the removal amount. It is suggested that a less removal amount is propitious to restraint of PVe. Figure 7b shows the PVn value as a function of scanning pitch. PVn increases as the scanning pitch is increased. It is noticeable that PVn increases slowly until the scanning pitch reaches ~1.1 mm, and then increases sharply. It is revealed that while correcting the surface form of optics by sub-aperture polishing, it is desired to adopt a smaller tool-path pitch for restraint of residual ripple.

Development of the multi-pitch tool path

Correction of the form error by CCOS processes aims to polish every region to a desired plane of absolute flatness, which is commonly located at the lowest point on the surface as shown in Fig. 8. In fact, a lower plane has to be selected due to the maximum motion speed of the tool. Such a removal map introducing extra removal isn't propitious to restraint of residual ripples as revealed above.

If the desired plane selected at the lowest point, the desired removal amount at the point would be zero. As

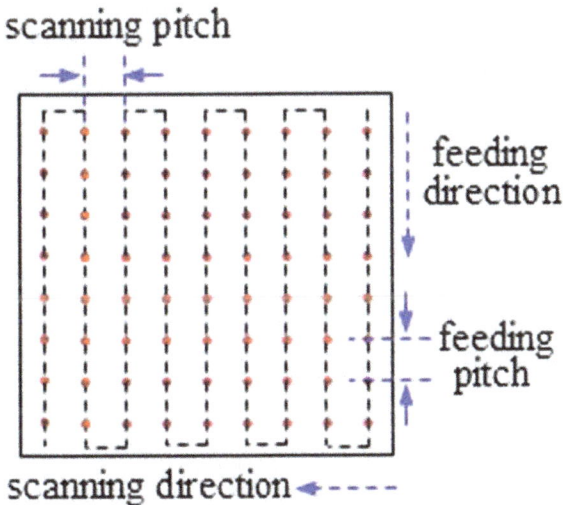

Fig. 3 Uniform pitch raster path for the modeling

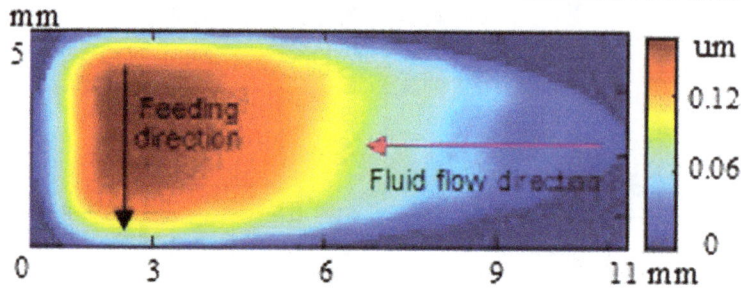

Fig. 4 MRF TIF chosen for the following simulation and experimental

the tool traverses across the point, it inevitably removes material deteriorating figure convergence, thus the tool are commonly driven with a most velocity allowable for the machine. Furthermore, the path pitch within lowest regions should be as large as possible so as to introduce less over-removal, but in uniform pitch tool path, a large pitch would deteriorate the residual ripple errors. Therefore, we develop a multi-pitch tool path which has a large pitch in less removal regions reducing over-removal and small pitch in more removal regions so as to decrease the residual ripples while guaranteeing the figure convergence.

The polishing procedure with the multi-pitch tool path is showed in Fig. 9. First, we should generate the removal map according to the actual surface figure and the desired surface figure. Then, the removal map is divided into several subregions based on the removal variance. After that we calculate the scanning path pitch and generate the path for each subregion. The spacing between adjacent dwell points along each path line, i.e. the feeding pitch, is also determined. The feeding pitch can be adopted within a wide range, yet value of the scanning pitch is recommended. After determination of the scanning and feeding pitches, we then acquire the dwell points on the whole surface. The polishing time at each dwell point (i.e. the dwell map) can be solved with various algorithms such as discrete convolution model, the linear equation model and so forth [17]. Finally, the CNC code can be generated

according to the dwell point on the path and the dwell time map.

Determination of the path pitch in each subregion

While generating multi-pitch tool path, we first divide the optic surface into several subregions according to the removal map. The whole material removal scope within the maximum and minimum removals is divided into several ranges, and then each removal range determines the corresponding subregions. The number of the removal ranges or subregions depends on the whole removal scope; the larger the removal, the more the ranges or subregions. Generally, 3 ~ 6 removal ranges or subregions are appropriate for most cases. Assuming a removal map has a maximum removal of r and a minimum removal of 0, it is divided into m subregions and the removal variance in each subregion has the same value dr, then the removal in each subregion can be derived by Eqs.4–5.

$$(k-1)\cdot\Delta r \le r_k < k\cdot\Delta r, \quad k=1,...,m. \tag{4}$$

$$\Delta r = r/m \tag{5}$$

While determining the path pitch in a subregion, a dwell point P which has a removal of h and covers a square area in the subregion is considered, as shown in Fig. 10. The removal is almost uniformly distributed in the tiny square area, and then the correlation among the

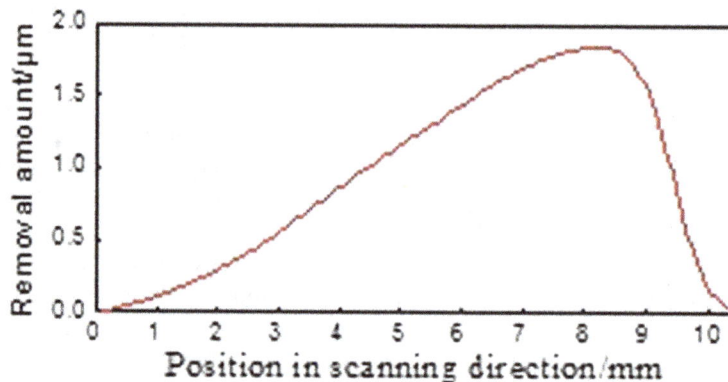

Fig. 5 Removal amount by a single path in the scanning direction

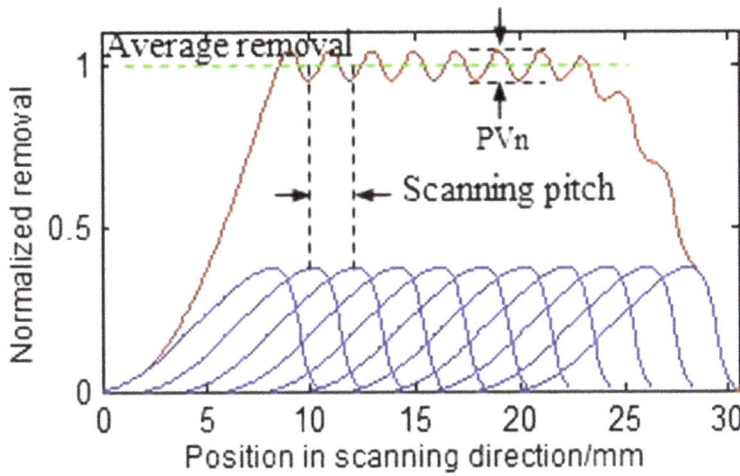

Fig. 6 Removal distribution in the scanning direction

removal depth (r), path pitch (d) and feeding velocity (v) can be obtained by Eq.6. It is revealed that a certain d can be calculated for a given s, r and v_{max}, as shown in Eq.7.

$$s = r \cdot d \cdot v \tag{6}$$

$$d = s/(r \cdot v_{max}) \tag{7}$$

In Eqs.6-7 h represents the feeding pitch, v_{max} the largest feeding velocity allowed by the machine, and s is the volume removal rate of the TIF, which can be derived by Eq.8:

$$s = p^2 \cdot \sum_{j=1}^{l} \sum_{i=1}^{k} R_{i,j} \tag{8}$$

Where p (unit in mm) is the pixel size of the TIF, and $R_{i,j}$ (unit in um) is the TIF removal rate.

As revealed in the previous section, minimum path pitches are desired for restraint of residual ripple errors. Eq.7 indicates that the feeding velocity is inversely proportional to the pitch; thus, we can adopt a maximum feeding velocity allowed by the machine so as to decrease the pitch.

However, increasing feeding velocity has a significant impact on the stability of TIF. A too large feeding velocity will result in alteration of TIF, and hence deteriorate efficiency of figure correction as well as MSF errors. Further, the machine imposes restrictions on the moving velocity and acceleration of every movable component. Hence, there is a favorable maximum velocity allowed for each polishing machine. Herein, the largest feeding velocity (v_{max}) allowed by the machine can be adopted in practice so as to reduce the pitch and hence the PVe.

As each subregion is determined within a material removal range, we adopt the minimum removal depth in each region for calculation of the corresponding path pitch (see Eq.9), which will prevent the feeding velocity exceeding the specified maximum value. Then, the pitch in each region can be obtained by Eq.10.

$$r_1 = 0.5 \cdot dr, r_k = (k-1) \cdot dr, \quad k = 2, ..., n. \tag{9}$$

$$d_k = s/(r_k \cdot v_{max}) \tag{10}$$

In CCOS processes, the scanning path pitch should be restricted within a range in practice. The minimum value of the pitch is determined by the positioning & moving precision of the polishing machine. The maximum one is

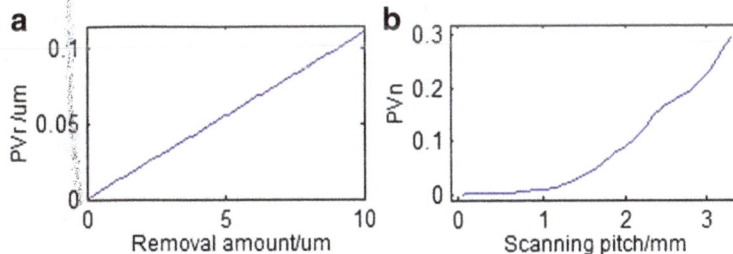

Fig. 7 Residual error as a function of (**a**) removal amount and (**b**) scanning pitch

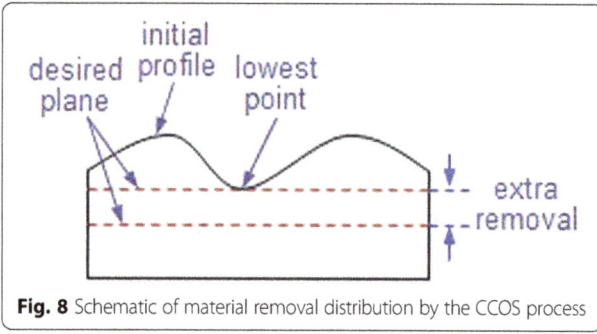

Fig. 8 Schematic of material removal distribution by the CCOS process

primarily dependent on the TIF size (i.e. <1/6 size). Further, a too large path pitch isn't propitious to correcting the form error and restraining the ripple error.

Solution and implementation of dwell time map

Dwell time map in terms of the polishing time at each dwell point provides the time that the tool dwells on the corresponding position to obtain desired removal. In the multi-pitch tool path, the dwell points are allocated at each path line with a feeding pitch. The

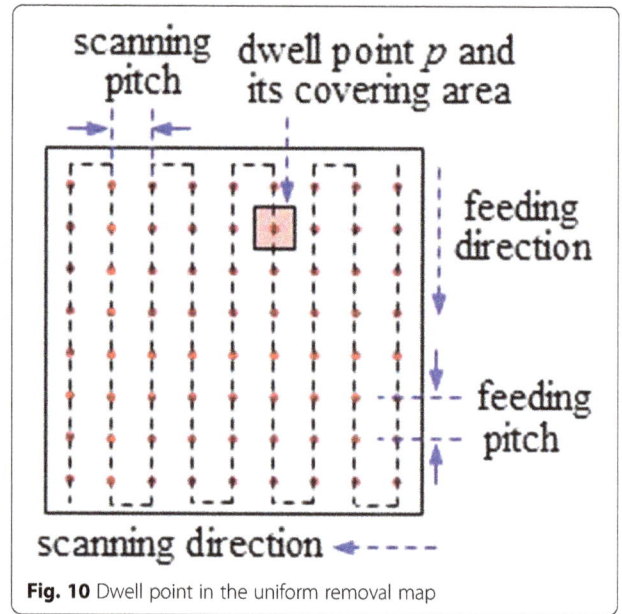

Fig. 10 Dwell point in the uniform removal map

feeding pitch can be specified according to the scanning pitch. Then, the dwell time map is solved by any developed algorithms, such as the discrete convolution method, linear equation method and so forth. The local feeding velocity (v_f) can be derived from the pitch and the local removal (r_f) at the corresponding point, as revealed in Eqs.11–12. In the multi-pitch tool path, the removal variance in each region is greatly decreased compared to the conventional tool path with a constant pitch on the whole optic surface. As the tool scans the path lines in any subregion, the path pitch is decreased as much as possible in every region, which is prone to improving the implementation precision of the dwell-time.

$$s \cdot t = r_f \cdot d \cdot v_f \cdot t \tag{11}$$

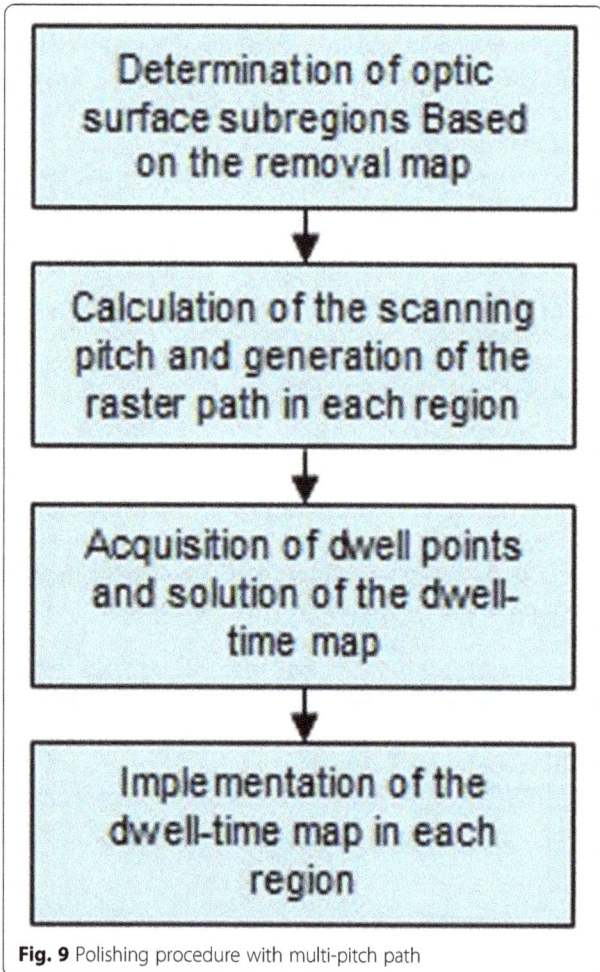

Fig. 9 Polishing procedure with multi-pitch path

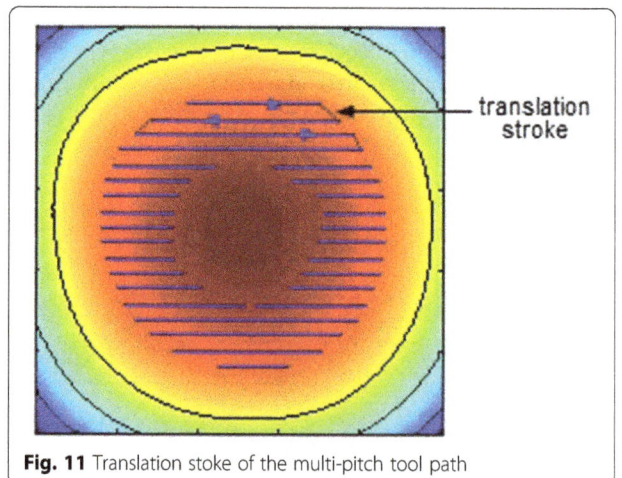

Fig. 11 Translation stoke of the multi-pitch tool path

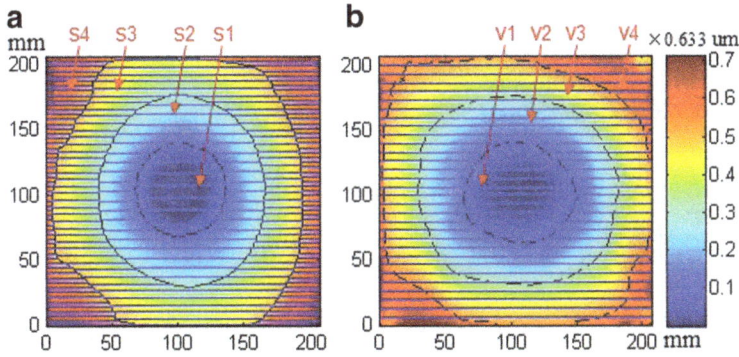

Fig. 12 Initial figures of the optics polished with (**a**) multi-pitch and (**b**) uniform pitch tool path

$$v_f = s/(d \cdot r_f) \qquad (12)$$

During generation of multi-pitch tool path, the optic surface is divided into several regions. In each region, the tool scans a raster path with a featured constant pitch. The pitch is dependent on the removal in the region, and the larger the removal, the smaller the pitch. During implementation of the dwell time map, the tool will traverse all the paths that generated covering the whole surface. Herein, we suggest that each region be scanned individually. In each region, adjacent path lines can be interconnected at the ends during implementation of the dwell-time map. As the tool traverses a path line and reaches the end, it translates to the nearby end of the next line and traverses this line (see Fig. 11). The translation stroke from one line to another maybe introduces extra dwell time, which will cause undesired removal and deteriorate the convergence rate of the surface form. If the tool lifts up after completing the last feeding segments in each path line, it will inevitably introduce extra removal during the lifting process. It is suggested that the tool lifts up while traversing the last feeding segment within a period longer than the determined dwell time. At this condition, the increased actual dwell time will compensate the decreased removal function achieving approximately the desired removal. Similarly, the tool descends while traversing the first feeding segment of the next path line. After the tool has covered one subregion, it also lifts off the optic and translates above to the first dwell point of another subregion. Then it descends to accomplish the subsequent dwell time. The lifting of the tool during the translation process wouldn't bring with extra removal.

Results and discussion

Herein, we utilize the multi-pitch tool path and regular uniform pitch tool path for figure correction with MRF process. The two paths are compared through simulation and experiments.

MRF process is a typical CCOS process characterized by the stable TIF and deterministic figuring procedure. The MRF machine has x, y, z axes for translation motions, C axis for rotation motion and A axis for swing motion. The maximum translating velocity of the x, y, z axes allow for 50 mm/s. The diameter of the wheel is 300 mm. The spotting and figuring processes are conducted under the condition: wheel speed 200 rpm, MR fluid ribbon height 1.6 mm and the penetration depth of the optic into the ribbon 0.4 mm. The magnetic field strength applied to the MR fluid ribbon is also stably controlled. The TIF obtained by spotting process is showed in Fig. 4.

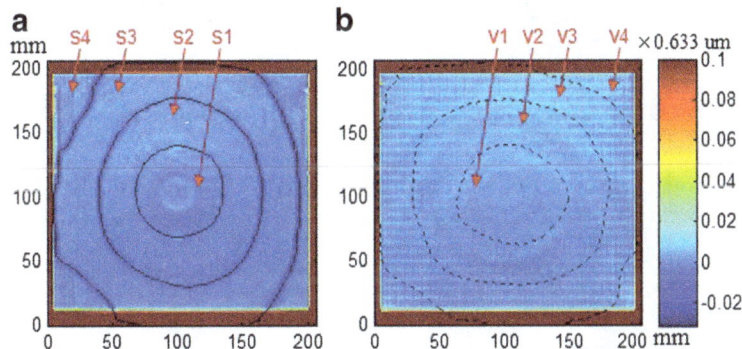

Fig. 13 Simulation results of the 1# and 2# optic figures with (**a**) multi-pitch and (**b**) uniform pitch tool paths. PVe of S1 ~ S4 are all smaller than 0.01 λ, while PVe of V1, V2, V3, V4 are approximately 0.008, 0.02, 0.04, 0.06λ

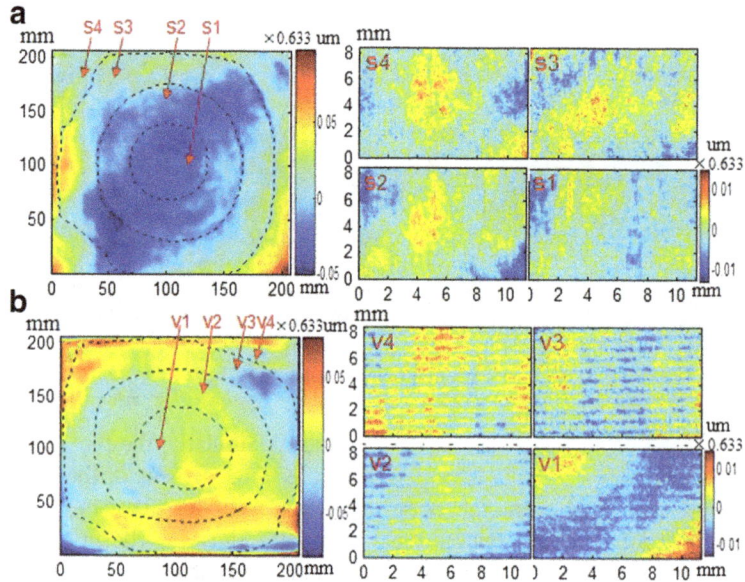

Fig. 14 Polishing results of the 1# and 2# optic figures with (**a**) multi-pitch and (**b**) uniform pitch tool pathss

We used two 200 mm × 200 mm sized optics (1#,2#). The optics are previously ground and polished with continuous polishing process. They both have a favorable initial MSF error specification because the continuous polishing has distinct advantage in restraint of MSF errors. Figures of the both are similarly distributed with a PV value of approximately 0.443um, as shown in Fig. 12. In the following, we employed the TIF to correct the optic figures respectively.

The practical feeding velocity is set to 50 mm/s for determining the pitches. We then calculate the desired pitch for each removal depth by Eq.7. Herein, 1# optic is polished with multi-pitch tool path, while 2# optic with uniform pitch tool path for

comparison. The removal map of 1# optic is divided into 4 regions, and the removal depths in the regions are as follows: 1) 0 ~ 0.0633um, 0.0633 ~ 0.190um, 0.190 ~ 0.316um, 0.315 ~ 0.443um, then the pitches in the regions can be obtained: 0.8, 0.395, 0.132, 0.099 mm. The pitch of 2# optic is set at 0.8 mm on the whole surface.

The dwell points are generated with a feeding pitch of 0.3 mm along every path, and the dwell time map is solved by common discrete convolution algorithm. Then the CNC code for controlling the kinematics of the MRF machine is generated based on the dwell point and dwell time. The simulation and experimental results are shown in Figs. 13, 14 and 15.

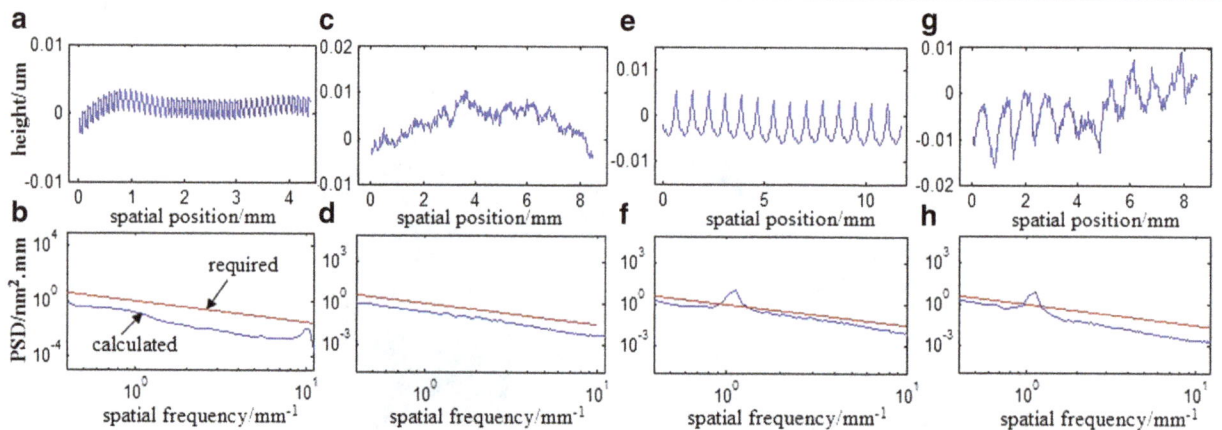

Fig. 15 Residual profiles and PSD errors of the simulation and polishing results, the sampling area is part of the subregions. **a** Residual profile of S4 simulation, **b** PSD error of S4 simulation, **c** residual profile of S4 polishing results, **d** PSD error of S4 polishing results, **e** residual profile of V4 simulation, **f** PSD error of V4 simulation, **g** residual profile of V4 polishing results, **h** PSD error of V4 polishing results

The both optics have a surface form of approximately 0.095um PV after polishing with multi-pitch and uniform pitch tool paths respectively in simulation and experiments. In the uniform pitch path, the residual ripples are fairly large and non-uniformly distributed depending on the local removal. The regions with more removal have larger residual ripples. In contrast, the multi-pitch polished optic exhibits superiority in restraining residual ripple. As the more removal regions with a much smaller pitch path, the residual ripples are restrained. It is noticeable that the optic polished with multi-pitch path has slight depression at the edge between adjacent regions due to that the tool translation stroke from one path line to another introduces extra removal. Although, the depression is so small that it has little effect on the figure error.

Conclusions

A multi-pitch tool path was developed for CCOS processes. With this tool path, the removal map is divided into several subregions, and the pitch in each subregion is set individually. In small removal subregions, the pitch is larger introducing less extra removal so that guarantee the convergence of the figure correction, while the large removal subregions the pitch is smaller so as to decrease the residual ripples. The multi-pitch tool path has been verified that it is beneficial to restraining the ripples while maintaining the convergence of the figure correction.

Abbreviations
CCOS: Computer-controlled optical surfacing; MRF: Magnetorheological Finishing; MSF: Middle spatial frequency; PSD: Power spectral density; PVe: Peak-to-valley value of the ripple; PVn: Normalized PV value of residual ripple; TIF: Tool influence function

Acknowledgements
Not applicable.

Funding
This work was supported by Science Challenge Project of China, No. JCKY2016212A506-0501.

Authors' contributions
DL and JH developed the multi-pitch tool path; HW assisted conducting the experiments. All authors read and approved the final manuscript.

Competing interests
The authors declare that they have no competing interests.

References
1. Betti, R., Hurricane, O.A.: Inertial-confinement fusion with lasers [J]. Nat. Phys. **12**(5), 435–448 (2016)
2. Pohl, M., Börret, R.: Simulation of mid-spatials from the grinding process [J]. J. Eur. Opt. Society-Rapid Publ. **11**, (2016)
3. Almeida, R., Börret, R., Rimkus, W., et al.: Polishing material removal correlation on PMMA–FEM simulation [J]. J. Eur. Opt. Society-Rapid Publ. **11**, (2016)
4. Wang, C.J., Cheung, C.F., Ho, L.T., et al.: A novel multi-jet polishing process and tool for high-efficiency polishing [J]. Int. J. Mach. Tools Manuf. **115**, 60–73 (2017)
5. Arnold, T., Boehm, G., Paetzelt, H.: New freeform manufacturing chain based on atmospheric plasma jet machining [J]. J. Eur. Opt. Society-Rapid Publ. **11**, (2016)
6. Tamkin, J.M., Milster, T.D.: Effects of structured mid-spatial frequency surface errors on image performance [J]. Appl. Opt. **49**(33), 6522–6536 (2010)
7. Hu, H., Dai, Y., Peng, X.: Restraint of tool path ripple based on surface error distribution and process parameters in deterministic finishing [J]. Opt. Express. **18**(22), 22973–22981 (2010)
8. Wang, C., Yang, W., Ye, S., et al.: Restraint of tool path ripple based on the optimization of tool step size for sub-aperture deterministic polishing [J]. Int. J. Adv. Manuf. Technol. **75**(9–12), 1431–1438 (2014)
9. Dai, Y.F., Shi, F., Peng, X.Q., et al.: Restraint of mid-spatial frequency error in magneto-rheological finishing (MRF) process by maximum entropy method [J]. Sci. China Ser. E: Technol. Sci. **52**(10), 3092–3097 (2009)
10. Wang, C., Wang, Z., Xu, Q.: Unicursal random maze tool path for computer-controlled optical surfacing. Appl. Opt. **54**(34), 10128–10136 (2015 Dec 1)
11. Yu, G., Li, H., Walker, D.: Removal of mid spatial-frequency features in mirror segments [J]. J. Eur. Opt. Society-Rapid Publ. **6**, (2011)
12. Dunn, C.R., Walker, D.D.: Pseudo-random tool paths for CNC sub-aperture polishing and other applications [J]. Opt. Express. **16**(23), 18942–18949 (2008)
13. Walker, D.D., Yu, G., Bibby, M., et al.: Robotic automation in computer controlled polishing [J]. J. Eur. Opt. Society-Rapid Publ. **11**, (2016)
14. Zhang, X., Yu, J., Zhang, Z., et al.: Analysis of residual fabrication errors for computer controlled polishing aspherical mirrors [J]. Opt. Eng. **36**(12), 3386–3391 (1997)
15. Cheng, H.B.: Independent variables for optical surfacing systems [M], p. 76. Springer-Verlag, Berlin (2014)
16. Spaeth, M.L., Manes, K.R., Widmayer, C.C., et al.: The National Ignition Facility wavefront requirements and optical architecture [C]. SPIE. **5341**, 25–42 (2004)
17. Wang, C., Yang, W., Wang, Z., et al.: Dwell-time algorithm for polishing large optics [J]. Appl. Opt. **53**(21), 4752–4760 (2014)

25

Research on edge-control methods in CNC polishing

Guoyu Yu[1]* (iD), David Walker[1,2,3], Hongyu Li[1,4], Xiao Zheng[1] and Anthony Beaucamp[3,5]

Abstract

Background: We have developed edge-control for the Precessions TM process suitable for fast fabrication of large mirror segments, and other applications sensitive to edge mis-figure. This has been applied to processing of European extremely large telescope (E-ELT) prototype mirror-segments, meeting the specification on maximum edge mis-figure. However we have observed residuals that have proved impossible to correct with this approach, being in part the legacy of asymmetries in the input edge-profiles.

Methods: We have therefore compared different proposed methods experimentally and theoretically and report here on a new edge-rectification step, which operates locally on edges, does not disturb the completed bulk area.

Results: A new toolpath has been developed and experiments have been carried out to demonstrate that local edge rectification can be carried out.

Conclusions: With this method, the residue error on edges can be removed separately and has potential to reduce total process time.

Keywords: Segment mirror, Optical fabrication, Telescopes, Polishing

Background

Segmented mirrors were adopted for 10 m–class telescopes and are being extended for the forthcoming 30-40 m class [1, 2]. This concept has found applications in other areas [3]. One important requirement of mirror segments is achieving adequate control of edge mis-figure, as this can deflect stray light or infrared emissivity into the science beam, reducing contrast and signal-to-noise ratio [4]. For polishing in the semiconductor sector it can be important because of depth-of-focus limitations of photolithography, and the need to maximize silicon useful real-estate. There are published theoretical reports on modelling edge-effects in polishing, one studying the parametric tool influence functions [5]. Other reports [6, 7] considered a "skin model" related to the Preston equation. Further reports [8, 9] have presented experimental data showing variation of tool influence functions when encroaching an edge, delivering improved edge performance.

The specified maximum edge mis-figure per edge on each E-ELT prototype segment is 200 nm PVq surface, and the average per edge over the prototypes is 100 nm. We have described in a previous paper [10] an end-to-end process-chain for mirror segments, targeted at the E-ELT, as well as other applications such as semiconductor polishing. This work has provided evidence that a fast, cost-effective process for polishing of the bulk surface and edges, directly on precision-ground aspheric hexagons, is achievable. This is based on bonnet-polishing of the entire segment with a raster tool-path (Fig. 1). The tool-lift algorithm progressively reduces spot-size of the near-Gaussian influence function, or 'IF') towards the outer extreme of the edge-zone (which we define as the peripheral zone one full-spot-size wide). This leaves a controllable edge up-stand. The process is followed by pitch-button polishing of the entire segment, to smooth the global surface and lower the raised edges. The maximum allowable mis-figure can be reached, but the average is more challenging.

The main reasons for this are as follows. First is print-through of edge-asymmetries from CNC grinding into

* Correspondence: g.yu@hud.ac.uk
[1]National Facility for Ultra Precision Surfaces, OpTIC Centre, University of Huddersfield, St. Asaph Business Park, Ffordd William Morgan, North Wales, St Asaph LL17 0JD, UK
Full list of author information is available at the end of the article

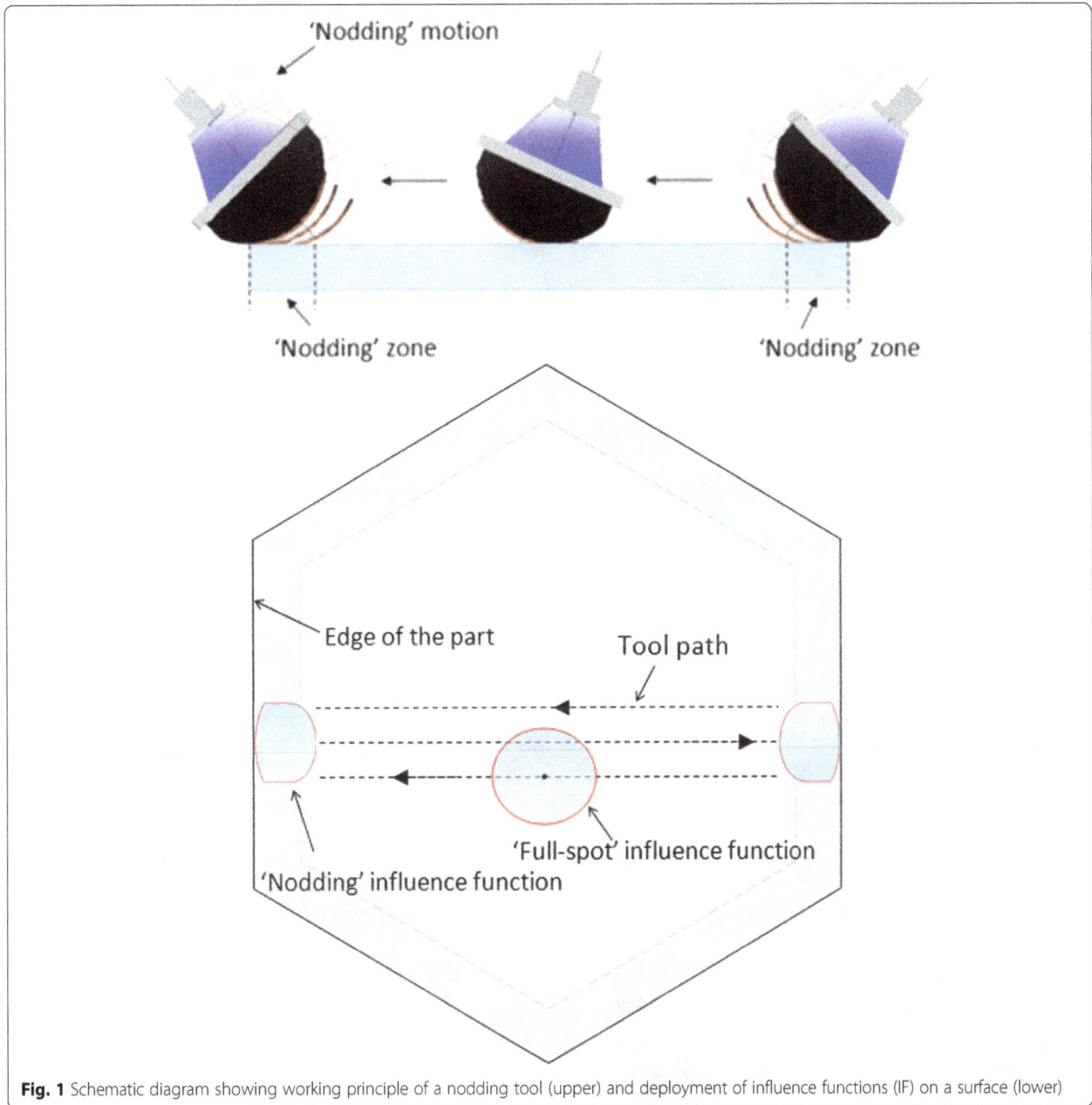

Fig. 1 Schematic diagram showing working principle of a nodding tool (upper) and deployment of influence functions (IF) on a surface (lower)

the final result. The second reason relates to the pitch processing stage. If the process continues sufficiently to eliminate the edge and corner up-stands completely, it tends to disturb the global form and turn down the extreme edge. Furthermore, it can lead to a trench at the interface between the tool-lift zone and the bulk surface i.e. ineffective blending.

We report in this paper on different methodologies we have been researching on this issue, supported by theoretical modelling and experimental results. In particular, we report on bonnet 'nodding', tool infill factor and local edge rectification.

Methods
Edge control with nodding bonnet tool

When a compliant bonnet-tool overlaps the edge of the part, excessive material is removed in the edge-zone due to increased local pressure. In 'nodding', as the IF moves along the tool-path and meets the edge of the part, the precess-angle is progressively increased. This maintains registration of the edge of the disk of polishing cloth with that of the part, truncating the IF, and avoiding edge-overlap (Fig. 1),

A bonnet for nodding needs first to be machined true to the machine virtual pivot (intersection of A, B axes), as is

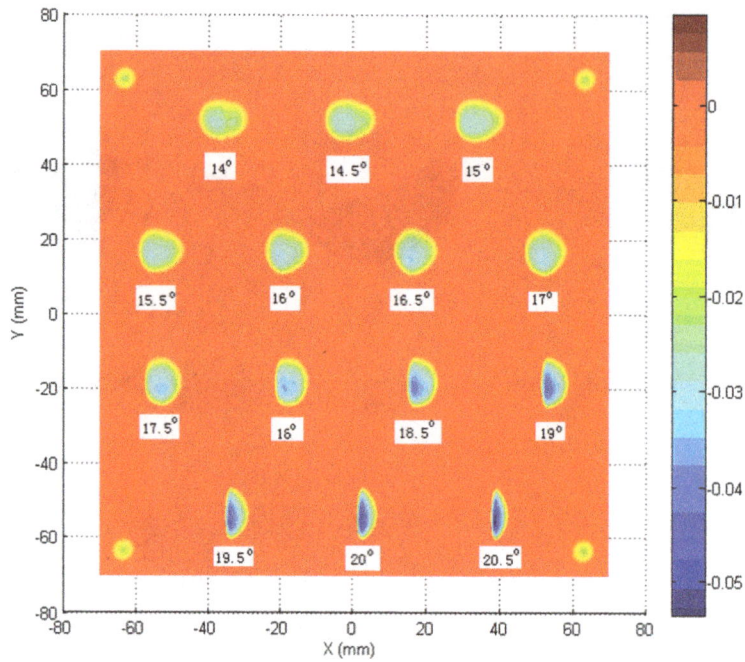

Fig. 2 Generated influence functions with varied precession angle

usual practice. The nominally circular disk of polishing cloth is molded to the bonnet radius and cemented to the bonnet. The periphery of the cloth is then machined, using a sharp tool, to be precisely circular and run true. This requires caution in order not to damage the bonnet. It is not necessary to machine right through the cloth; a step is sufficient to deliver an influence function with a sharp edge.

Successful application of nodding for edge control relies on two aspects. First, a set of IFs in the edge zone requires stability and accuracy, including both removal rate and shape. Second, a nodding motion is required such that the truncated edge of the influence function is tangential to the edge of the part under polishing, as shown in Fig. 2.

An interferogram and Form Talysurf scan (edge to edge) of a surface processed with nodding, are shown in Fig. 3. It can be seen that a feature is left on the surface throughout the nodding zone. However, there

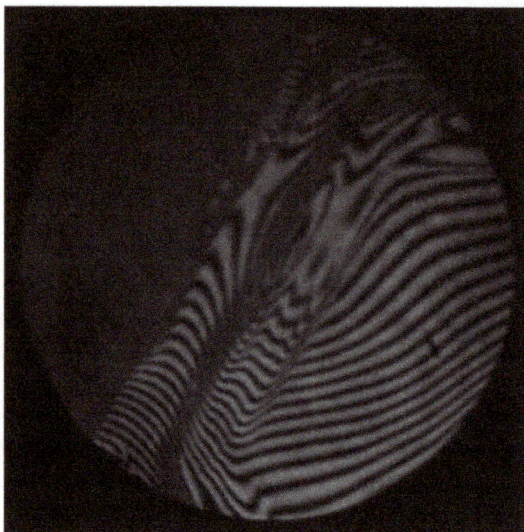

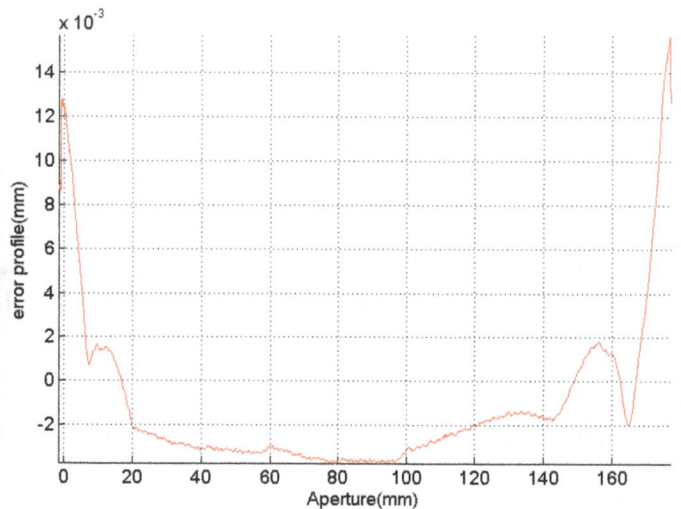

Fig. 3 Processed surface edge with nodding method

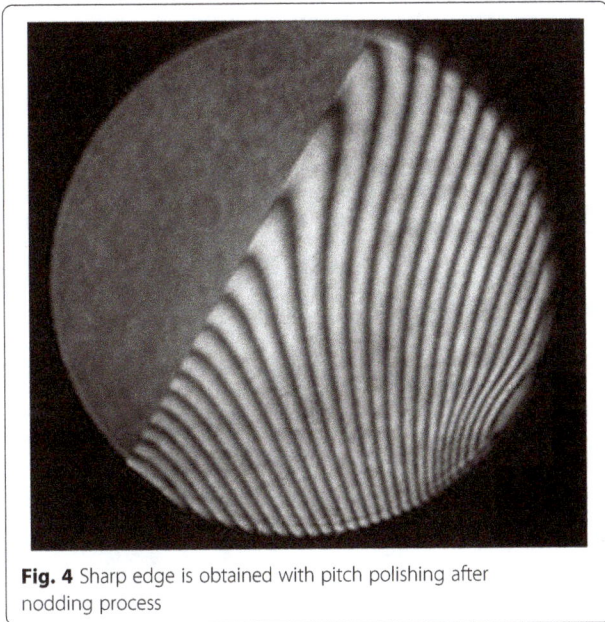

Fig. 4 Sharp edge is obtained with pitch polishing after nodding process

is no edge turn-down. This demonstrates that the process is fundamentally sound. The up-standing edge is about 7 mm wide. This can be flattened by a pitch polishing process, as shown in Fig. 4.

Edge control with non-uniform treatment-tools with different infill factors

Complementing bonnets, rigid tools can be used to rectify mid-spatial features and assist edge control. Their size is limited by aspheric misfit, leading to compliant (visco-elastic or non-Newtonian) materials often being incorporated. Traditionally, tools either 'float' on the part by gravity, or under additional spring-force. It is fundamentally not possible to achieve uniform removal throughout the bulk and edge zones. This would require the tool-path and tool to leave the part completely;

impossible, because the tool would rock on the edge. Stopping the tool-path short leaves an excess of material, which cannot be fully rectified simply by changing process variables. This is exacerbated by other local boundary-condition issues, such as differences in slurry mobility on edges and bulk. To explore some of these effects, simulation work has been carried out to optimize process parameters to achieve optimal edge form.

A simulation has been conducted in the MatLab environment to predict the surface profile with rigid tool working on a flat workpiece surface based on Preston's Law. The influence function of the rotating tool is computed (Fig. 5a) and the material removals then integrated over the tool path (Fig. 5b).

A variable we haven't explored previously is the tool 'in-fill factor', applied to a rotating rigid or semi-rigid tool. We define this as the proportion of the circumference of each radial zone from tool-centre to tool-edge, which is occupied by a physical pad. In craft polishing this is represented by 'petal laps'. To improve the relative removal ability on the edge, the infill factor on the edge of the part is fixed to be 1 and reduced as it comes closer to the centre, followed by Eq. 1:

$$Infill\ Factor = \left(\frac{r}{R}\right)^n \tag{1}$$

where R is the radius of the tool, r is the distance of concentric circles to the centre of the tool and n is the infill factor power of the equation.

Some examples generated based on Eq. 1 with different powers have been modelled (shown in Fig. 6), where the coloured part represents the raised pad in contact with the part, and the white part the spaces between. It is expected that the tool with higher infill factor power can relatively remove more material on the edge than the bulk.

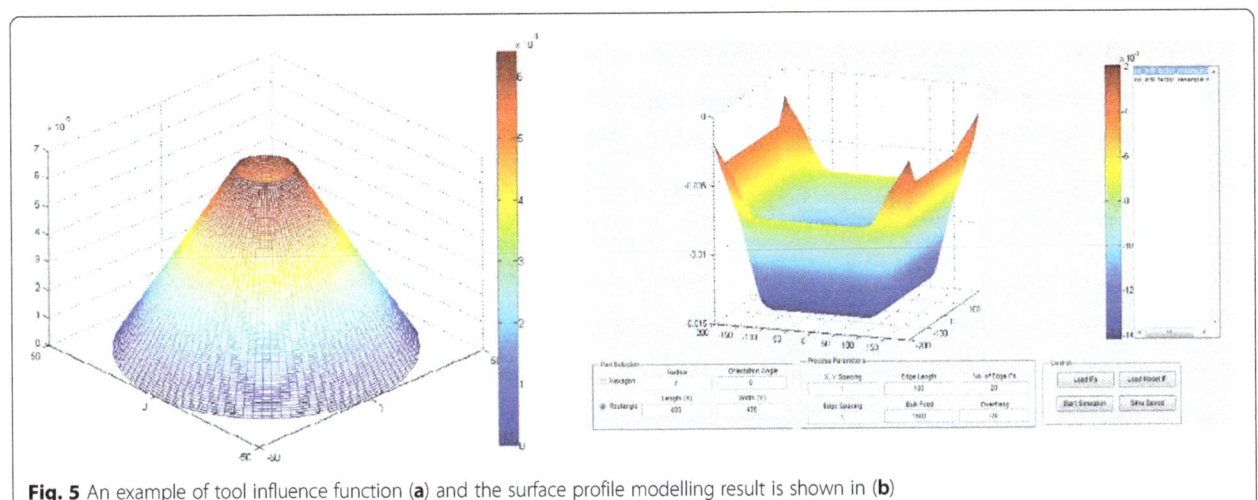

Fig. 5 An example of tool influence function (**a**) and the surface profile modelling result is shown in (**b**)

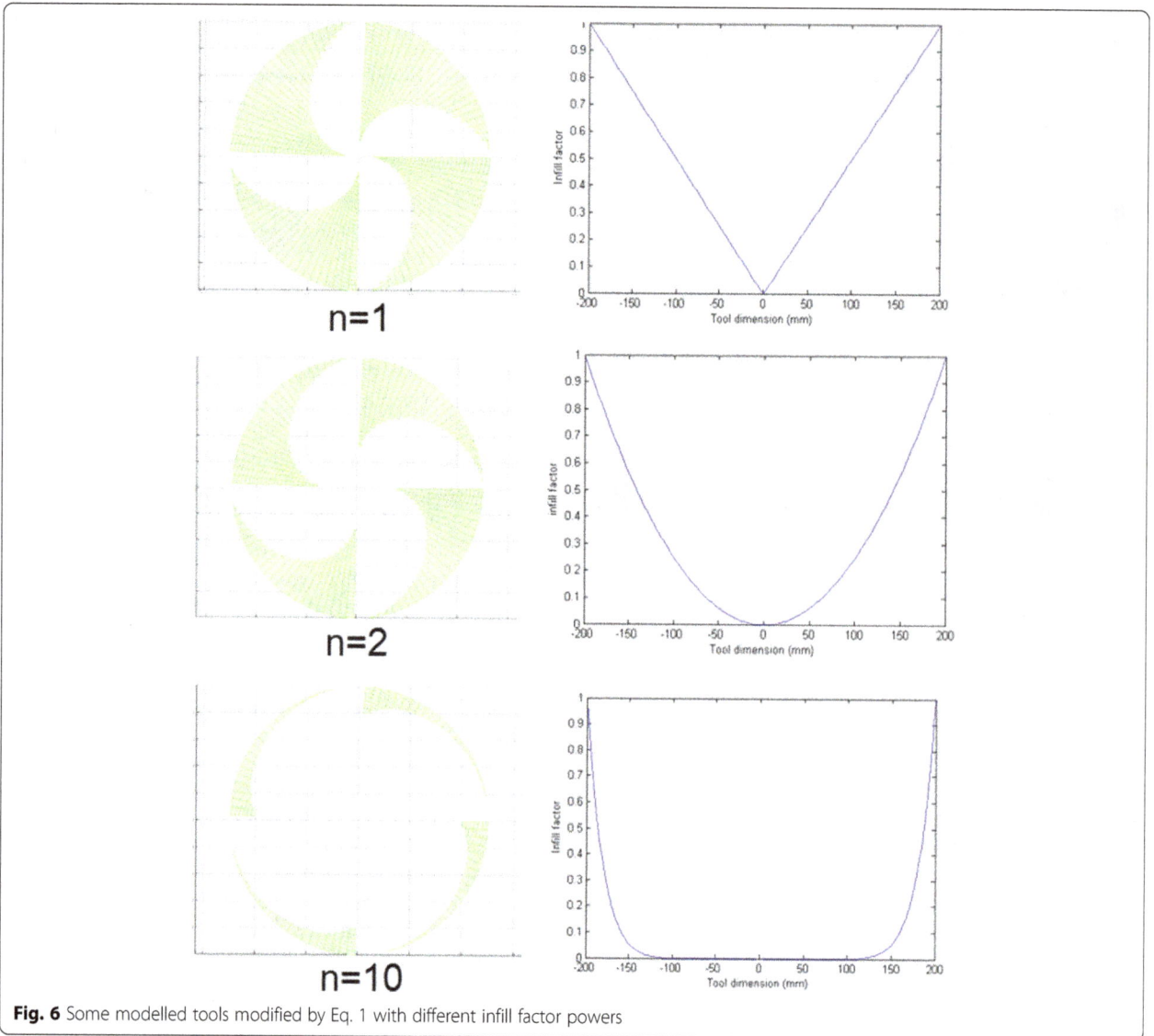

Fig. 6 Some modelled tools modified by Eq. 1 with different infill factor powers

The effect of such tools with different infill factor power has been simulated by integrating the corresponding influence function. Cross sections of the simulated surface profile is shown in Fig. 7. Interestingly, the modelling has shown that the edge-profile is *insensitive* to changing the in-fill factor in the manner shown, and therefore we have not progressed this idea further.

Edge control with local rectification

Local edge rectification means the polishing tool addresses the edge zones specifically, as shown in the illustrative hexagonal-spiral tool-path (Fig. 8). Preliminary work has been reported on modelling edge-zone correction using IFs generated where the polishing spot overlaps the edge. We report in this paper on experimental results, following

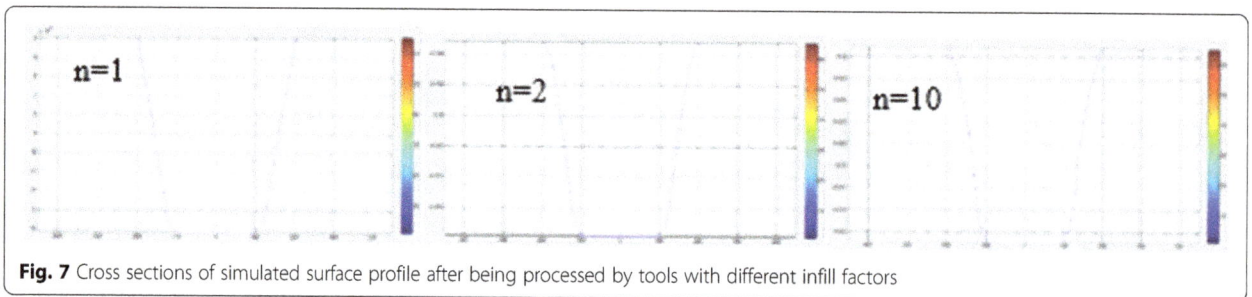

Fig. 7 Cross sections of simulated surface profile after being processed by tools with different infill factors

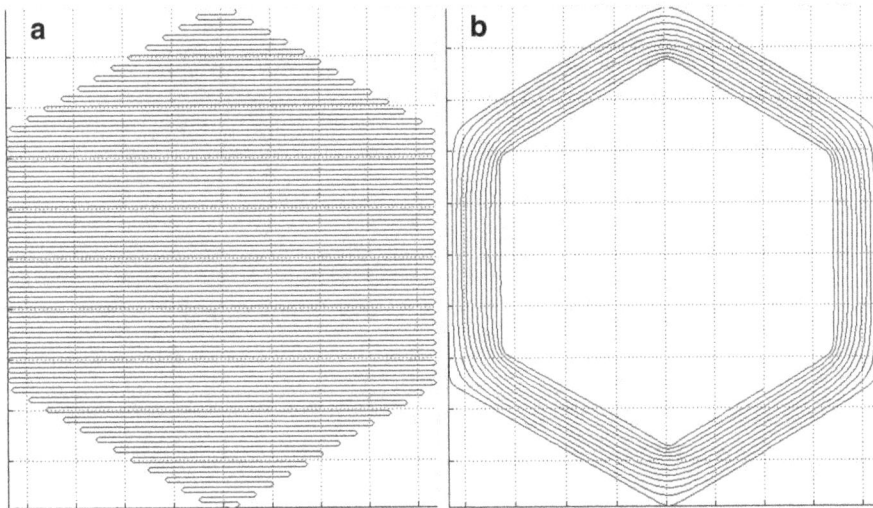

Fig. 8 Schematic diagrams of tool paths (**a**) Raster tool path for bulk-surface. **b** Hex-Hex tool path for local-rectification (not to scale)

development of a hexagonal tool-path, and software to model removal within the edge zone.

Correction of the edge-zone residuals may then proceed as follows. Bulk polishing is programmed to leave an upstanding edge-zone. An R20 bonnet is selected, as its small footprint gives greater sensitivity for local rectification than the larger bonnets used for bulk polishing. The R20 cores-out most of the up-stand using a hexagonal-spiral tool-path, constrained to lie within the edge-zone. Tool-lift is deployed towards the *inner* boundary of the edge-zone, to avoid creating a trench. Finally, a rotating pitch tool is used to 'blend' the edge zone into the bulk area.

Local rectification can improve edges of uneven height, and corners that are higher than the edges. Bonnet polishing tools used for the main bulk deliver spots too large for local correction at corners and edges. In principle, any bonnet can provide spot-sizes tapering to zero, but with larger bonnets the sensitivity to Z-offset errors becomes extreme. For example, assume a typical edge-correction spot-size of 6.9 mm. The corresponding Z-offsets for R20, R80 and R160 bonnets are 0.3, 0.07, 0.037 mm respectively. An error (increase) of a nominal 0.1 mm in Z offset then increases the area of each of the spots by 31%, 127% and 266% respectively. On the other hand, the process time will be unacceptable if the whole surface is processed with a small tool optimized for edges and corners.

Technical challenges and solutions

An error (increase) of a nominal 0.1 mm in Z offset then increases the area of each of the spots by 31%, 127% and 266% respectively. On the other hand, the process time will be unacceptable if the whole surface is processed with a small tool optimized for edges and corners.

The input quality for a local rectification step is a part such as a hexagon i) with the bulk form to specification, and ii) with unacceptable edge residuals that are all upstanding with respect to the extrapolated bulk form. The required fidelity of the mapping of the metrology-data coordinate-frame onto the part's surface, and onto the machine's CNC coordinate-frame is $\leq \sim 100$ μm; significantly more demanding than the ~ 200 μm for the prior corrective polishing operation. This demonstrates one of the key advantages of performing such corrections, with on-machine metrology, on a dynamically-stiff Cartesian CNC platform of the quality of the Zeeko machines. A robot platform would hardly be competitive.

In the following section, various issues are discussed relating to metrology and polishing. The solutions provided have been experimentally verified.

Overshooting When polishing and correcting bulk-form by dwell-time moderation, a significant offset or pedestal is required for two reasons i) to remove the overall surface and sub-surface damage layer from prior grinding, and ii) to avoid the infinite traverse speeds and accelerations that would be needed to 'skip over' the surface to allow zero local removal. Corrective polishing is then *differential* i.e. proportional with respect to the offset. Drifts in removal rate will affect the correction, but the principle of leaving edges always turned up provides a contingency.

For local edge-rectification of the up-stand, the material removal has to be controlled *absolutely*, so that the correction never overshoots i.e. it never causes the local surface to project below the extrapolated bulk-form, as this can be rectified only by re-working the entire *global* surface. Upward residuals, in contrast, can always be rectified

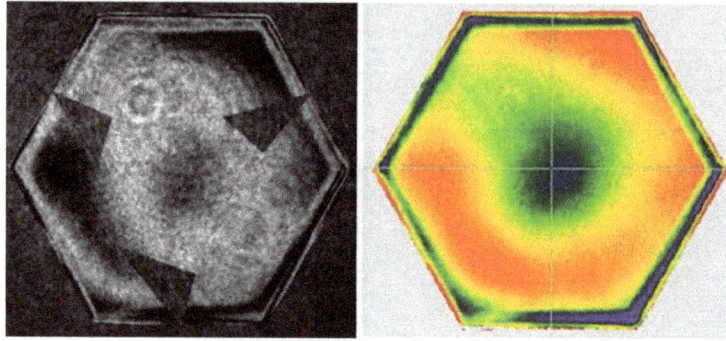

Fig. 9 Overshoot due to touch-on error

with another *local* pass. There are several factors that can lead to overshoot, described below.

In the Zeeko process, the bonnet tooling is advanced towards the part prior to polishing, and first-contact is determined using feedback from a load cell in the polishing head. This gives the zero datum of bonnet compression. The tool is then advanced further towards the part ('Z-offset') to compress the bonnet, and so create a contact-spot (influence function) of the desired size. Any error in establishing first-contact directly disturbs the spot-size, and so affects volumetric removal rate by approximately a square law (see Section 3). This effect is therefore very sensitive, and is dominant in overshoot. It can be caused by movement of the part in its fixturing, thermal growth in the machine, residual hysteresis in the bonnet material, electrical noise in the load cell signal, or drift in the DC signal amplifier. The principal impact is an error in polishing-rate *between* acquiring influence function data and starting polishing.

Figure 9 gives a typical example, where such errors have led to excessive removal, creating a depressed trench inboard of the edge, of 270 nm in depth. The triangular masks visible in the interferogram are to identify the true physical edges (start of bevel), rather than the end of the visible fringes.

The solution to touch-on errors has been independent calibration, both immediately prior to acquiring influence function data, and immediately before polishing. The procedure is to perform touch-on, then back-off the tool through a known distance under software control, and manually check the distance by inserting and withdrawing a shim of known thickness. Any discrepancies can then be compensated in software through modifying the Z-offset.

A second cause of overshoot results from the definition of the depth-of-error to be removed. As mentioned above, skipping over areas is precluded by machine acceleration/speed limits. Thus, if an attempt is made fully to correct the edge-zone error, this will result in overshoot somewhere. This risk is mitigated by invoking the ability to change spot-size 'on the fly' in the *Precessions* process, by varying Z-offset. However, this is most sensitive to error when Z-offset and

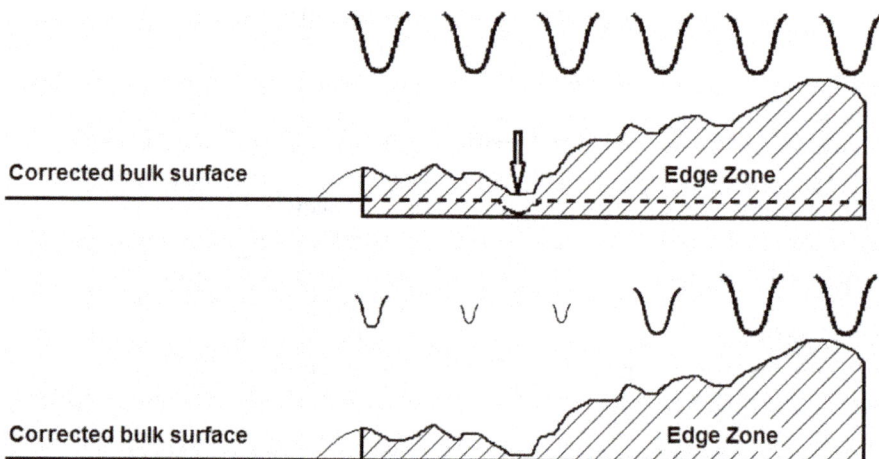

Fig. 10 Principle of reducing process time. (Upper) Using uniform IFs requires a larger amount of material removal. (Lower) Deploy different size of IFs requires fewer material removal

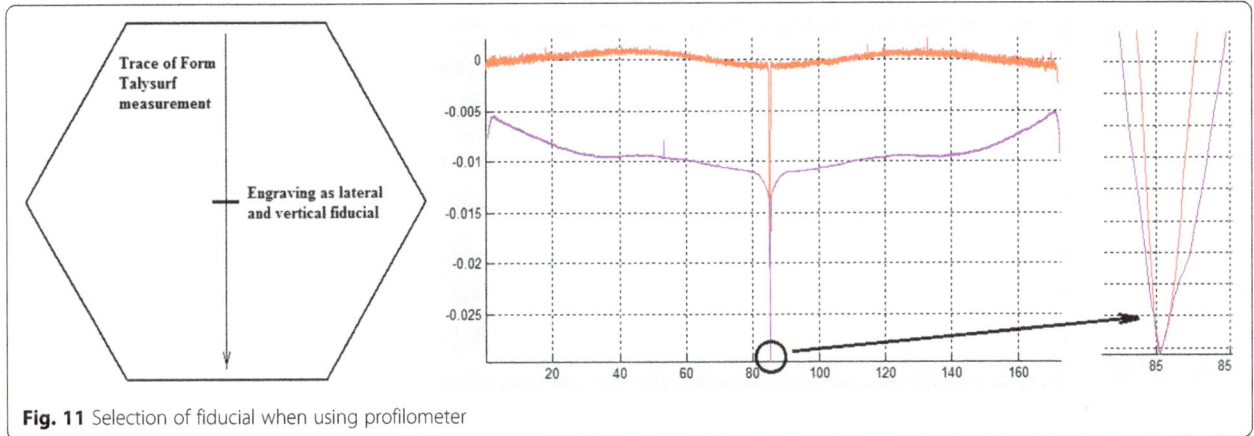

Fig. 11 Selection of fiducial when using profilometer

spot-size both approach zero i.e. around the area(s) of zero target removal. Therefore, a small positive residual is deliberately retained as a contingency, defined to be within the edge mis-figure specification.

A third cause of overshoot arises from any tilt of the part in the machine coordinate frame. The Z-offset variation will then be most at the edge zones, where the effect on volumetric removal rate is greatest. The standard Zeeko 'Non Linear Correction' algorithm performs touch-on at various locations over the surface, and the result of a numerical fit automatically corrects the CNC file. This in turn corrects the Z-offset along the tool-path, for any distortions or global tilts of the part on the polishing support system. However, the numerical fit is weighted in favour of the bulk area, by virtue of the predominance of sample points therein, and this can lead to errors in the edge-zone. A modified Non Linear Correction has been implemented, where samples are acquired only around the periphery of the part.

An additional factor is variation in specific gravity, temperature and/or pH of the slurry, between acquiring influence functions and polishing the part. On the IRP1600 machine used for full-size segment fabrication, this has been mitigated by increasing the slurry-tank volume to 150 l, increasing the slurry-delivery to the part to 30 l/min, and improving slurry agitation. Digital monitoring and archiving of slurry conditions has also been implemented.

Process time The aim of the local edge-rectification is to reduce both the residual edge mis-figure and the total process time, with respect to the standard process-chain that rasters the entire surface. In this standard process, the pitch-button time is dominated by the need to lower the upstanding edges. However, this also removes an unnecessarily large DC level from the polished bulk area, and global form tends to regress, as shown in Fig. 10 (upper). With local rectification, the volume of the edge up-stand is reduced by typically 80% using a small bonnet/spot-size, operating *only* in the limited area around the edge (Fig. 10 lower). This in turn increases the effectiveness, and reduces the

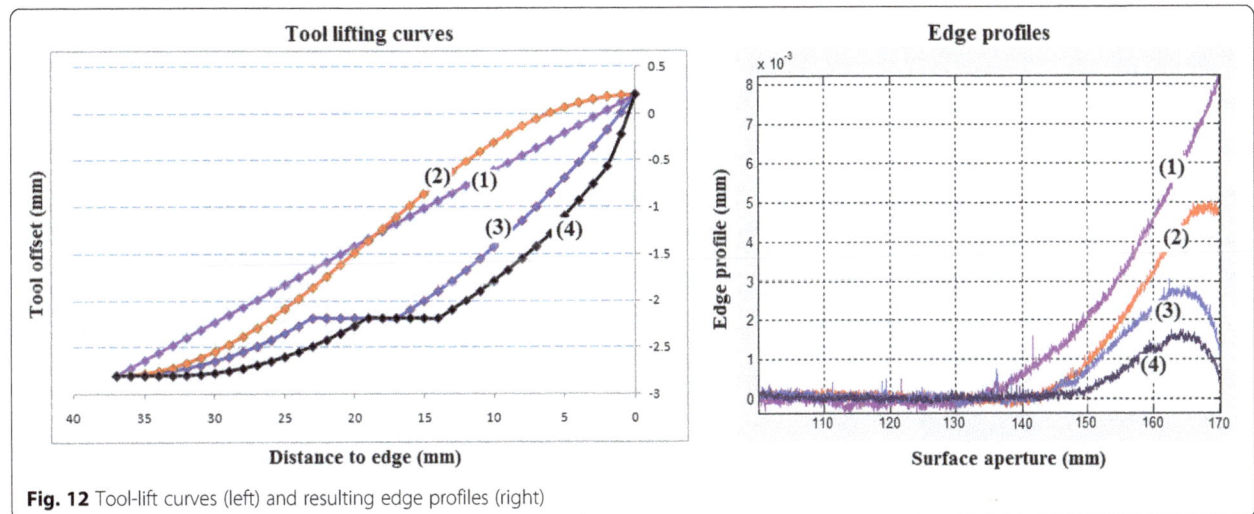

Fig. 12 Tool-lift curves (left) and resulting edge profiles (right)

Table 1 Process parameters of large tool polishing

Parameter	Value	Unit
Surface feed	2000	mm/min
Head speed	800	RPM
Precess angle	14	Degree
Tool offset	2.8	mm
Tool overhang	0	mm
Track spacing	3	mm

time, for the final pitch-button step. On the experimental sample, the total process time saved is 165 min which is 22% of the total process time. On a real segment, this saving would increase due to fact that the time spent on the bulk area will increase by a square law, whereas the time increase on edges will be of linear.

Optimising the bulk surface polishing for edges The first process step is to pre-polish the surface with a large polishing tool. This is to remove subsurface damage left by the prior generating process and to improve surface roughness so that global form can be measured by full-aperture interferometry. The polishing tool in the work on witness parts reported here is an R200 bonnet, with

2.8 mm Z-offset, delivering a nominal 60 mm diameter polishing spot. The target removal was 10 μm P-V depth. Edge control at this stage aimed to leave an up-stand, so that this can be locally corrected afterwards. Accurate metrology of the edge profile is very important in debugging the process, and so a fiducial was engraved on the surface of the witness-part surface to act as:-

1. A datum to measure absolute material-removal by comparing surface profile measurements across the engraved mark before and after polishing, shown in Fig. 11.
2. A fiducial to register the lateral position of profile measurements so that true edge positions were known.

Figure 12 (left) shows examples of several tool lifting schemes we have investigated. The horizontal axis represents the distance from the centre of the polishing spot, to the extreme edge of the part. The vertical axis comprises the Z-offsets on the surface with respect to touch-on. Different tool lifting parameters have produced different edge profiles, as shown in Fig. 12 (right), where the extreme right of the graph represents the true edge of the surface.

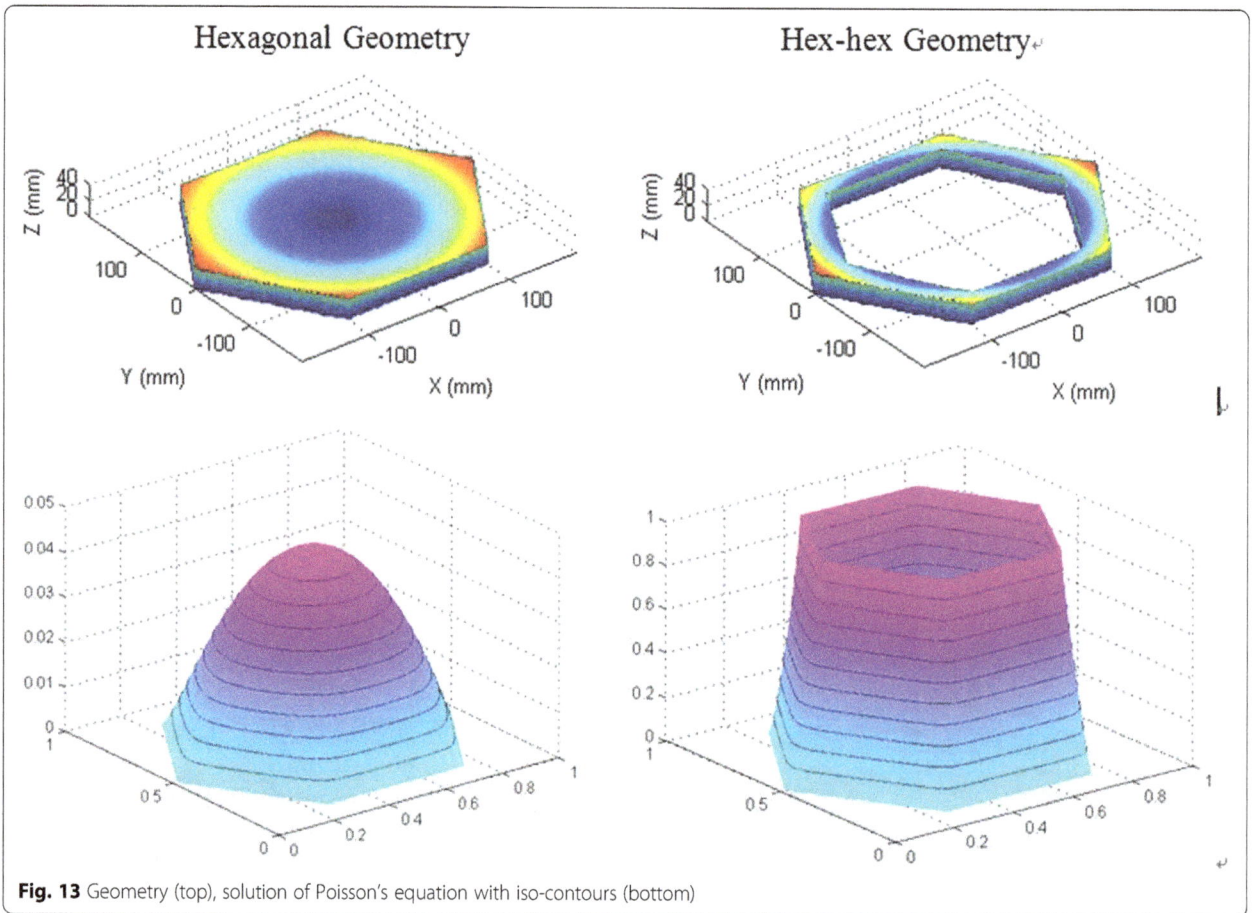

Fig. 13 Geometry (top), solution of Poisson's equation with iso-contours (bottom)

There are several parameters that can be explored, such as precess-angle, local dwell time, Z-offset, and maximum overhang of the polishing spot at the edge. Typical process parameters are listed below in Table 1. With these parameters, a total of 10 μm can be removed through 3 polishing runs. The polishing should start from every other corner to create a symmetric up-stand.

Mathematical basis of hexagonal spirals We have developed two new forms of spiral tool-path; starting from the original circular-spiral, to the 'adaptive spiral' and 'hex-hex' spiral, which are reported below.

The adaptive spiral starts by following the edge of the part (e.g. hexagonal), and morphs progressively to a circular spiral at the centre. The hex-hex spiral does not morph, but stays (e.g.) hexagonal between the outer boundary of the part, and a defined inner boundary on the surface. In both cases, if the traverse were slowed to zero at the corner, then accelerated to follow the direction of the new edge, a depression at the corner would result. The corners must therefore be rounded.

The adaptive spiral has previously been successfully applied to cutting process such as high speed routing http://www.wseas.us/e-library/transactions/mechanics/2008/27-159.pdf and 5-axis milling http://www.sciencedirect.com/science/article/pii/S0168927403000394. One method to compute adaptive spirals consists of computing the solution to Poisson's equation:

$$-div(grad(u)) = 1 \qquad (2)$$

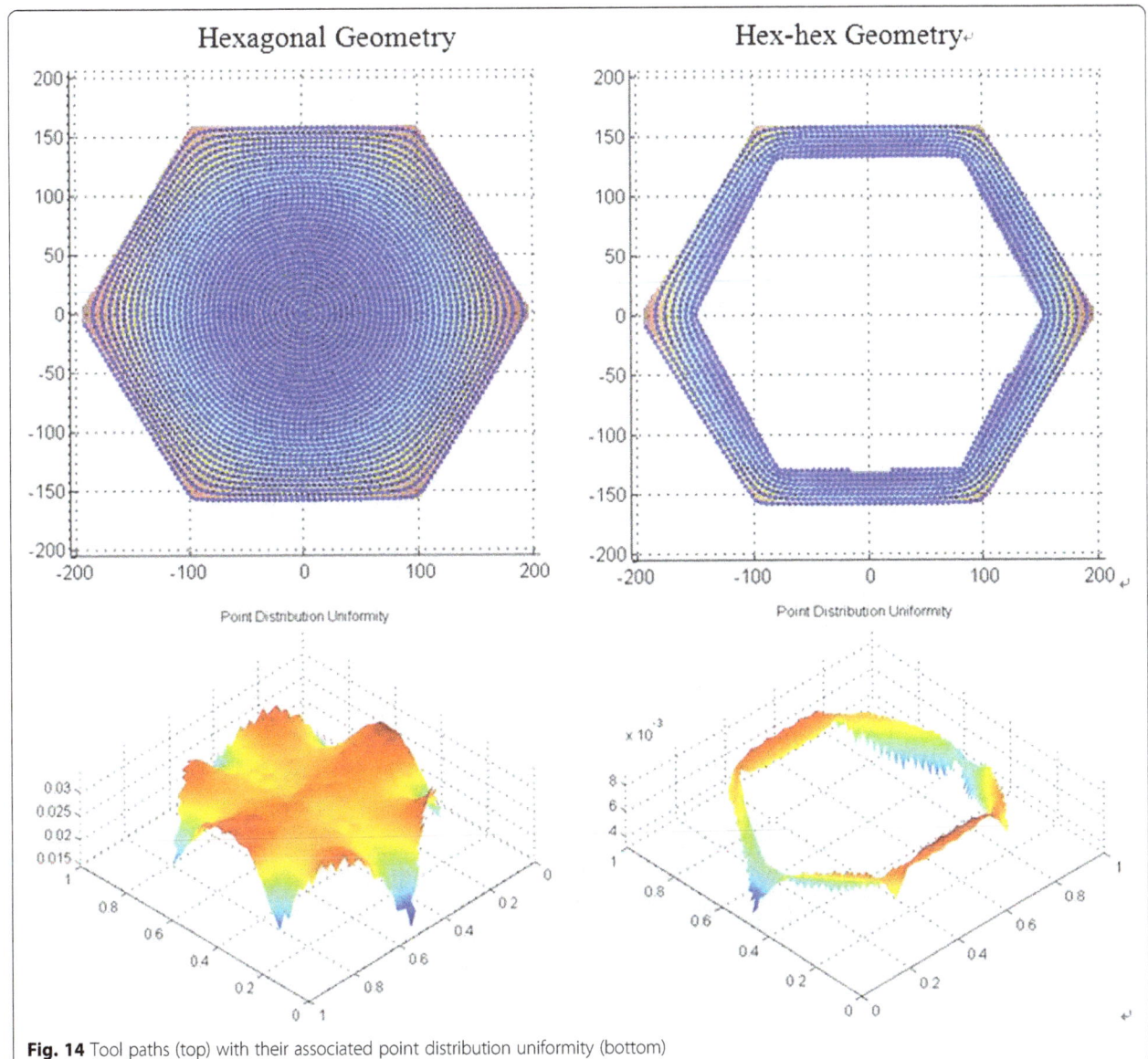

Fig. 14 Tool paths (top) with their associated point distribution uniformity (bottom)

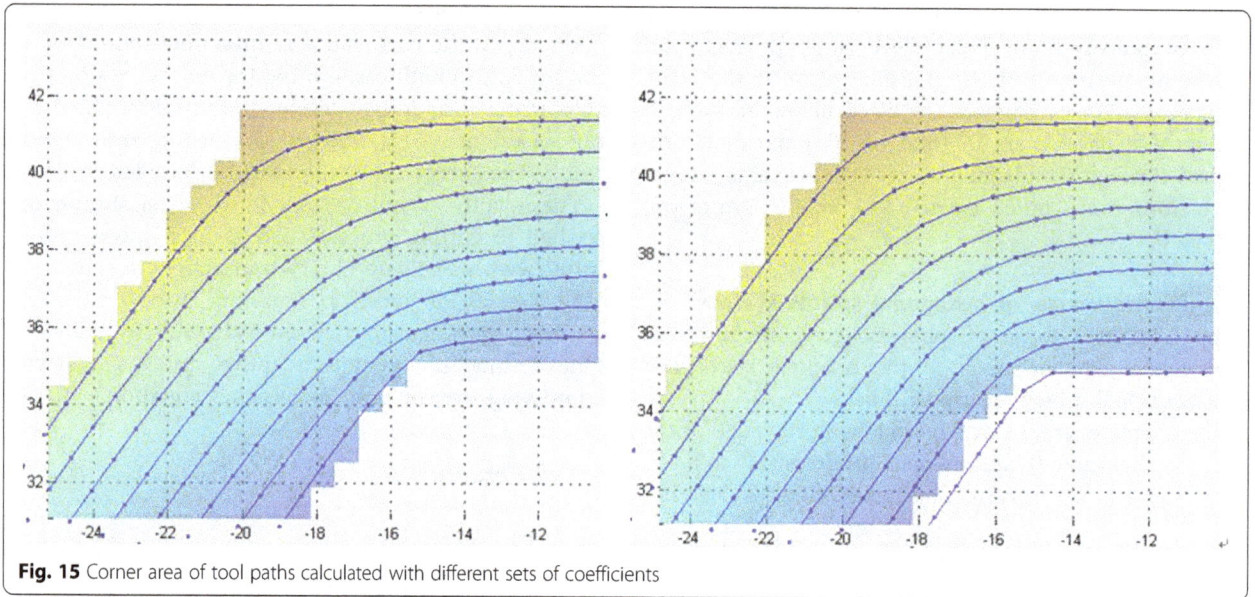

Fig. 15 Corner area of tool paths calculated with different sets of coefficients

with Dirichlet conditions applied to the boundaries of the surface:

$$\begin{cases} u = 0 & \textbf{\textit{for outer boundaries}} \\ u = 1 & \textbf{\textit{for inner boundaries}} \end{cases} \qquad (3)$$

A hex-to-hex surface (upper right) and associated solution of Poisson's equation (bottom left) are shown in Fig. 13. The tool path is then generated by following iso-contours on the solution to Poisson's equation (bottom right).

While this method can generate hex-hex and adaptive tool paths featuring smooth cornering, a drawback is that the track-spacing changes progressively across the tool path. Since bonnet polishing is a time and space dependent sub-aperture process, a change in track-spacing introduces local variations in removal depth. To compensate for this effect, it is necessary to compute the relative density of tool-path points across the surface, and use this information to moderate the feed rate of the polishing spot along the tool path. This

density can be computed by convoluting the dataset with a Gaussian filter:

$$g(x, y) = \frac{1}{2\pi\sigma^2} \cdot e^{-\frac{x^2 + y^2}{2\sigma^2}} \qquad (4)$$

where σ relates to the width of the spot size used during subsequent polishing. Examples of tool paths with their resulting point distribution uniformity are shown in Fig. 14.

Finally, it is possible to balance tool path smoothness against the uniformity of point distribution by introducing extra coefficients in Poisson's equation:-

$$-\textbf{\textit{div}}(\textbf{\textit{coef}}1^*\textbf{\textit{grad}}(\textbf{\textit{u}})) + \textbf{\textit{coef}}2^*\textbf{\textit{u}} = \textbf{\textit{coef}}3 \qquad (5)$$

The corner area of tool paths calculated for different coefficients on the same hex-hex surface are shown in Fig. 15.

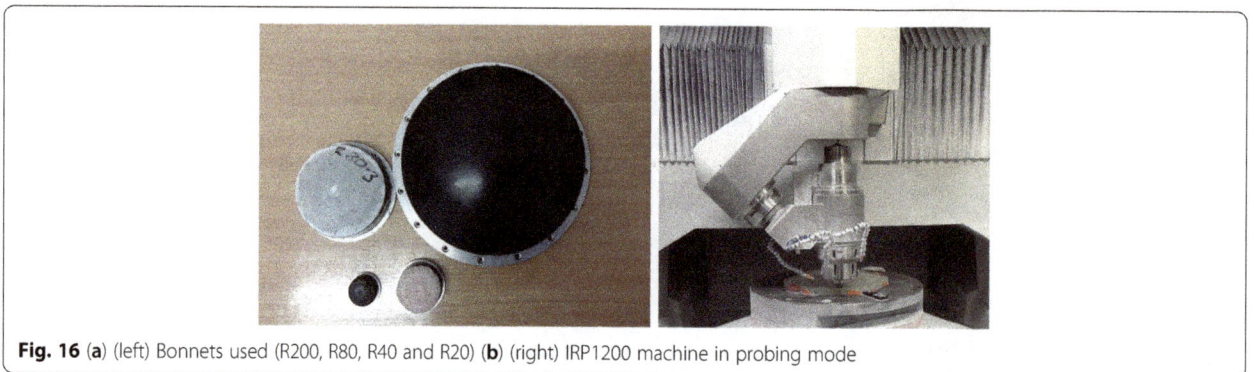

Fig. 16 (a) (left) Bonnets used (R200, R80, R40 and R20) **(b)** (right) IRP1200 machine in probing mode

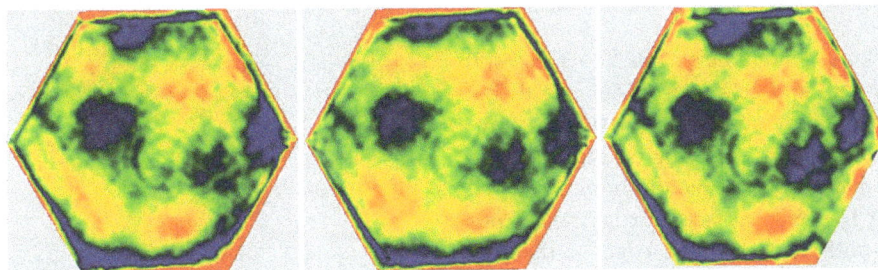

Fig. 17 a Interferometry after form corrections. **b** After1st and (**c**) 2nd local edge rectification. The average edge mis-figures after the 3 stages were 108 nm, 91 nm and 72 nm respectively

Results and discussion

Process optimization was carried out on a Zeeko IRP1200 machine. The demonstration part used was a borosilicate hexagon of 390 mm corner-to-corner. The surface was spherical with a radius of curvature 3.0 m for ease of testing. An interferometer was located on a test tower above the Zeeko polishing machine to provide in-situ metrology.

The part was pre-polished using an R200 bonnet shown in Fig. 16 (a). This removed subsurface damage from generating the spherical form, and improved surface quality so that full-aperture interferometry could be deployed. Smaller tools, such as R80, were then used for form-correction and during these stages, process parameters were optimized for minimum edge-zone up-stand. Local edge rectification then used a R20 bonnet, delivering spot-sizes up to 6.9 mm. These small spot sizes were well-matched to the spatial frequencies in the up-stand. The use of Uninap polishing cloth from Universal Photonics provided a benign removal characteristic contributing to the precision of removal.

The surface after form correction is shown in Fig. 17 (a). The errors of the entire surface were 27 nm RMS and 182 nm PVq (95%). The average edge zone mis-figure was 108 nm PVq. For the first local edge rectification, 30% of the total edge error was targeted and the polishing time was 49 min. The average error mis-figure was 91 nm after this correction, as can be seen

in Fig. 17 (b). The second run target 70% of the total error in edge zone. This was to avoid over-polishing of certain areas due to residual polishing spot-variation, even after surface tilt compensation. The average error mis-figure was 72 nm after this run, as can be seen in Fig. 17 (c).

The final process was blending of the edge zone into the bulk area using a pitch-button tool. The diameter of 50 mm was selected for compatibility with processing a full-size segment, to meet the criterion that the aspheric misfit would be well below the slurry particle size [11]. This step also helped to remove certain mid-spatial frequency features resulting from small-tool corrections. The edge mis-figure after this process, averaged over the six individual PVqs, was 68 nm PVq, as shown in Fig. 18.

Conclusions

We have previously reported on our edge-control work, in regard to a process-chain that operates on the entire segment surface using raster tool-paths. This has resulted in a full-size segment being completed to specification. In this paper, we report on a new extension to the process, where we have treated the edge-zone separately with small tools, without disturbing the finished bulk area. We have pointed out that this requires the edge-zone residuals to be high with respect to the

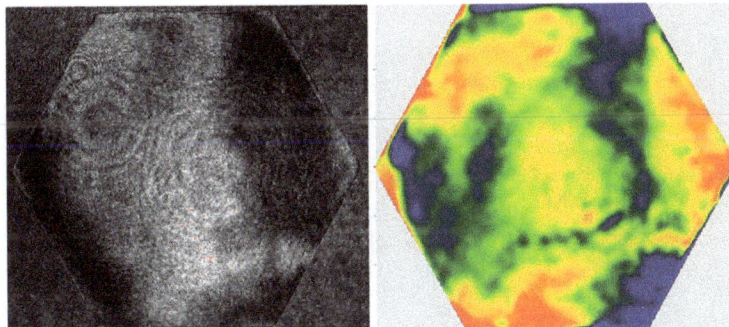

Fig. 18 Surface fringes and phase map after final blending

extrapolated bulk surface, which the tool-lift method can deliver.

We have identified specific issues related to successful local edge rectification, and reported on our simulation and tool-path software-development. In particular, simulation has demonstrated that superior edge-quality can be achieved with small-tool local rectification, compared with processing of the bulk surface. Furthermore, constraining the tool-path to follow the edge of the part drastically reduces rectification time compared with using the same tool to raster-polish the entire surface. New software functions have been developed to execute a hexagonal tool path, modified in the corners to accommodate acceleration/ deceleration limits of the machine. Modified probing software has also been implemented, so that probing is constrained to the edge-zone, in order to improve precision of determining tip/tilt of the part.

Further work to be conducted will focus on understanding the trade-off between edge-quality and total process-time. We have confirmed that edge-quality improves with smaller tool-sizes, and that total process time is reduced if rectification is constrained to the edge-zone alone. If tools smaller than R20 are invoked for finer edge rectification, total process-time will start to increase. Nevertheless, this may confer advantages in some applications if more stringent edge-specifications have to be met.

Abbreviations
CNC: Computer numerically controlled; E-ELT: European extremely large telescope; IF: Influence function; PV: Peak-to-valley

Acknowledgements
We gratefully acknowledge Glyndŵr University for the commitment and financial support to the National Facility for Ultra Precision Surfaces in general, and to the E-ELT prototype project in particular. Thanks are also due to the European Southern Observatory for awarding the contract for fabrication of prototype segments. We are also grateful to Zeeko Ltd. for software development that has enabled the local edge rectification work to proceed. Research grants from the UK EPSRC and STFC, from a NASA-SBIR grant, and from the Welsh Government, are also gratefully acknowledged.

Funding
Declared at acknowledgements.

Authors' contributions
GY: section 2.3 and 3. DW: Section 1 and 2.1. HL: Experimental work on section 2.1 and 3. XZ: Section 2.2. AB: Mathematical basis of hexagonal spirals. All authors read and approved the final manuscript.

Competing interests
The authors declare that they have no competing interests.

Author details
[1]National Facility for Ultra Precision Surfaces, OpTIC Centre, University of Huddersfield, St. Asaph Business Park, Ffordd William Morgan, North Wales, St Asaph LL17 0JD, UK. [2]Department of Physics and Astronomy, University College, Gower St, London, WC1E 6BT, UK. [3]Zeeko Ltd, 4 Vulcan Court, Vulcan Way, Coalville, Leicestershire LE67 3FW, UK. [4]Researh Center for Space Optical Engineering, Harbin Institute of Technology, Harbin 150001, China. [5]Kyoto University, C3 Bldg. Kyotodaigaku-Katsura, Nishikyo-ku, Kyoto 615-8540, Japan.

References
1. Nelson, J, Mast, TS: Construction of the keck observatory. Proc. SPIE 1236. 47–55 (1990)
2. Alvareza, P, Rodríguez Espinosab, JM, Sánchez, F: The gran Telescopio Canarias (GTC) project. New Astron. Rev. 42,553–42,556 (1998)
3. Moses, E, Dunne, M: Generating laser energy. http://www.ingenia.org.uk/ingenia/issues/issue33/Moses_Dunne.pdf
4. Li, H, Walker, D, Yu, G, Sayle, A, Messelink, W, Evans, R, Beaucamp, A: Edge control in CNC polishing, paper 2: simulation and validation of tool influence functions on edges. Opt. Express. 21, 370–381 (2013)
5. Kim, DW, Park, WH, Kim, SW, Burge, J: Parametric modelling of edge effects for polishing tool influence functions. Opt. Express. 17, 5656–5665 (2009)
6. Hu, H, Dai, Y, Peng, X, Wang, J: Research on reducing the edge effect in magnetorheological finishing. Appl. Opt. 50, 1220–1226 (2011)
7. Li, H, Yu, G, Walker, D, Evans, R: Modelling and measurement of polishing tool influence functions for edge control. J Eur. Opt. Soc. Rapid. Publ. 6, 1104801–1104806 (2011)
8. Jing, H, King, C, Walker, DD: Measurement of influence function using swing arm profilometer and laser tracker. Opt. Express. 18(5), 5271–5281 (2010)
9. Walker, DD, Beaucamp, A, Evans, R, Fox-Leonard, T, Fairhurst, N, Gray, C, Hamidi, S, Li, H, Messelink, W, Mitchell, J, Rees, P, Yu, G: Edge-control and surface-smoothness in sub-aperture polishing of mirror segments. Proc. SPIE 8450, 84502A-1–A-9, (2012)
10. Walker, DD, Yu, G, Li, H, Messelink, W, Rob, E, Anthony, B: Edges in CNC polishing: from mirror-segments, towards semiconductors paper 1: edges on processing the global surface. Opt. Express. 20, 19787–19798 (2012)
11. Song, C, Walker, D, Yu, G: Misfit of rigid tools and interferometer subapertures on off-axis aspheric mirror segments. Opt. Eng. 50, 073401 https://doi.org/10.1117/1.3597328 (2011)

26

Controlling the optical properties of a laser pulse at $\lambda = 1.55\mu m$ in InGaAs\InP double coupled quantum well nanostructure

Jalil Shiri[1*] and Abdollah Malakzadeh[2]

Abstract

Background: The transient and steady-state behaviour of the absorption and the dispersion of a probe field propagating at $\lambda = 1.55\mu m$ through an InGaAs\InP double coupled quantum well are studied. The effect of terahertz signal excitation, electron tunnelling and incoherent pumping on the optical properties of the probe field is discussed.

Methods: The linear dynamical properties of the double coupled quantum well by means of perturbation theory and density matrix method are discussed.

Results: We show that the group velocity of a light pulse can be controlled from superluminal to subluminal or vice versa by controlling the rates of incoherent pumping field, terahertz signal and tunnelling between the quantum wells. The required switching time is calculated and we find it between 3 to 15 ps.

Conclusions: In the terahertz (30 ~ 300 μm or 1 ~ 10THz) intersubband transition, the incoming photon energy is (4 ~ 41mev) and maybe in the order of electron thermal broadening (KT ~ 6 meV-25 meV for 77 K -300 K). Therefore in the conventional structure, the incoming photon can directly excite the ground state electrons to higher energy levels. It is shown that the absorption and the dispersion of the probe field can be controlled by the intensity of terahertz signal and incoherent pumping field.

Keywords: Electro-optical switching, Dispersion and absorption, Group velocity, Terahertz signal, Tunnelling effects, Incoherent pumping field

Background

It is known that the absorption and the dispersion properties of a weak probe field can be modified effectively by atomic coherence and quantum interference [1–5]. Atomic coherence can be achieved by the strong coupling fields, the spontaneous emission and incoherent pumping fields. It is known that atomic coherence due to the coherent laser field has essential roles for modifying the optical properties of atomic systems such as spontaneously generated coherence (SGC) [1], lasing without inversion [2], modifying spontaneous emission [3], coherent population trapping (CPT) [5], optical bistability [6–10] and so on [11–16]. furthermore, it

has been shown that quantum interference arising from SGC [8] and incoherent pumping field [17] can be used for analyse of some interesting phenomena such as lasing without population inversion [4], optical bistability [17], and superluminal/subluminal light propagation [18]. Similar phenomena involving quantum coherence in solid state systems such as semiconductor quantum wells (QWs) and quantum dots (QDs) [19], can also be occurred [20, 21]. In the past decade, there has been an increasing interest in optical properties of quantum dot molecules (QDMs) and quantum wells (QWs), due to important role in optoelectronic devices. Recently, investigators have examined the effects of an external field and inter-dot tunnel coupling on the optical properties of QDs and QWs [22–28]. Quantum well semiconductors were

* Correspondence: jalil.shiri@chmail.ir
[1]Young Researchers and Elite Club, North Tehran Branch, Islamic Azad University, Tehran, Iran
Full list of author information is available at the end of the article

chosen because of their advantage in flexible design, controllable interference strength, long dephasing times [29, 30], large dephasing rates [~10 ps-1] [31] and large electric dipole moment which make them suitable for application in the optoelectronic devices. Quantum coherence in a QW structure can be induced by electron tunnelling or applying a laser field [32, 33]. Coherence induced by incoherent field and tunnel coupling in the QW system plays an important role in light–matter interaction and has found numerous implementations in semiconductor optics. On the other hand, coherent control [34–37] over the dispersive and absorptive properties of solid-state media such as photonic crystals and semiconductors has recently attracted a lot of attention [38–42]. Several proposals for quantum coherence and interference in QWs have been performed and analysed. To utilize the tunnelling effect, an electron is excited by a laser field, then tunnels to the second QW by controlling the external voltage between the wells [43]. An interesting application of QWs is modification of light pulse to make a fast electro-optical switch by controlling the propagation of a weak light pulse in a semiconductor system, which depends on the dispersive properties of the medium.

In this paper, we introduce a compact four level quantum wells system composed of two QWs. Then, we investigate effect of terahertz signal, incoherent pumping field and tunnelling between QWs on the absorption, dispersion and the group velocity of a weak probe field. The required switching time when propagation of light changes from subluminal to superluminal and vice versa is also discussed. We find that the dispersion/absorption spectra of the probe pulse can be changed via the effect of terahertz signal, incoherent pumping field and tunnelling effect.

Methods

In Fig. 1 we consider a compact double coupled quantum well nanostructure which is fabricated using InGaAs/InP nanostructures in material grown by an attractive growth technique i.e. organometallic vapor phase epitaxy (OMVPE). The QWs consist of two periods of alternating 10 nm InGaAs and 10 nm InP layers. The sample can be grown in a horizontal OMVPE system at atmospheric pressure. The growth chamber should contain a system that allows the growth of a QW as narrow as 10 A° with an average roughness of half the lattice constant of InGaAs. Typical growth rates are 10 A° /s for lnGaAs and 5 A° /s for InP [44]. An incoherent pumping field and weak probe field are applied to first quantum well (QW$_1$). According to the band gap difference between In.47Ga0.53As (0.7 eV) and InP (1.35 eV), the wave length of the incoherent

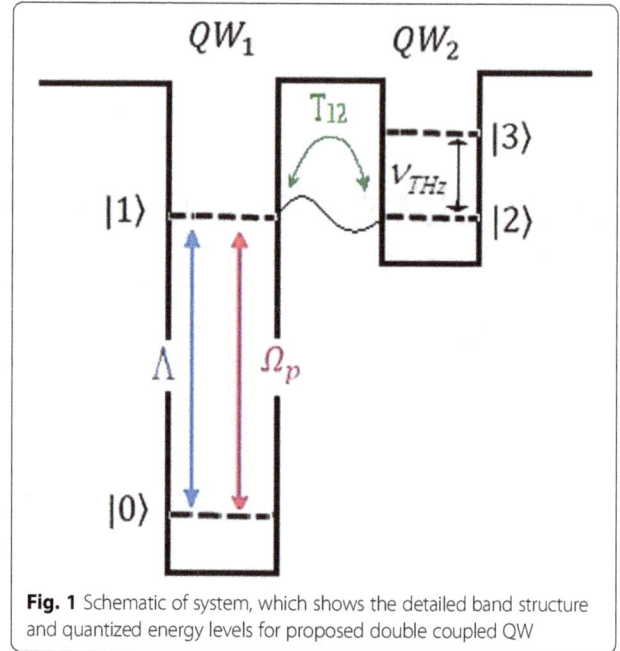

Fig. 1 Schematic of system, which shows the detailed band structure and quantized energy levels for proposed double coupled QW

pump field can be in the range of $\lambda = 1 - 2\mu m$ [44]. For controlling of the tunnelling rate between QWs, the system is placed between two connected electrodes, as electrodes are in contact to this system. By applying independently tuneable gate voltages, electron tunnelling can easily be accomplished between QW$_1$ and QW$_2$. The range of the applied voltage to the electrodes varies as V ≃ 0 −30 mV. For more details we refer to [44–46]. Figure 1 shows the detailed band structure and energy levels of the system. Lower level $|0\rangle$ and upper level $|1\rangle$ are conducting band levels of QW$_1$. Level $|2\rangle$ and level $|3\rangle$ are the excited conducting levels of the QW$_2$ of the right of QW$_1$. It is assumed that the energy difference of three excited levels and the lower level is large, so their tunnelling couplings can be ignored. By applying a gate voltage the level $|2\rangle$ and the level $|3\rangle$ get closer to the level $|1\rangle$. A weak tuneable probe field of the frequency ω_p with Rabi frequency $\Omega_p = \frac{\vec{E}.\vec{\wp}}{2\hbar}$ and an incoherent pumping field Λ are applied to the transition $|0\rangle \rightarrow |1\rangle$. Here, $\vec{\wp}$ is electric dipole moment and E is amplitude of the probe field laser. By forming the resonant coupling of the probe field with the QW$_1$, an electron is excited from the $|0\rangle$ band to the $|1\rangle$ band of the QW$_1$. By providing the tunnelling conditions the electron can be transferred to level $|2\rangle$ in QW2 and the Ω_{THz} (terahertz signal) prompt the electron from the level $|2\rangle$ to the level $|3\rangle$. The total Hamiltonian in the rotating-wave approximation method [47, 48], which represents the interaction of the probe laser field, terahertz signal and incoherent pumping field with the double coupled QWs system, can be expressed in the form of

$$H = \sum_{j=0}^{3} E_j |j\rangle\langle j| + \left[(\Omega_p e^{-i\omega_p t}|0\rangle\langle 1| + \wp\varepsilon|0\rangle\langle 1| + T_{12}|1\rangle\langle 2| + \Omega_{THz} e^{-i\omega_{THz}t}|2\rangle\langle 3| + H.C. \right]$$

$$(1)$$

Where $E_j = \hbar\omega_j$ denotes the energy of state $|i\rangle$. $\wp$ is the dipole moment of the atomic transition corresponding to the pumping of the electrons from level $|0\rangle$ to level $|1\rangle$, and the electric field ε implies the electrical amplitude of the incoherent pumping field. T_{12} correspond to tunnelling between "QW$_1$" and "QW$_2$". The tunnelling can be described by perturbation theory which can be given by Bardeen's approach [49]. According to this approach, the tunnelling probability of an electron in state Ψ with energy E_Ψ from the first QW to state Φ with energy E_Φ in the second QW is given by Fermi's golden rule [50]

$$W = \frac{2\pi}{\hbar}|T_e|^2\delta(E_\Psi - E_\Phi).$$

$$(2)$$

The tunnelling matrix elements can then be acquired by an integral over a surface in the barrier region lying between the QWs

$$T_e = \frac{\hbar}{2m}\int_{z=z_0}\left(\Phi^*\frac{\partial\Psi}{\partial z} - \Psi\frac{\partial\Phi^*}{\partial z}\right)dS,$$

$$(3)$$

where z_0 lies in the barrier, and m is the effective mass of the electron. Applying a bias voltage V, the current is

$$I = \frac{4\pi e}{\hbar}\int_0^{eV}\rho_1(E_F - eV + \varepsilon_0)\rho_2(E_F + \varepsilon_0)|T_e|^2 d\varepsilon_0,$$

$$(4)$$

where ε_o is the energy difference between two discrete states in two wells. The current corresponds to the local density state of each QW (ρ_0, ρ_1) of the Fermi energy (E_F). The magnitude of coupling between two QWs can be adjusted by the bias voltage applied to the wells. Note that T_e is relevant to the applied bias to the molecule. For $T_e = T_{12} \neq 0$ some interaction terms should be appeared in total Hamiltonian as depicted in Eq. (1).

The density-matrix approach given by

$$\frac{\partial\rho}{\partial t} = -\frac{i}{\hbar}[H, \rho],$$

$$(5)$$

Can be used to obtaining the density operator in an arbitrary multilevel QWs system. Substituting Eq. (1) in Eq. (2), the density matrix equations of motion can be expressed as

$$\dot{\rho}_{01} = (i\delta_p - \Gamma_{01} - \Lambda)\rho_{01} + iT_{12}\rho_{02} - i\Omega_p(\rho_{00} - \rho_{11}),$$
$$\dot{\rho}_{02} = iT_{12}\rho_{01} + (i(\delta_p + \omega_{12}) - \Gamma_{02} - \Lambda/2)\rho_{02} + i\Omega_{THz}\rho_{03} + i\Omega_p\rho_{12},$$
$$\dot{\rho}_{03} = i\Omega_{THz}\rho_{02} + (i(\delta_p + \omega_{12} + \omega_{23}) - \Gamma_{03} - \Lambda/2)\rho_{03} + i\Omega_p\rho_{13},$$
$$\dot{\rho}_{12} = i\Omega_p\rho_{02} - (i\omega_{12} + \Gamma_{12} + \Lambda/2)\rho_{12} + i\Omega_{THz}\rho_{13} + iT_{12}(\rho_{11} - \rho_{22}),$$
$$\dot{\rho}_{13} = i\Omega_p\rho_{03} + i\Omega_{THz}\rho_{12} - (i(\omega_{12} + \omega_{23}) + \Gamma_{13} + \Lambda/2)\rho_{13} - iT_{12}\rho_{23},$$
$$\dot{\rho}_{23} = -iT_{12}\rho_{13} - i(\omega_{23} + \Omega_{THz})\rho_{23} + i\Omega_{THz}(\rho_{22} - \rho_{33}),$$
$$\dot{\rho}_{00} = -i\Omega_p(\rho_{01} - \rho_{10}) - \Lambda\rho_{00} + (\gamma_{10} + \Lambda)\rho_{11} + \gamma_{20}\rho_{22} + \gamma_{30}\rho_{33},$$
$$\dot{\rho}_{11} = i\Omega_p(\rho_{01} - \rho_{10}) + \Lambda\rho_{00} - (\gamma_{10} + \Lambda)\rho_{11} + iT_{12}(\rho_{12} - \rho_{21}),$$
$$\dot{\rho}_{22} = -iT_{12}(\rho_{12} - \rho_{21}) + i\Omega_{THz}(\rho_{23} - \rho_{32}) - \gamma_{20}\rho_{22},$$
$$\dot{\rho}_{33} = -i\Omega_{THz}(\rho_{23} - \rho_{32}) - \gamma_{30}\rho_{33},$$

$$(6)$$

Where $\rho_{mn} = |m\rangle\langle n|(m, n = 0, 1, 2, 3)$ and $\rho_{mm} = |m\rangle\langle m|(m = 0, 1, 2, 3)$ [51] represent the coherent terms and the population operators for the QWs, respectively. We get $\omega_{12} = \omega_{10} - \omega_{20}$ and $\omega_{23} = \omega_{20} - \omega_{30}$. The probe field detuning with respect to the QW transition frequencies is $\delta_p = \omega_{10} - \omega_p$. The term $\Lambda = 2(\wp^2/\hbar^2)\Gamma_p$ is the incoherent pumping rate. Note that the incoherent pumping process can also take place in unspecified auxiliary levels. So we assume that the electric field has a broad frequency spectrum or effectively δ-like correlation, i.e., $\langle\varepsilon^*(t)\varepsilon(t)\rangle = \Gamma_p\delta(t - t')$. The spontaneous emission rates for sub band $|i\rangle$, denoted by γ_{10}, are due primarily to longitudinal optical (LO) phonon emission events at low temperature. The total decay rates $\Gamma_{ij}(i \neq j)$ are given by $\Gamma_{0n} = \gamma_{n0}/2 + \gamma_{n0}^{dph}$, $\Gamma_{mn} = (\gamma_{n0} + \gamma_{m0})/2 + \gamma_{mn}^{dph}$, $m, n = 1, 2, 3$ and $m \neq n$, here γ_{mn}^{dph}, are the dephasing rates of the quantum coherence of the $|i\rangle \leftrightarrow |j\rangle$ pathway and determined by electron–electron, interface roughness, and phonon scattering processes. Usually, γ_{mn}^{dph} is the dominant mechanism in a semiconductor solid-state system. Equation (3) can be solved to obtain the steady state response of the medium. The susceptibility of the compact double QWs to the weak probe field is determined by coherence term ρ_{01}

$$\chi = \frac{2N\wp}{E\varepsilon_0}\rho_{01},$$

$$(7)$$

Where N is the carrier density in the proposed QWs system. Ssusceptibility comprise two parts, real and imaginary $(\chi = \chi' + i\chi'')$. Note that the real part of the susceptibility χ' correspond to the dispersion and imaginary part χ'' correspond to absorption. The dispersion slope of the probe field has a major role in the group velocity. The group velocity v_g of the light, which is propagates in the medium, given by [52]:

$$v_g = \frac{c}{1 + 2\pi\chi'(\omega_p) + 2\pi\omega_p(\partial\chi'(\omega_p)/\partial\omega_p)},$$

$$(8)$$

Equation (8) implies that for a negligible real part of susceptibility, the light propagation can be superluminal as a negative slope of dispersion, on the other hand, for positive dispersion slope, the light propagation in the medium can be subluminal.

Results and discussion

Now, we analyse the numerical results of the above equations and discuss the transient and the steady-state behaviour of the absorption and the dispersion. It is assumed that the system is initially in the ground state, i.e. $\rho_{00}(0) = 1$ and $\rho_{ij}(0) = 0$ (i, j = 0, 1, 2, 3). We take typically spontaneous emission $\gamma_{10} = 1 THz$ [45] and other relevant parameters by the factor of these rates. Introduced rates are equivalent to dephasing times in the order of picoseconds. Here the according to the Eq. (8), the positive and negative dispersion slope are representing the propagation of light subluminal and superluminal respectively.

Figure 2(a) shows the dispersion (dashed) and absorption (solid) properties of a probe field versus wavelength, in the absence of incoherent pumping field Λ and tunnelling effect. We observe that the absorption peak accompanies by a negative dispersion. Thus, superluminal light with large absorption propagates through the medium. In Fig. 2(b) we show at the steady state behaviour of the probe dispersion (dashed) and absorption (solid) in the presence of the first tunnelling effect T_{12} and absence of an incoherent pumping field. It can easily be seen that the absorption of the probe field reduced around $\lambda = 1.55 \mu m$ just by applying tunnelling rate $T_{12} = 1\gamma$. We find that the slope of dispersion is very sensitive to the tunnelling effect. When we increase T_{12} from 0 to 1γ, slope of dispersion changes from negative to positive. Figure 2(c) Shows the effect of the terahertz signal on the system, when keeping all other parameters fixed in Fig. 2(b), and with applying the terahertz signal, the dispersion slope does not change, but tow windows transparency created that is accompanied by three absorption peaks.

In Fig. 3. We investigate the incoherent pumping field effect on the system. In Fig. 3(a) we applied an incoherent pumping filed to condition of Fig. 2(a) with values $\Lambda = 8\gamma$. By increasing the incoherent pumping field Λ from 0 to 8γ, the absorption peak of the probe field becomes broaden, while the slope of the dispersion is still negative around the $\lambda = 1.55 \mu m$. Physically, by increasing the incoherent pump rate, the upper levels are populated, for this condition the probe absorption will reduce in transition $|0\rangle \rightarrow |1\rangle$, thus the peak maximum of the probe field absorption gets reduce. In Fig. 3(b), by applying the tunnelling rate T_{12} mid incoherent pumping field, one absorption peak of the probe field is crated at $\lambda = 1.55 \mu m$. By applying the incoherent pumping field, when keeping all other parameters fixed in Fig. 2(b), the slope of dispersion changes from positive to negative that is

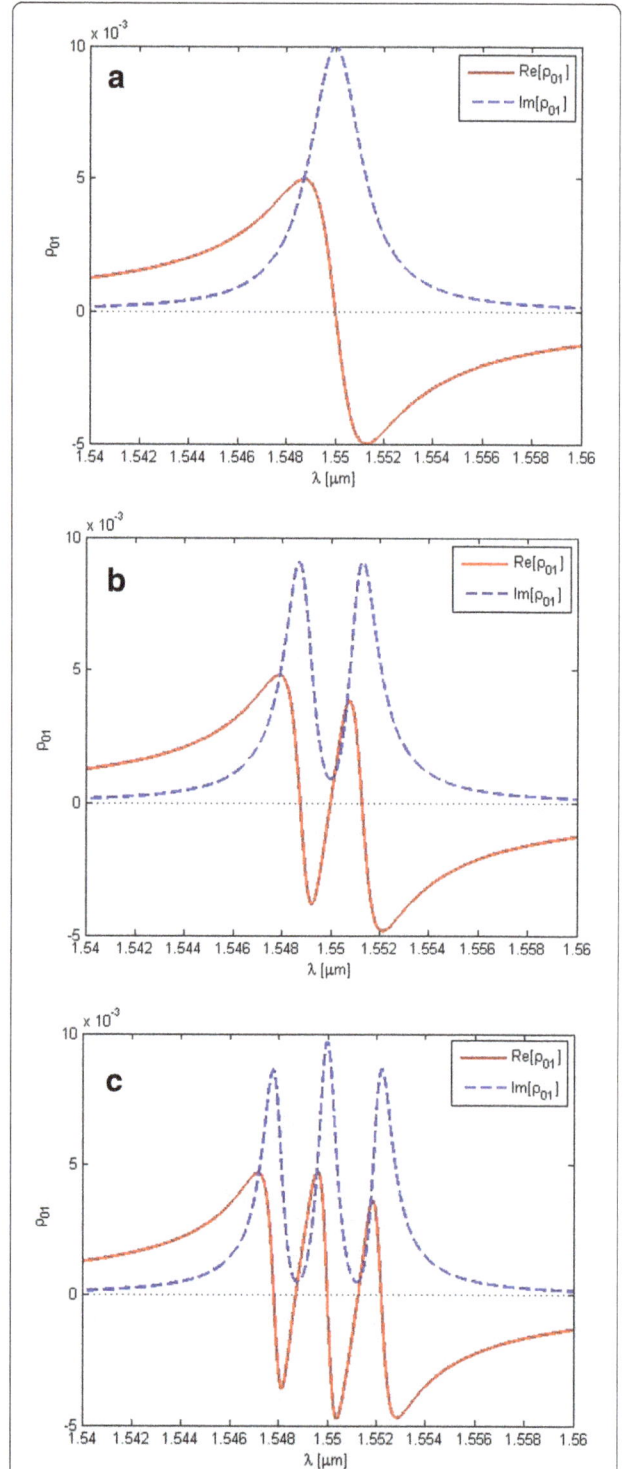

Fig. 2 Real (*dashed*) and imaginary (*solid*) parts of susceptibility as a probe field wavelength for (**a**). $T_{12} = 0$, THz signal = of, $\Lambda = 0$, (**b**) $T_{12} = 1 \gamma$, THz signal = of, $\Lambda = 0$, (**c**) $T_{12} = 1 \gamma$, THz signal = on $\Lambda = 0$, Other parameters are $\omega_{12} = \omega_{23} = 1\gamma$, $\Gamma_{01} = \gamma = 1$ THz, $\Gamma_{02} = 0.1 \Gamma_{01}$, $\Gamma_{03} = 0.01 \Gamma_{01}$, $\Gamma_{12} = 0.05 \Gamma_{01}$, $\Gamma_{13} = 0.025 \Gamma_{01}$, $\Gamma_{23} = 0.05 \Gamma_{01}$, $\gamma_{10} = 0.6$ THz, $\gamma_{20} = 0.1 \gamma_{10}$, $\gamma_{30} = 0.01 \gamma_{10} \Omega_p = 0.01 \gamma$

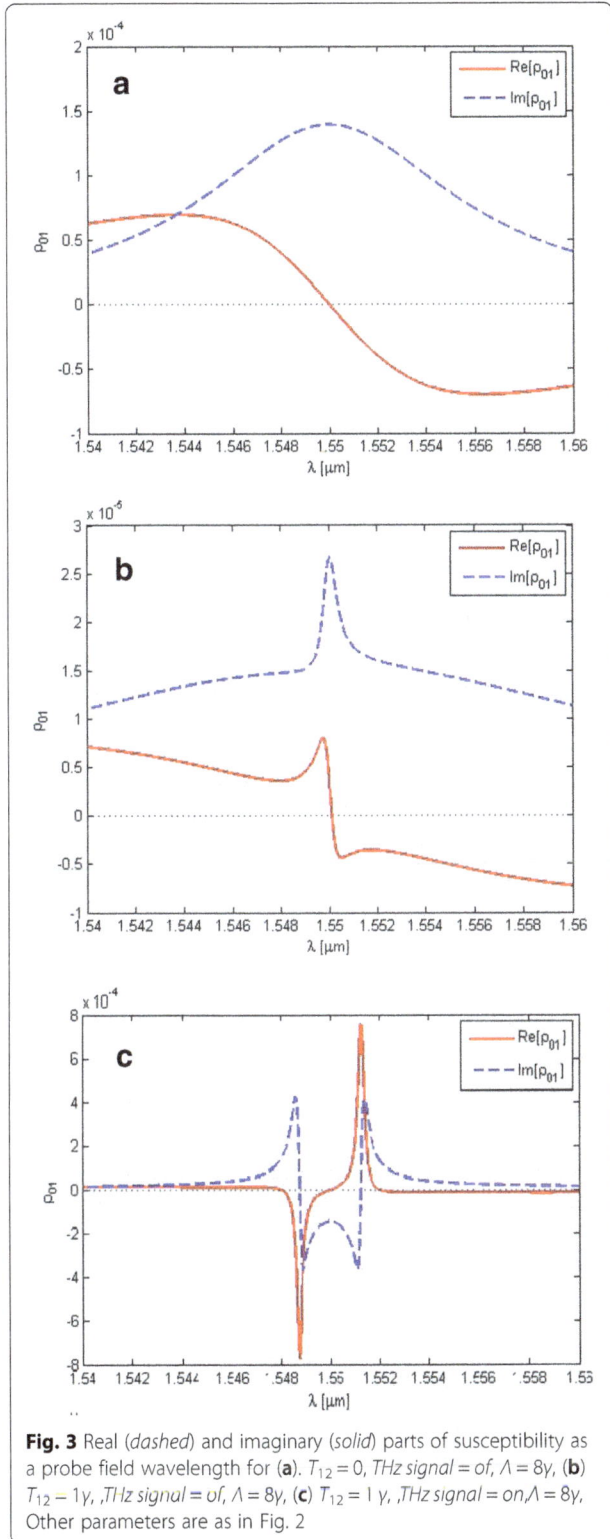

Fig. 3 Real (*dashed*) and imaginary (*solid*) parts of susceptibility as a probe field wavelength for (**a**). $T_{12} = 0$, *THz signal = of*, $\Lambda = 8\gamma$, (**b**) $T_{12} - 1\gamma$, *,THz signal = of*, $\Lambda = 8\gamma$, (**c**) $T_{12} = 1\gamma$, *,THz signal = on,* $\Lambda = 8\gamma$, Other parameters are as in Fig. 2

QW_2 and the terahertz signal, while the incoherent pump rate is still $\Lambda = 8\gamma$. Figure 3(c) show the probe dispersion (solid) and absorption (dashed) in the presence of the terahertz signsl and tunnelling effect. It can easily be seen that the slope of dispersion changes from negative to positive as the presence of terahertz signal. Here, by taking into account the incoherent pumping field, tunnelling effect and terahertz signal, the probe gain is appeared around $\lambda = 1.55\mu m$ in the spectrum. Physically, by taking into account the inter-dot tunnel coupling, the coherence in the system is created. So, the electromagnetically induced transparency is established by the inter-dot tunnel coupling. But in the presence of the incoherent pump field, the upper levels are populated, for this condition the probe absorption will reduce in transition. Then, by applying the terahertz signal there will be a population inversion between level $|1>$ and level $|0>$, which may lead to a probe field amplification. This probe amplification (or gain) might be remarkable for laser applications. Also, it is clear that the dispersion around the probe wavelength $\lambda = 1.55\mu m$ is positive, corresponding to the superluminal light propagation. In fact, the group velocity of the probe field strongly sensitive to the rate of the incoherent pump and the tunnelling effect. This

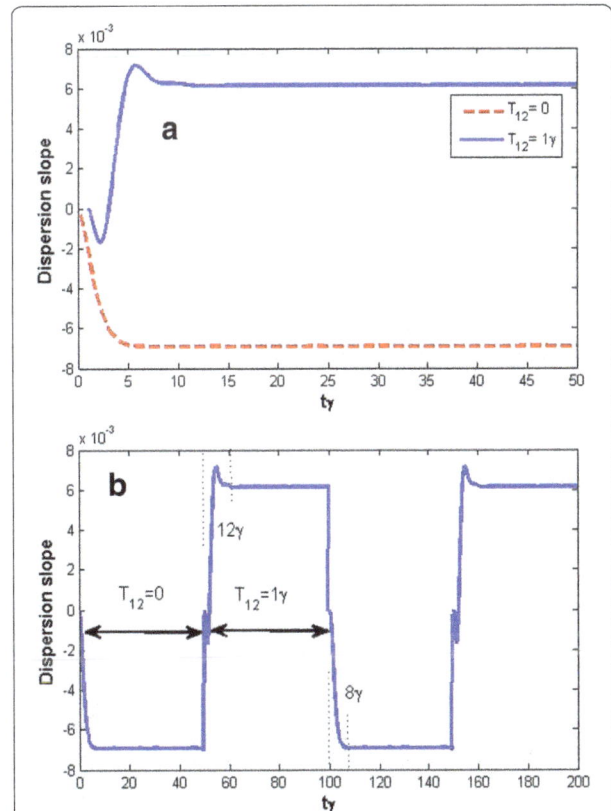

Fig. 4 Dynamical behaviour (**a**) and switching process (**b**) of dispersion slope for $T_{12} = (0, 1\gamma)$, *THz signal = of*, $\Lambda = 0$. Other parameters are as in Fig. 2

shown in Figs. 2(b) and 3(b). This indicates that group velocity of light propagates, increasing from subluminal to superluminal through the medium. In Fig. 3(c), we simultaneously apply the tunnelling rate T_{12} between QW_1 and is an important mechanism in which the probe field absorption and dispersion can be controlled by the incoherent pump field in the quantum well.

Now, we are interested in the dynamical behaviour of the dispersion slope due to this properties can be used as an electro-optical switch of group velocity from subluminal to superluminal or vice versa. The dispersion slope is variation of the dispersion (dashed line in Figs. 2 and 3) around the $\lambda = 1.55\mu m$, which is controllable from negative to positive or vis versa by controlling of the parameters. We are interested in the required switching time for change of the light propagation from subluminal to superluminal or vice versa. In fact, this system can be used as an optical switch, in which the propagation of a laser pulse can be controlled with tunnelling effect, incoherent pumping field and terahertz signsl. The transient behaviour of dispersion slope of probe field at $\lambda = 1.55\mu m$ as the controllable parameters are shown. Switching time for subluminal/superluminal light propagation is defined as the time to reach a steady-state from the superluminal state to the subluminal state and vice versa. Figure 4(a), shows the transition behaviour of the dispersion slope by consecutively switching the tunnelling rate from $T_{12} = 0$ to $T_{12} = 1\gamma$ and vice versa. By increasing the normalized time

$t\gamma$ ($\gamma = \Gamma_{10} = 1\ THz$), the dispersion slope takes a steady negative values for $T_{12} = 0$ corresponding to superluminal light propagation, while it changed to negative on applying the terahertz signal which is corresponding to subluminal light propagation. In Fig. 4(b), we plot the switching diagram of the dispersion slope for two different values of T_{12}. The required switching time for propagation of the light from subluminal to superluminal is about 8 ps and from superluminal to subluminal is about 12 ps. The transient behaviour of the dispersion slope is displayed for two various rate of terahertz signal, while keeping $T_{12} = 1\gamma$ and $\Lambda = 8\gamma$ fixed in Fig. 4. We are looking for the required switching time for changing the group velocity from subluminal to superluminal or vice versa by proper manipulating the tunnelling rate T_{12} and terahertz signal. Effect of the terahertz signal leads to superluminal propagation of light in the medium (Fig. 5 (a)). Figure 5(b) shows that the required switching time from subluminal propagation light to superluminal propagation or vice versa is about 7γ. Now, we investigate the effect of the incoherent pumping rate by consecutively switching incoherent pumping rate from $\Lambda = 0$ to $\Lambda = 8\gamma$ and vice versa, while the tunnelling rate $T_{12} = 1\gamma$ is fixed in Fig. 4(a). It is clearly find that in the Fig. 6(a) the slope of the dispersion changes from positive to negative by adjusting the

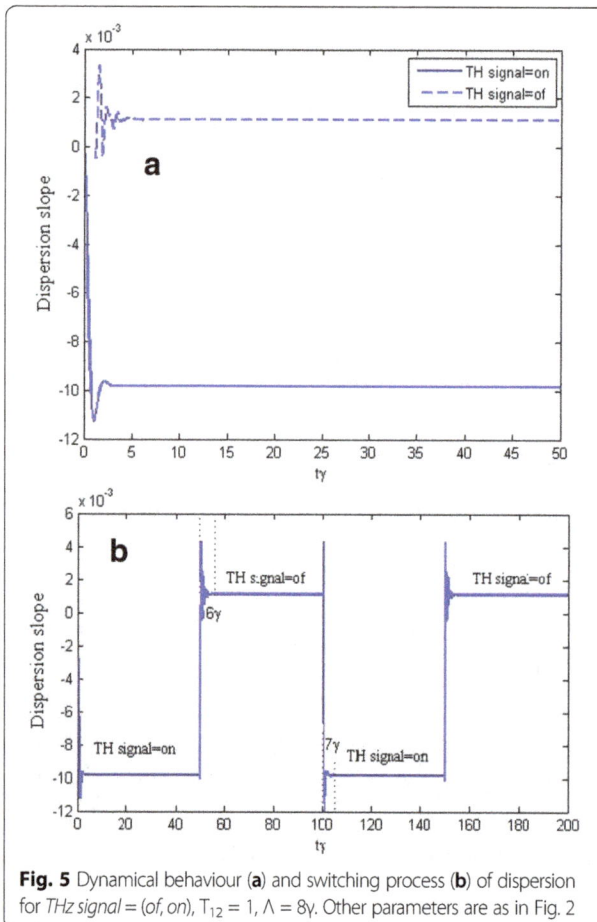

Fig. 5 Dynamical behaviour (a) and switching process (b) of dispersion for *THz signal* = (of, on), $T_{12} = 1$, $\Lambda = 8\gamma$. Other parameters are as in Fig. 2

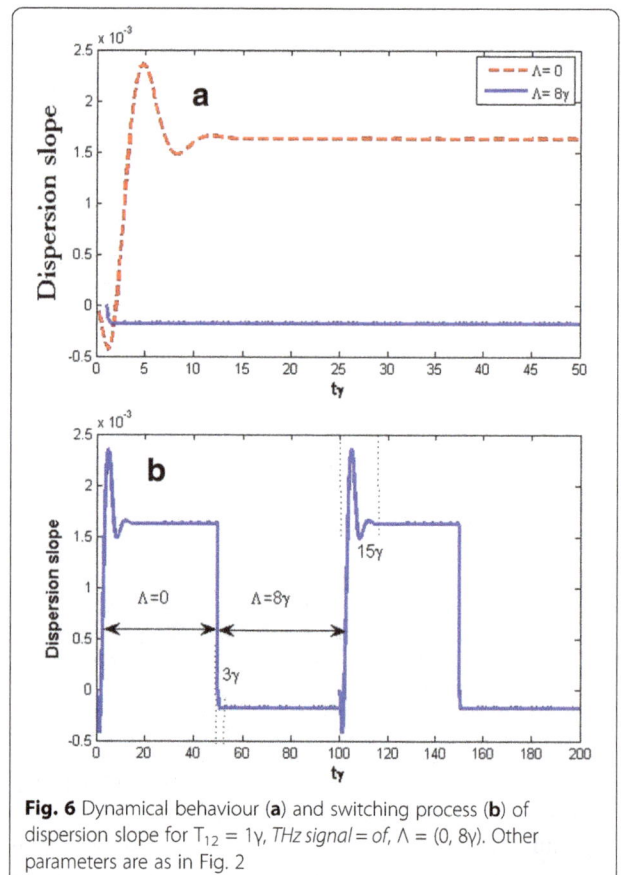

Fig. 6 Dynamical behaviour (a) and switching process (b) of dispersion slope for $T_{12} = 1\gamma$, *THz signal* = of, $\Lambda = (0, 8\gamma)$. Other parameters are as in Fig. 2

incoherent field. For $\Lambda = 0$ the slope of the dispersion is positive corresponding to subluminal light propagation, while it changes to negative as the incoherent pumping changes from $\Lambda = 0$ to $\Lambda = 8\gamma$ corresponding to superluminal light propagation. The required switching time for change of the propagation light from subluminal to superluminal is about 3 ps, and vice versa is 15 ps Fig. 6(b). This approach can be utilized to produce a switch operating only by controlling the tunnelling rate T_{12} and incoherent rate Λ.

Conclusion

We investigated the transient and the steady-state behaviour of a weak probe field at $\lambda = 1.55\mu m$ in a compact double coupled QWs system with applying the tunnelling between QWs, terahertz signal and one incoherent pumping field. It is shown that the absorption and the dispersion of the probe field can be controlled by applying the tunnelling between QWs, terahertz signal and incoherent pumping fields. It has also been shown that the medium can be used as an optical switch in which the propagation of the laser pulse can be controlled with tunnelling between QWs and the incoherent pumping field. We obtained the switching time, between 3 to 15 ps as a high-speed optical switch of group velocity from subluminal to superluminal or vice versa.

Abbreviations
CPT: coherent population trapping; OMVPE: organometallic vapour phase epitaxy; QD: quantum dot; QDMs: quantum dot molecules; QW: Quantum well; SGC: spontaneously generated coherence

Funding
This work was not supported by any specific funding.

Authors' contributions
JSh in the development of the mathematical model and carried out the simulation, contributed in the analysis of the results. AM conceived of the study and finalized the manuscript. Both authors helped to draft the manuscript. Both authors have read and approved the final manuscript.

Competing interests
The authors declare that they have no competing interests.

Author details
[1]Young Researchers and Elite Club, North Tehran Branch, Islamic Azad University, Tehran, Iran. [2]Centre for laser and optics, Basic Sciences Department, Imam Hussein University, Tehran, Iran.

References
1. Macovei, M. A. and Evers, J.: Opt. Commun. **240**, 379 (2004)
2. Javanainen, J.: Europhys. Lett. (EPL). **17**, 407 (1992)
3. Hu, X., and Zhang, J.-P.: J. Phys. B Atomic Mol. Opt. Phys. **37**, 345 (2004).
4. Fleischhauer, M., Keitel C. H., Scully, M. O., and Su, C.: Opt. Commun. **87**, 109 (1992)
5. Bullock, D., Evers, J., and Keitel, C. H.: Phys. Lett. A. **307**, 8 (2003)
6. Wu, J., Lü, X.-Y., and Zheng, L.-L.: J. Phys. B Atomic Mol. Opt. Phys. **43**, 161003 (2010)
7. Wang, Z., and Xu, M.: Opt. Commun. **282**, 1574 (2009)
8. Osman, K. I., and Joshi, A.: Opt. Commun. **293**, 86 (2013)
9. Joshi, A., Yang, W., and Xiao, M.: Phys. Rev. A. **68**, 015806 (2003)
10. Hossein Asadpour, S., and Eslami-Majd, A.: J. Lumin. **132**, 1477 (2012)
11. Antón, M. A., Carreño, F., Calderón, O. G., and Melle, S.: Opt. Commun. **281**, 3301 (2008)
12. Chen, A.: Opt. Express. **22**, 26991 (2014)
13. Si, L.-G., Yang, W.-X., and Yang, X.: J. Opt. Soc. Am. B. **26**, 478 (2009)
14. Yang, W.-X., Chen, A.-X., Lee, R.-K., and Wu, Y.: Phys Rev. A. **84**, 013835 (2011)
15. Yang, W.-X., Hou, J.-M., Lin, Y., and Lee, R.-K.: Phys. Rev. A. **79**, 033825 (2009)
16. Yannopapas, V., Paspalakis, E., and Vitanov, N. V.: Phys. Rev. Lett. **103**, 063602 (2009)
17. Mahmoudi, M., Sahrai, M., and Tajalli, H.: Phys. Lett. A. **357**, 66 (2006)
18. Sahrai, M., Asadpour, S. H., Mahrami, H., and Sadighi-Bonabi, R.: J. Lumin. **131**, 1682 (2011)
19. Shiri, J.: Laser Phys. **26**, 056202 (2016)
20. Su, H. and Chuang, S. L.: Opt. Lett. **31**, 271 (2006)
21. Palinginis, P., Sedgwick, F., Crankshaw, S., Moewe, M., and Chang-Hasnain, C. J.: Opt. Express. **13**, 9909 (2005)
22. Schmidt, K. D. M. H., Nikonov, D. E., Campman, K. L., and Gossard, A. I. lu A. C.: Laser Phys. **9**, 797 (1999)
23. Chang-Hasnain, C. J., Ku, Pei-Cheng, Kim, Jungho, and Chuang, Shun-Lien, Proc. IEEE. **9**, 1884 (2003)
24. Che, C., Han, Q., Ma, J., Zhou, Y., Yu, S., and Tan, L.: Laser Phys. **22**, 1317 (2012)
25. Kosionis, S. G., Terzis, A. F., and Paspalakis, E.: Phys. Rev. B. **75**, 193305 (2007)
26. Li, J., Liu, J., and Yang, X.: Superlattice. Microst. **44**, 166 (2008)
27. Tsukanov, A. V. and Openov, L. A.: Semiconductors. **38**, 91 (2004)
28. Wang, Z.: Ann. Phys. **326**, 340 (2011)
29. Joshi, A. and Xiao, M.: Appl. Phys. B Lasers Opt. **79**, 65 (2004)
30. Wijewardane, H. O. and Ullrich, C. A.: Appl. Phys. Lett. **84**, 3984 (2004)
31. Serapiglia, G. B., Paspalakis, E., Sirtori, C., Vodopyanov, K. L., and Phillips, C. C.: Phys. Rev. Lett. **84**, 1019 (2000)
32. Sahrai, M., Mahmoudi, M., and Kheradmand, R.: Phys. Lett. A. **367**, 408 (2007)
33. Dynes, J. F., Frogley, M. D., Rodger, J., and Phillips, C. C.: Phys. Rev. B. **72**, 085323 (2005)
34. Kash, M. M., Sautenkov, V. A., Zibrov, A. S., Hollberg, L., Welch, G. R., Lukin, M. D., Rostovtsev, Y., Fry, E. S., and Scully, M. O.: Phys. Rev. Lett. **82**, 5229 (1999)
35. J. Shiri and A. Malakzadeh, Laser Physics **27**, 016201 (2017)
36. Hau, L. V., Harris, S. E., Dutton, Z., and Behroozi, C. H.: Nature. **397**, 594 (1999)
37. Ghulghazaryan, R. G. and Malakyan, Y. P.: Phys. Rev. A. **67**, 063806 (2003)
38. Borri, P., Langbein, W., Schneider, S., Woggon, U., Sellin, R. L., Ouyang, D., and Bimberg, D.: Phys. Rev. B. **66**, 081306 (2002)
39. Cole, B. E., Williams, J. B., King, B. T., Sherwin, M. S., and Stanley, C. R.: Nature. **410**, 60 (2001)
40. Michler, P.: Science. **290**, 2282 (2000)
41. Pelton, M., Santori, C., Vučković J., Zhang B., Solomon G. S., Plant, J., and Yamamoto, Y.: Physical Rev. Lett. **89**, 233602 (2002)
42. Yamamoto, Y., Kim, J., Benson, O., and Kan, H.: Nature. **397**, 500 (1999)
43. Ginzburg, P. and Orenstein, M.: Opt. Express. **14**, 12467 (2006)
44. I. Bar-Joseph, C. Klingshirn, D. A. B. Miller, D. S. Chemla, U. Koren, and B. I. Miller,Appl Phys Lett. **50**, 1010 (1987)
45. H. Sattari, M. Sahrai, and S. Ebadollahi-Bakhtevar, Applied Optics **54**, 2461 (2015)
46. Alexander, M. G. W. and RUhle, W. W.: Appl. Phys. Lett. **55**, 885 (1989)
47. Mehmannavaz, M. R., Nasehi, R., Sattari, H., and Mahmoudi, M.: Superlattice. Microstruct. 75, 27 (2014)
48. Kapale, K. T., Scully, M. O., Zhu, S.-Y., and Zubairy, M. S., Phys. Rev. A. **67**, 023804 (2003)
49. Gardiner, C. W. and Collett, M. J.: Phys. Rev. A. **31**, 3761 (1985)
50. Reittu, H. J.: Am. J. Phys. **63**, 940 (1998)
51. Sahrai, V. T. A. B. M., Arzhang, B., Taherkhani, D.: Physica E. **67**, 121 (2015)
52. Bardeen, J., Phys. Rev. Lett. **6**, 57 (1961)

Permissions

The contributors of this book come from diverse backgrounds, making this book a truly international effort. This book will bring forth new frontiers with its revolutionizing research information and detailed analysis of the nascent developments around the world.

We would like to thank all the contributing authors for lending their expertise to make the book truly unique. They have played a crucial role in the development of this book. Without their invaluable contributions this book wouldn't have been possible. They have made vital efforts to compile up to date information on the varied aspects of this subject to make this book a valuable addition to the collection of many professionals and students.

This book was conceptualized with the vision of imparting up-to-date information and advanced data in this field. To ensure the same, a matchless editorial board was set up. Every individual on the board went through rigorous rounds of assessment to prove their worth. After which they invested a large part of their time researching and compiling the most relevant data for our readers.

The editorial board has been involved in producing this book since its inception. They have spent rigorous hours researching and exploring the diverse topics which have resulted in the successful publishing of this book. They have passed on their knowledge of decades through this book. To expedite this challenging task, the publisher supported the team at every step. A small team of assistant editors was also appointed to further simplify the editing procedure and attain best results for the readers.

Apart from the editorial board, the designing team has also invested a significant amount of their time in understanding the subject and creating the most relevant covers. They scrutinized every image to scout for the most suitable representation of the subject and create an appropriate cover for the book.

The publishing team has been an ardent support to the editorial, designing and production team. Their endless efforts to recruit the best for this project, has resulted in the accomplishment of this book. They are a veteran in the field of academics and their pool of knowledge is as vast as their experience in printing. Their expertise and guidance has proved useful at every step. Their uncompromising quality standards have made this book an exceptional effort. Their encouragement from time to time has been an inspiration for everyone.

The publisher and the editorial board hope that this book will prove to be a valuable piece of knowledge for researchers, students, practitioners and scholars across the globe.

List of Contributors

Ping An, Fu-zhong Bai, Zhen Liu and Xiao-juan Gao
College of Mechanical Engineering, Inner Mongolia University of Technology, Huhhot 010051, China

Xiao-qiang Wang
College of Information Engineering, Inner Mongolia University of Technology, Hohhot 010080, China

Taner Oguzer
Department Electrical and Electronics Engineering, Dokuz Eylul University,
Buca, 35160 Izmir, Turkey

Ayhan Altintas
Department Electrical and Electronics Engineering, Bilkent University, 06800 Ankara, Turkey

Alexander I. Nosich
Laboratory of Micro and Nano Optics, Institute of Radio-Physics and Electronics NASU, Kharkiv 61085, Ukraine

M. A. A. Rosli
Department of Computer and Communication Systems Engineering, Faculty of Engineering, Universiti Putra Malaysia, 43400UPM Serdang, Selangor, Malaysia

A. S. M. Noor
Department of Computer and Communication Systems Engineering, Faculty of Engineering, Universiti Putra Malaysia, 43400UPM Serdang, Selangor, Malaysia
Research Centre of Excellence for Wireless and Photonic Network, Faculty of Engineering, Universiti Putra Malaysia, 43400UPM, Serdang, Selangor, Malaysia

P. T. Arasu
Communication Technology Section, Universiti Kuala Lumpur-British Malaysia Institute, 53100 GOMBAK, Kuala Lumpur, Malaysia

H. N. Lim
Department of Chemistry, Faculty of Science, Universiti Putra Malaysia, 43400UPM, Serdang, Selangor, Malaysia

N. M. Huang
Physics Department, Low Dimensional Materials Research Centre, University of Malaya, 50603 Kuala Lumpur, Malaysia

K. Esakki Muthu
University VOC College of Engineering, University VOC, Thoothukudi, Tamilnadu 630003, India

A. Sivanantha Raja
A.C.College of Engineering and Technology, Karaikudi, Tamilnadu 630004, India

M. Lukovic and V. Lukovic
University of Kragujevac, Faculty of Technical Sciences, Cacak 32000, Serbia

I. Belca, B. Kasalica and M. Vicic
University of Belgrade, Faculty of Physics, Belgrade 11000, Serbia

I. Stanimirovic
University of Nis, Faculty of Science and Mathematics, Nis 18000, Serbia

A. Ghaffar and M. M. Hussan
Department of Physics, University of Agriculture, Faisalabad, Pakistan

Majeed A. S. Alkanhal and Sajjad ur Rehman
Department of Electrical Engineering, King Saud University, Riyadh, Saudi Arabia

Ricardo I. Álvarez-Tamayo, José G. Aguilar-Soto, Baldemar Ibarra-Escamilla and Evgeny A. Kuzin
Optics Department, Instituto Nacional de Astrofísica, Óptica y Electrónica, Luis Enrique Erro 1, Puebla 72824, Mexico

Manuel Durán-Sánchez
Optics Department, Instituto Nacional de Astrofísica, Óptica y Electrónica, Luis Enrique Erro 1, Puebla 72824, Mexico
CONACyT Research Fellow - Instituto Nacional de Astrofísica, Óptica y Electrónica, Luis Enrique Erro 1, Puebla 72824, Mexico

José E. Antonio-López
CREOL, The College of Optics and Photonics, University of Central Florida, Orlando, FL 32816-2700, USA

Mohamed Bouhadda, Mustapha Serhani and Ali Boutoulout
Moulay Ismail University, MACS Laboratory, Meknès, Morocco

Fouad Mohamed Abbou and Fouad Chaatit
Al Akhawayen University, School of Sciences and Enginnering, Ifrane, Morocco

Fabrice Kwefeu Mbakop and Noël Djongyang
Department of Renewable Energy, The Higher Institute of the Sahel, University of Maroua, PO Box 46 Maroua, Cameroon

Danwé Raïdandi
Department of Mechanical Engineering, National Advanced Polytechnic School, University of Yaounde I, PO Box 8390 Yaounde, Cameroon

Benjamin Vest
Laboratoire Charles Fabry, 2 Avenue Augustin Fresnel, 91127 Palaiseau, France

Baptiste Fix, Julien Jaeck and Riad Haïdar
ONERA, Chemin de la Hunière, 91761 Palaiseau Cedex, France

Masaki Misawa
Theranostic Device Research Group, Health Research Institute, National Institute of Advanced Industrial Science and Technology (AIST), 1-2-1 Namiki, Tsukuba, Ibaraki 305-8564, Japan

Kiyofumi Matsuda
Theranostic Device Research Group, Health Research Institute, National Institute of Advanced Industrial Science and Technology (AIST), 1-2-1 Namiki, Tsukuba, Ibaraki 305-8564, Japan
The graduate School for the Creation of New Photonics Industries, 1955-1 Kurematsu, Nishi-ku, Hamamatsu, Shizuoka 431-1202, Japan

Juan C. Aguilar
Instituto Nacional de Astrofísica,, Óptica y Electrónica, Luis Enrique , Tonantzintla, Puebla, Mexico

Masato Yasumoto
Research Institute for Measurement and Analytical Instrumentation, NMIJ, National Institute of Advance Industrial Science and Technology, Tsukuba 305-8568, Ibaraki,
Japan
Japan Synchrotron Radiation Research Institute, SPring-8, Sayo, Hyogo 679-5198, Japan

Yoshio Suzuki and Akihisa Takeuchi
Japan Synchrotron Radiation Research Institute, SPring-8, Sayo, Hyogo 679-5198, Japan

Shakil Rehman
Singapore-MIT Alliance for Research and Technology (SMART) Centre1 CREATE Way #09-03, CREATE Tower, Singapore 138602, Singapore

Ilpo Niskanen
Faculty of Technology, University of Oulu, PO Box 73009014 Oulu, Finland

Abdelghafour Messaadi and Ignacio Moreno
Departamento de Ciencia de Materiales, Óptica y Tecnología Electrónica, Universidad Miguel Hernández, 03202 Elche, Spain

María del Mar Sánchez-López
Departamento de Físicay Arquitectura de Computadores, Instituto de Bioingeniería, Universidad Miguel Hernández, 03202 Elche, Spain

Pascuala García-Martínez
Departament d'Òptica, Universitat de València, 46100 Burjassot, Spain

Asticio Vargas
Departamento de Ciencias Físicas, Universidad de La Frontera, Temuco, Chile

Lucas Alber, Martin Fischer, Marianne Bader and Markus Sondermann
Max-Planck-Institute for the Science of Light, Staudtstr. 2, 91058 Erlangen, Germany
Department of Physics, Friedrich-Alexander University Erlangen-Nürnberg (FAU), Staudtstraße 7/B2, 91058 Erlangen, Germany

Klaus Mantel
Max-Planck-Institute for the Science of Light, Staudtstr. 2, 91058 Erlangen, Germany

Gerd Leuchs
Max-Planck-Institute for the Science of Light, Staudtstr. 2, 91058 Erlangen, Germany
Department of Physics, Friedrich-Alexander University Erlangen-Nürnberg (FAU), Staudtstraße 7/B2, 91058 Erlangen, Germany
Department of Physics, University of Ottawa, 75 Laurier Avenue East, ON K1N 6N5 Ottawa, Canada

Linning Peng
Institute of Information Science and Engineering, Southeast University, No.2 SiPaiLou, 210096 Nanjing, China

Ming Liu
School of Computer and Information Technology, Beijing Jiaotao University, No.3 ShangYuanCun, 100044 Beijing, China

Maryline Hélard and Sylvain Haese
IETR (Institute of Electronic and Telecommunications in Rennes), INSA-Rennes (Institut National des Sciences Appliquées de Rennes), 20 Avenue des Buttes de Coësmes, 35708 Rennes, France

Mengjun Li
Institute of Modern Optics, College of Electronic Information and Optical Engineering, Nankai University, Tianjin 300071, China

Hui Fang, Xiaoming Li and Xiaocong Yuan
Nanophotonics Research Centre & Key Laboratory of Optoelectronic Devices and Systems of Ministry of Education and Guangdong Province, College of Optoelectronic Engineering, Shenzhen University, Shenzhen 518060, China

Jaromír Pištora
Nanotechnology Centre, VSB – Technical University of Ostrava, 17. Listopadu 15/2172, 708 33 Ostrava, Poruba, Czech Republic

Jan Chochol
Nanotechnology Centre, VSB – Technical University of Ostrava, 17. Listopadu 15/2172, 708 33 Ostrava, Poruba, Czech Republic
Department of Electrical and Computer Engineering, Dalhousie University, 6299 South St, Halifax NS B3H 4R2, Canada

Lukáš Halagačka
Nanotechnology Centre, VSB – Technical University of Ostrava, 17. Listopadu 15/2172, 708 33 Ostrava, Poruba, Czech Republic
Department of Physics, VSB – Technical University of Ostrava, 17. listopadu 15/2172, 708 33 Ostrava, Poruba, Czech Republic

Martin Mičica
Nanotechnology Centre, VSB – Technical University of Ostrava, 17. Listopadu 15/2172, 708 33 Ostrava, Poruba, Czech Republic
Institut d'Electronique, de Microélectronique et de Nanotechnologie, UMR CNRS 8520, Avenue Poincaré, F-59652 Villeneuve d'Ascq cedex, France

Michael Čada
Department of Electrical and Computer Engineering, Dalhousie University, 6299 South St, Halifax NS B3H 4R2, Canada.

Kamil Postava
Department of Physics, VSB – Technical University of Ostrava, 17. listopadu 15/2172, 708 33 Ostrava, Poruba, Czech Republic

Mathias Vanwolleghem and Jean-François Lampin
Institut d'Electronique, de Microélectronique et de Nanotechnologie, UMR CNRS 8520, Avenue Poincaré, F-59652 Villeneuve d'Ascq cedex, France

M Gargouri
Unité de l'état solide, Faculté des Sciences de Sfax, Route Soukra, 3018 Sfax, Tunisie

M Karray
Unité de l'état solide, Faculté des Sciences de Sfax, Route Soukra, 3018 Sfax, Tunisie
LAUM UMR CNRS 6613, Université du Maine, Av. O. Messiaen, 72085 Le Mans, France

P Picart
LAUM UMR CNRS 6613, Université du Maine, Av. O. Messiaen, 72085 Le Mans, France

C Poilane
CIMAP UMR6252, Université de Caen, 6 Boulevard du Maréchal Juin, 14050 Caen cedex 4, France

Guoyu Yu and Xiao Zheng
National Facility for Ultra Precision Surfaces, OpTIC Centre, University of Huddersfield, St. Asaph Business Park, Ffordd William Morgan, North Wales, St Asaph LL17 0JD, UK

David Walker
National Facility for Ultra Precision Surfaces, OpTIC Centre, University of Huddersfield, St. Asaph Business Park, Ffordd William Morgan, North Wales, St Asaph LL17 0JD, UK
Department of Physics and Astronomy, University College, Gower St, London, WC1E 6BT, UK
Zeeko Ltd, 4 Vulcan Court, Vulcan Way, Coalville, Leicestershire LE67 3FW, UK

Li Yang and Li Jiu-Sheng

Margarita L. Shendeleva

Petr Křen

Fahimeh Abrinaei

Ying Yuan, Xiongxiong Wu, Xiaorui Wang and Yan Zhang

Yilin Jiang, Qi Tong, Haiyan Wang and Qingbo Ji

Hongyu Li
National Facility for Ultra Precision Surfaces, OpTIC Centre, University of Huddersfield, St. Asaph Business Park, Ffordd William Morgan, North Wales, St Asaph LL17 0JD, UK
Researh Center for Space Optical Engineering, Harbin Institute of Technology, Harbin 150001, China

Anthony Beaucamp
Zeeko Ltd, 4 Vulcan Court, Vulcan Way, Coalville, Leicestershire LE67 3FW, UK
Kyoto University, C3 Bldg. Kyotodaigaku-Katsura, Nishikyo-ku, Kyoto 615-8540, Japan

Jalil Shiri
Young Researchers and Elite Club, North Tehran Branch, Islamic Azad University, Tehran, Iran

Abdollah Malakzadeh
Centre for laser and optics, Basic Sciences Department, Imam Hussein University, Tehran, Iran

Index

www.ingramcontent.com/pod-product-compliance
Lightning Source LLC
Chambersburg PA
CBHW082036190326
41458CB00010B/3379